Beyond-CMOS Technologies for Next Generation Computer Design

Rasit O. Topaloglu • H.-S. Philip Wong
Editors

# Beyond-CMOS Technologies for Next Generation Computer Design

Springer

*Editors*
Rasit O. Topaloglu
IBM
Hopewell Junction, NY, USA

H.-S. Philip Wong
Department of Electrical Engineering
Stanford University
Stanford, CA, USA

ISBN 978-3-030-08003-7 ISBN 978-3-319-90385-9 (eBook)
https://doi.org/10.1007/978-3-319-90385-9



Printed on acid-free paper

This Springer imprint is published by the registered company Springer International Publishing AG part of Springer Nature.
The registered company address is: Gewerbestrasse 11, 6330 Cham, Switzerland

# Foreword

This book surveys and summarizes recent research aimed at new devices, circuits, and architectures for computing. Much of the impetus for this research stems from a remarkable development in information technology that played out in the brief period between 2003 and 2005. After decades of rapid exponentially compounding improvement, microprocessor clock frequencies abruptly plateaued—a stunning break from a long-established and highly desirable trend in computing performance. The proximate cause was the increasing difficulty and cost of powering and removing the waste heat from ever denser and faster transistor circuits. The root cause—the reason for excessive heat generation—was the inability, for fundamental reasons, to reduce transistor threshold voltage in proportion to reductions in power supply voltage. As a result, standby (or passive) power had grown exponentially from technology generation to technology generation until it equaled or exceeded active power, which was also increasing. Further increases would have driven unacceptable costs for power and cooling across many product categories. Instead, development teams pivoted and began to optimize each new technology generation for operation in this new power-constrained environment. While many had foreseen the need and developed strategies for highly power-constrained device-, circuit-, and system-level design, the net outcome of the power-performance trade-offs at all levels of the design hierarchy was difficult to predict. Thus the complete and abrupt cessation of advances in clock frequency came as a surprise.

This event sent ripples throughout the worldwide microelectronics industry. Since 2005, integration density and cost per device have continued to improve, and manufacturers have emphasized the increasing number of processors and the amount of memory they can place on a single die. However, with clock frequencies stagnant, the resulting performance gains have been muted compared to those of the previous decades. The return on investment for development of each new generation of ever-smaller transistors has therefore been reduced, and the number of companies making that investment has declined. To be clear, the total effort remains enormous by the standards of any industry, with the vast majority of R&D dollars going toward further advancement of silicon CMOS field effect transistor technology and, increasingly, toward advancement of circuit and system architectures.

But the experience of 2003–2005 also sparked a bold new industry initiative. In 2005, the Nanoelectronics Research Initiative (NRI) was chartered by a consortium of Semiconductor Industry Association (SIA) member companies to develop and administer a university-based research program to address the increasingly evident limitations of the field effect transistor. In partnership with the National Science Foundation (NSF), NRI would fund university research to "Demonstrate novel computing devices capable of replacing the CMOS FET as a logic switch in the 2020 timeframe." In 2007, the National Institute of Standards and Technology (NIST) joined the private-public partnership, resulting in the creation of four multi-university, multidisciplinary research centers. NRI's bold and clearly articulated research goals caught the attention of funding agencies in Europe and Asia and helped to spark new initiatives in those geographies. In 2013 the Defense Advanced Research Projects Agency (DARPA) joined with industry to fund STARnet, further focusing US university researchers on the exploration of post-CMOS devices. As the NRI and STARnet programs evolved, the interest of the industrial sponsors shifted from exploration of isolated devices toward co-development of new devices, circuits, and architectures. This was made explicit with new programs announced with NSF in 2016 and with NIST and DARPA in 2017.

The research initiatives and results described in *Beyond-CMOS Technologies for Next Generation Computer Design* reflect and address this broad industry need for new approaches to energy-efficient computing. Indeed, much of the work was funded to a greater or lesser extent through NRI, STARnet, or programs with closely related goals. The research ranges from new materials and the devices they enable, to novel circuits and architectures for computing. In many cases, the results span two or more levels in this hierarchy. For example, Subhasish Mitra describes the novel fabrication processes that made it possible to build a simple computer from transistors based on carbon nanotubes. Xueqing Li and coauthors tell us why and how the negative capacitance field effect transistor (NCFET) and other "steep slope" devices are poised to open a new circuit design space for ultra-low-power electronics. Two sets of authors provide perspectives on the interplay between emerging nonvolatile memory devices, 3D integration schemes, and "compute in memory" architectures. Looking beyond conventional FETs and the traditional computing architecture, it seems there is still a *lot* to explore!

Department of Electrical Engineering, Columbia University Thomas N. Theis
New York, NY, USA
April 15, 2018

# Preface

Advances of traditional CMOS devices may be hitting a bottleneck soon due to electrostatic control, power, device density, and variability limitations. It may be necessary to complement silicon transistors with beyond-CMOS counterparts in integrated circuits. Yet, a straightforward replacement may not yield optimal architecture and system response. Hence, circuits need to be redesigned in the context of beyond-CMOS devices. This book in particular targets to bridge the gap between device availability and architecture/system considerations. With this book, readers should be able to understand:

- Why we need to consider beyond-CMOS devices,
- What are the challenges of beyond-CMOS options,
- How should architecture and systems be designed differently,
- How would designs take advantage of beyond-CMOS benefits.

The book consists of the following seven chapters from distinguished authors:

Hills, Mitra, and Wong focus on carbon nanotube transistors. They further analyze a monolithic 3D integration with carbon nanotube transistors. A new device integration enabled by carbon nanotube transistors would lead to three orders of magnitude energy delay product improvement.

Resta, Gaillardon, and de Micheli discuss a novel device (MIG-FET) with intrinsic doping where the device type is not fixed at manufacture but is adjustable using inputs to the gate. The authors analyze this functionality-enhanced MIG-FET device.

Nourbakhsh, Yu, Lin, Hempel, Shiue, Englund, and Palacios study devices of 2D layered materials that have weak van der Waals forces between the layers. They discuss not only electrical but also optoelectrical and biological applications in their chapter.

Khwa, Lu, Dou, and Chang discuss nonvolatile memories including resistive RAM (ReRAM), phase change memory, and spin-torque transfer magnetic RAM (STT-RAM), and their circuit implementations such as nonvolatile SRAM.

Ghose, Hsieh, Boroumand, Ausavarungnirun, and Mutlu study processing-in-memory to avoid CPU to memory transfers. They propose and discuss an in-memory accelerator for pointer chasing and a data coherence support mechanism.

Li, Kim, George, Aziz, Jerry, Shukla, Sampson, Gupta, Datta, and Narayanan investigate tunneling FET (TFET), negative capacitance FET (NCFET), and HyperFET as steep-slope device candidates to achieve low power consumption.

Finally, Zografos, Vaysset, Soree, and Raghavan analyze spin-wave devices and spin-torque majority gates including circuit benchmarking against silicon devices.

Hopewell Junction, NY, USA Rasit O. Topaloglu
Stanford, CA, USA H.-S. Philip Wong

# Contents

# Chapter 1
# Beyond-Silicon Devices: Considerations for Circuits and Architectures

**Gage Hills, H.-S. Philip Wong, and Subhasish Mitra**

## 1.1 Introduction

While beyond-silicon devices promise improved performance at the device level, leveraging their unique properties to realize novel circuits and architectures provides additional benefits. In fact, the benefits afforded by the new architectures that beyond-silicon devices enable can far exceed the benefits any improved device by itself could achieve. As a case study, we provide an overview of carbon nanotube (CNT) technologies, and highlight the importance of understanding—and leveraging—the unique properties of CNTs to realize improved devices, circuits, and architectures. In this chapter, we begin by reviewing state-of-the-art CNT technologies and summarizing their benefits. We then discuss the obstacles facing CNT technologies and the solutions for overcoming these challenges, while highlighting their circuit-level implications. We end by illustrating how CNTs can impact computing architectures, and the considerations that must be taken into account to fully realize the benefits of this emerging nanotechnology.

G. Hills (✉) · H.-S. P. Wong
Department of Electrical Engineering, Stanford University, Stanford, CA, USA
e-mail: ghills@stanford.edu

S. Mitra
Department of Electrical Engineering, Stanford University, Stanford, CA, USA

Department of Computer Science, Stanford University, Stanford, CA, USA



R. O. Topaloglu, H.-S. P. Wong (eds.), *Beyond-CMOS Technologies for Next Generation Computer Design*, https://doi.org/10.1007/978-3-319-90385-9_1

## 1.2 Carbon Nanotube Field-Effect Transistors

For decades, improvements in computing performance and energy efficiency (characterized by the energy-delay product, EDP, of very-large-scale-integration (VLSI) digital systems) have relied on physical and equivalent scaling of silicon-based field-effect transistors (FETs). This "equivalent scaling" path included strained silicon, high-k gate dielectric and metal gate, and advanced device geometries (e.g., FinFETs, and potentially nanowire FETs). However, continued scaling is becoming increasingly challenging, spurring the search for beyond-silicon emerging nanotechnologies to supplement—and 1 day supplant—silicon CMOS. One such promising emerging nanotechnology for VLSI digital systems is carbon nanotube field-effect transistors (CNFETs). CNFETs are excellent candidates for continuing to improve both the performance and energy efficiency of digital VLSI systems, as CNFETs are expected to improve digital VLSI EDP by and order of magnitude compared to silicon-CMOS (at the same technology node for both CNFETs and silicon-CMOS). Moreover, CNFETs are projected to scale beyond the limitations of silicon-CMOS, providing an additional opportunity for further EDP benefits [1]. A full description of the device-level advantages of CNTs falls beyond the scope of this chapter, but we summarize some of the key benefits below (note that, the following list is not exhaustive):

1. CNTs achieve ultrahigh carrier transport (e.g., mobility and velocity) even with ultrathin ($\sim$1 nm) bodies. In contrast, when bulk materials, such as silicon, are scaled below sub-10 nm dimensions, the carrier transport (e.g., mobility) degrades dramatically, resulting in reduced drive current (not to mention the challenges of robust manufacturing of sub-10 nm thin silicon). In contrast, a CNT naturally has an ultrathin body of $\sim$1 nm, dictated by the diameter of the CNT, while still achieving ultrahigh carrier mobility. This enables CNFETs to achieve high drive current even with an ultrathin body.
2. Ultrathin CNTs for CNFET channels result in improved electrostatic control, which is necessary for controlled off-state leakage current and steep subthreshold slope (*SS*). CNFETs can therefore maintain controlled off-state leakage current and steep *SS* due to their thin body while simultaneously maintaining high drive current (discussed above). In contrast, silicon channels incur a fundamental trade-off: thin bodies for improved electrostatic control, but thicker bodies for improved drive current.
3. Leveraging a planar device structure for CNFETs (in contrast to today's three-dimensional silicon FinFETs or stacked nanowire FETs) results in both reduced gate-to-channel capacitance and also reduced parasitic capacitance, improving both circuit speed and energy consumption.

A schematic of a CNFET is shown in Fig. 1.1. Multiple CNTs compose the transistor channel, whose conductance is modulated by the gate, as with a conventional metal-oxide-semiconductor FET (MOSFET). The gate, source, and drain are defined using conventional photolithography, while the doping of the CNT

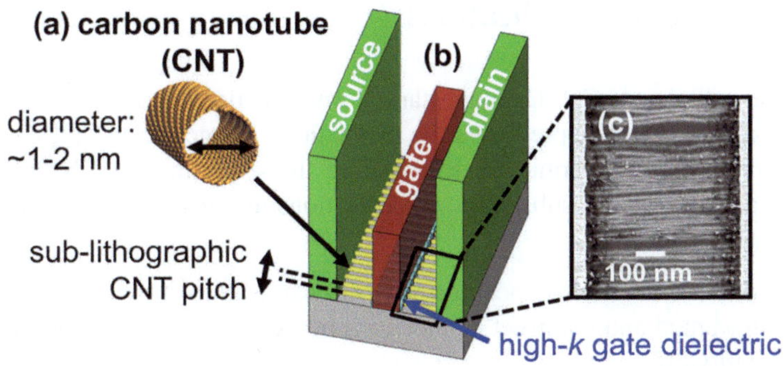


**Fig. 1.1** Carbon nanotube FET (CNFET) schematic. (**a**) Carbon nanotube (CNT), indicating ultrathin ∼1–2 nm CNT diameter; (**b**) CNFET, with multiple parallel CNTs comprising the CNFET channel; (**c**) scanning electron microscopy (SEM) image of CNTs in the channel region

is typically controlled via electrostatic doping, instead of interstitial doping (as is typically the case for silicon CMOS). The inter-CNT spacing is determined by the CNT growth, and can therefore exceed the minimum lithographic pitch. For high drive current, the target inter-CNT spacing is 4–5 nm, corresponding to CNT density of ∼200–250 CNTs/μm [2].

There has been significant progress worldwide toward physically realizing a high-performance VLSI CNFET technology. Recent experimental demonstrations have shown: CNFETs with 5 nm gate lengths [3] while simultaneously maintaining strong electrostatic control of the channel (with subthreshold slope = 70 mV/decade for both PMOS and NMOS CNFETs), high-performance CNFETs with current densities both competing with and exceeding silicon-based FETs (simultaneously with high on/off current ratio) [3–8]), techniques to reduce hysteresis with high-*k* gate dielectrics (with nm-scale thickness, deposited at low temperatures through atomic layer deposition (ALD)) [9], and negative capacitance CNFETs with subthreshold slope = 55 mV/decade (exceeding the 60 mV/decade limit at 300°K) [10].

Importantly, CNFETs are unique among emerging nanotechnologies, as complete and complex digital systems fabricated entirely using CNFETs have been experimentally demonstrated (descriptions in Sect. 1.4 and in Fig. 1.9). The first complete digital subsystem, a digital implementation of a sensor/sensor-interface circuit, was demonstrated by Shulaker et al. [13]. Since then, increasingly complex systems, including a simple microprocessor built entirely using CNFETs [14], have been demonstrated. Furthermore, as discussed in Sect. 1.4 of this chapter, CNFETs have been exploited to realize new system architectures, such as monolithic three-dimensional (3D) integrated systems, where multiple vertical layers of CNFET circuits are fabricated directly overlapping one another, interleaved with layers of memory, resulting in even larger EDP benefits at the system level [1].

## 1.3 Circuit-Level Implications

Despite the promise of CNFETs, substantial imperfections and variations inherent with CNTs had previously prevented the realization of larger-scale CNFET circuits, and thus had to be overcome to demonstrate the experimental CNFET circuits described above [2]. The substantial imperfections and variations associated with CNFETs are

1. *Mis-positioned CNTs*: Mis-positioned CNTs can lead to stray conducting paths. These unwanted and incorrect connections in a circuit can cause incorrect logic functionality [15].
2. *Metallic CNTs (m-CNTs)*: Due to imprecise control over CNT properties, CNTs can be either semiconducting (s-CNT) or metallic (m-CNT); m-CNTs, which have little or no bandgap due to their chirality and diameter, lead to degraded (decreased) on/off ratio (drive current/off-state leakage current), increased leakage power, and incorrect logic functionality [16].
3. *CNT-specific variations*: In addition to variations that exist in conventional silicon CMOS circuits (such as channel length variation and oxide thickness variations), CNTs suffer from CNT-specific variations [2, 17]. These are discussed in detail later in this section.

To overcome these inherent CNT imperfections, researchers developed the imperfection-immune design paradigm [2], which relies on both understanding and leveraging CNT-specific circuit-design techniques to overcome the above imperfections and realize larger-scale CNFET circuits.

### *1.3.1 Overcoming Mis-Positioned CNTs*

It is currently impossible to guarantee exact alignment and positioning of all CNTs on a wafer, especially for VLSI CNFET circuits that potentially require billions of CNTs. The resulting mis-positioned CNTs introduce stray conducting paths, resulting in incorrect logic functionality. While improved CNT synthesis techniques to improve the CNT alignment have been developed, they remain insufficient. Therefore, the remaining mis-positioned CNTs must be dealt with through design and are a major consideration for circuit design.

As a first measure to address mis-positioned CNTs, wafer-scale aligned CNT growth is accomplished by growing the CNTs on a quartz crystalline substrate (Fig. 1.2a) [18]. The CNTs grow preferentially along the crystalline plane of the substrate, and >99.5% of CNTs are synthesized aligned [18]. Importantly, after growth, the CNTs are transferred to a traditional amorphous $SiO_2$/Si substrate, to remain silicon-CMOS compatible (Fig. 1.2b). However, as discussed above, 99.5% aligned CNTs are insufficient for digital VLSI systems. Thus, a circuit design technique can also be leveraged to overcome the remaining <0.5% mis-positioned

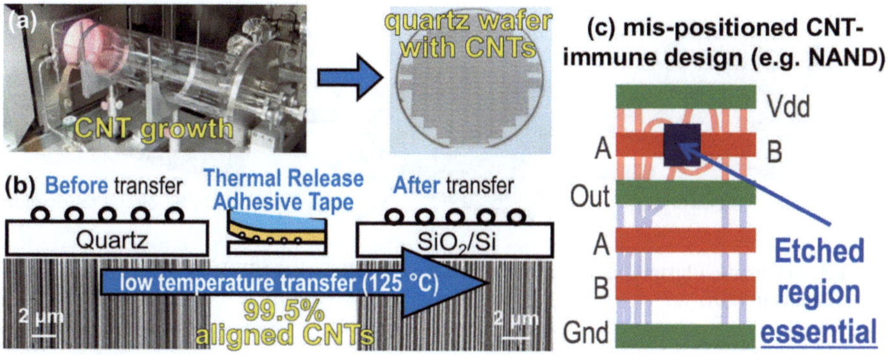


**Fig. 1.2** Overcoming mis-positioned CNTs. (**a**) Wafer-scale aligned CNT growth on crystalline quartz substrate, yielding 99.5% aligned CNTs; (**b**) low-temperature CNT transfer process from quartz onto wafer used for final circuit fabrication; (**c**) mis-positioned CNT-immune design (example shown for a NAND gate)

CNTs. Specifically, the mis-positioned CNT-immune design technique can be used, which ensures that the resulting circuit is immune to any mis-positioned CNTs [15] (Fig. 1.2c). Important points of consideration for circuit design techniques, which mis-positioned CNT-immune design satisfies, are that the design technique can be (1) applied to any arbitrary logic function, (2) is compatible with VLSI design flows (e.g., is implemented entirely within the standard cell and does not require die-specific customization), and (3) has minimal cost (in terms of area, power, and speed; mis-positioned CNT-immune design has significantly smaller impact compared with traditional redundancy-based defect- and fault-tolerance techniques).

### *1.3.2 Overcoming Metallic CNTs*

In addition to mis-positioned CNTs, m-CNTs are also inherent to every CNT synthesis, resulting in up to 50% m-CNTs depending on the growth conditions. While advancements in CNT synthesis can yield >99% semiconducting CNTs (s-CNTs), it is currently impossible to grow 100% s-CNTs. Thus, m-CNTs must be removed post growth. To meet digital VLSI requirements, 99.99% of all m-CNTs must be removed [19]. Several methods exist for m-CNT removal, such as solution-based sorting and single-device electrical breakdown, SDB. SDB—whereby a sufficiently large source–drain voltage is pulsed to break down m-CNTs through self-Joule heating (while the gate turns off all s-CNTs)—has shown the ability to remove the required 99.99% of m-CNTs for VLSI applications. However, while SDB can achieve such a high degree of m-CNT removal, it simultaneously poses several scalability challenges: It is infeasible to perform SDB on individual devices, due to both probing time and the inability to physically contact the source,

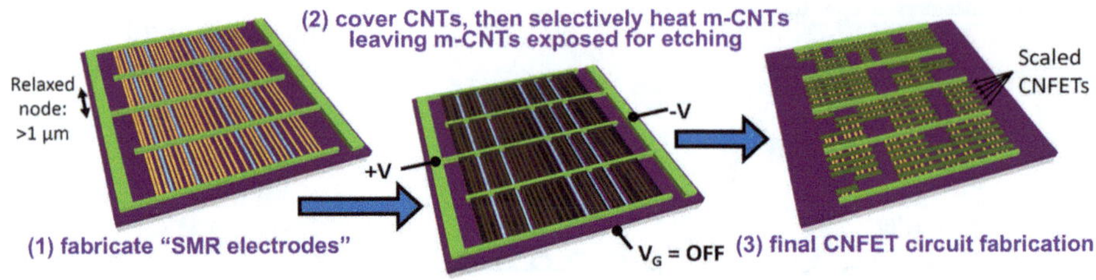


**Fig. 1.3** Schematic illustrations of Scalable Metallic CNT Removal (SMR) design and processing steps (details in [20])

drain, and gate of every transistor in a logic circuit, particularly those within logic gates where there does not exist a contact for each of the terminals, as can be the case with series transistors.

To perform electrical breakdown in a VLSI-compatible manner, a combined CNFET processing and circuit design technique can be used, called Scalable Metallic CNT Removal (SMR) [20], which selectively removes >99.99% of all m-CNTs across an entire wafer, all at once. Importantly, SMR meets the same three requirements described above for mis-positioned CNT-immune design: (1) It can be applied to any arbitrary logic function, (2) it is compatible with VLSI design flows, and (3) has minimal area, energy, and delay cost at the system level. SMR involves three steps, shown in Fig. 1.3: (1) Fabricate "SMR electrodes" for m-CNT removal; (2) cover all CNTs with a protective mask, then apply source–drain bias using the SMR electrodes at full wafer scale while turning off s-CNTs via transistor gates; this causes selective heating of m-CNTs (m-CNTs flow current since they do not turn "off") and the protective mask around the m-CNTs sublimates, leaving them exposed so they can be etched away from the wafer (extensive design and process details in [20]); (3) fabricate final CNFET circuits after m-CNT removal. This follows VLSI processing and design flows with no die-specific customization. Using SMR, 99.99% of m-CNTs can be removed selectively versus inadvertent removal of 1% of s-CNTs (Fig. 1.3) [20].

### *1.3.3 CNT-Specific Variations*

In addition to variations that exist in silicon CMOS circuits, CNTs are also subject to CNT-specific variations, including variations in CNT type (m-CNT or s-CNT), CNT density, diameter, alignment, and doping [2]. These CNT-specific variations can lead to significantly reduced circuit yield, increased susceptibility to noise, and large variations in CNFET circuit delays. Such variations are common for emerging nanotechnologies, owing to imprecise synthesis of nanomaterials today. One method to counteract these effects is to upsize all transistors in a circuit. However, such naïve upsizing incurs large energy and delay costs that diminish potential beyond-

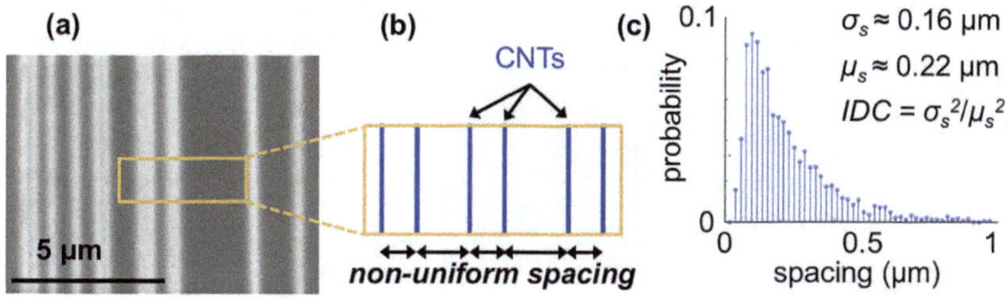


**Fig. 1.4** CNT density variations. (**a**) SEM of CNTs with nonuniform inter-CNT spacing. (**b**) Illustration of nonuniform inter-CNT spacing. (**c**) Experimentally extracted inter-CNT spacing distribution [21]

silicon technology benefits. Rather, various process improvement options, when combined with new circuit design techniques, provide an energy-efficient method of overcoming variations. As an example, without such strategies, CNT variations can degrade the potential speed benefits of CNFET circuits by ≥20% at sub-10 nm nodes, even for circuits with upsized CNFETs to achieve ≥99.9% yield [17]. By leveraging CNT process improvements, together with CNFET circuit design, the overall speed degradation can be limited to ≤5% with ≤5% energy cost while simultaneously meeting circuit-level noise margin and yield constraints [2, 17].

As an example, we summarize circuit design considerations for overcoming the dominant source of CNT variations. The dominant source of variations in CNFET circuit is due to CNT count variations, that is, variations in the number of CNTs per CNFET. CNT count variations lead to increased delay variations, reduced noise margin, and possible functional failure of devices (e.g., CNFETs with no s-CNTs in the channel). There are multiple sources of CNT count variations, including the probabilistic presence of m-CNTs in a CNFET, and the probabilistic removal of both m-CNTs and inadvertent s-CNT removal. Additionally, CNT count variations are caused by nonuniform inter-CNT spacing from the CNT growth (Fig. 1.4). This results in local density variations across a wafer. Therefore, CNFETs with a specific width will not always be comprised of a fixed number of CNTs.

As mentioned above, a naïve solution to overcoming functional failures is upsizing CNFETs. Increasing the width of a CNFET increases the average number of CNTs per CNFET, thus exponentially reducing the probability of CNFET functional failure [22]. Yet upsizing all CNFETs leads to significant energy penalties.

While naïve upsizing improves circuit yield, it overlooks the opportunity to improve yield through taking advantage of properties unique to CNTs. Specifically, due to the fact that CNTs are one-dimensional nanostructures with lengths typically much longer than the length of a CNFET, CNTs exhibit asymmetric correlations [22]. For instance, if the active region (area of channel which has CNTs) of multiple CNFETs is aligned perpendicular to the direction of CNT growth, the CNFETs are comprised of different and distinct CNTs. These CNFETs are thus uncorrelated. However, if the active regions of CNFETs are aligned along the direction of CNT growth, then all CNFETs are comprised of essentially the same set of CNTs,

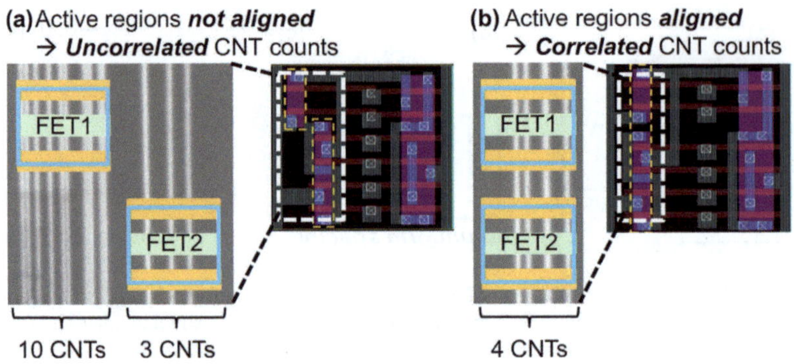


**Fig. 1.5** Aligned-active layouts illustration (example AOI222_X1 standard cell), with and without aligned-active layout [22]. (**a**) Without aligned-active layout. The CNT counts of FET1 and FET2 are uncorrelated since each FET is comprised of different CNTs. (**b**) With aligned-active layout. The CNT counts of FET1 and FET2 are correlated, reducing CNT count variations

and thus their electrical properties are highly correlated. This asymmetric CNT correlation provides a unique opportunity to improve yield otherwise limited by CNT count variations with only minimal upsizing, resulting in smaller energy penalty than naively upsizing all CNFETs in a circuit. Special layouts, called *aligned-active layouts* (illustrated in Fig. 1.5), constrain the active regions of the CNFETs within the standard cell to be aligned along the direction of CNT growth [22]. By aligning the active regions of the CNFETs, the probability of having the entire column of CNFETs function or fail is approximately at the probability of just a single CNFET functioning or failing, irrespective of the actual number of CNFETs in the column (for CNTs oriented in the vertical direction). It has been shown that aligned-active layouts and selective upsizing can improve (i.e., reduce) the probability of functional failures by multiple orders of magnitude at significantly reduced energy penalties associated versus naïve CNFET upsizing [22]. The costs of implementing aligned-active layouts at the standard cell level and at the system level are minimal (<5% energy cost). Additionally, the locations of I/O pins are mostly retained, minimizing the impact on intercell routing during circuit-level place-and-route. For a detailed description of the aligned-active layout design technique for large-scale designs (aligned-active regions inside each standard cell, aligned-active regions between different cells on the same set of CNTs, and selective upsizing to ensure high yield given that all standard cells cannot share the same set of CNTs in a large design), please refer to previous work in [22].

To further reduce the impact of CNT variations, CNFET processing should continue to be improved, and so a combination of improved CNT processing and CNFET circuit design should be leveraged simultaneously [2]. However, this raises two key questions: (1) which processing parameters should be improved? and (2) by how much? Without a systematic methodology to evaluate the circuit-level impact of CNT variations, one might blindly pursue difficult CNT processing paths with diminishing returns, while overlooking other processing parameters that enable larger performance gains. For example, much research has focused solely on improving the initial purity of the CNT synthesis (e.g., reducing the percentage of grown m-CNTs and preferentially grown s-CNTs). However, reducing the percentage of grown m-CNTs (for instance, beyond 1% m-CNTs) suffers from diminishing returns and can be insufficient to meet digital VLSI system design targets [2]. Previously, co-optimization of processing and design has been performed via a trial-and-error-based approach. However, such a brute-force approach can be prohibitively time-consuming, potentially requiring months of simulation time. Therefore, a framework and methodology that efficiently selects effective combinations of processing options and circuit design techniques to overcome variations is essential for emerging nanotechnologies.

An example of such a methodology is described in detail in [17]. To fully understand the importance of systematic searches for effective combinations of processing options and circuit design techniques for overcoming variations, we first describe a brute-force approach. A designer would iterate over many design points (design point: a combination of values for the processing parameters of a technology, and the transistor design parameters, such as upsizing). Each design point would be analyzed until a design point that satisfies a target delay penalty (i.e., the increase in critical path delay due to variations) with small energy cost is found. To analyze each design point, computationally expensive models to calculate delay penalties would be exploited. However, this has two major bottlenecks: (1) the time required to calculate delay penalties limits to a number of design points that can be explored, and (2) as the number of processing and design parameters increase, the number of required simulations can increase exponentially leveraging a brute-force search of all design points. In stark contrast, the methodology in [17], implemented for CNFETs, relies on a gradient descent search algorithm, based on key metrics such as delay and noise margin sensitivity information with respect to CNT processing parameters (i.e., parameters to quantify CNT count variations, e.g., due to variations in CNT spacing or due to the presence of m-CNTs), to systematically guide the exploration of design points (example illustration in Fig. 1.6). This drastically decreases the number of design points that need to be explored. Moreover, the delay penalties are calculated >100× faster that previous approaches by leveraging computation approximations and techniques (such as highly-efficient sampling methods and variation-aware timing models). This enables exploration of many more design points, while still maintaining sufficient accuracy to make correct design decisions. An important consequence of efficient search of vast design spaces is that it allows finding more than a single target design point that meets the required specifications. Such efficient search is critical, as it allows multiple acceptable design points to be found. Therefore, if processing constraints result in one design point becoming infeasible, an alternative design point that relaxes the constraint that is difficult to achieve can be chosen. Such a framework then guides experimental work, motivating and setting concrete processing targets to realize digital VLSI systems with emerging nanotechnologies.

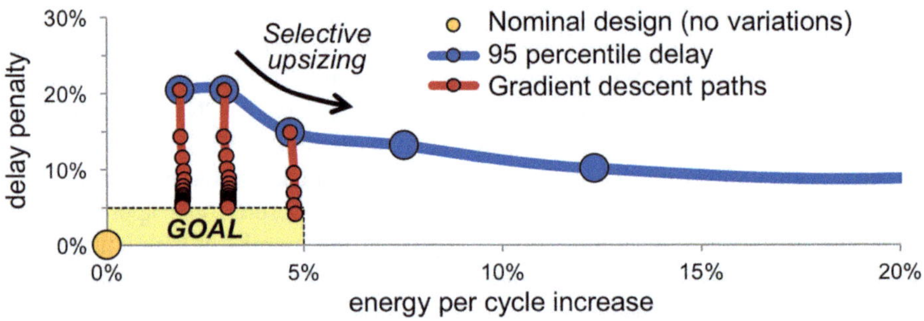


**Fig. 1.6** Gradient descent illustration to overcome CNT variations at the circuit level (by achieving 5% delay penalty with <5% energy increase) [17]. Multiple gradient descent paths (red) are initialized from the initial delay penalty vs. energy trade-off curve (blue) and descend until delay penalty ≤5% (extensive details in [17])

## 1.4 Architectural-Level Implications

While CNFETs promise improved devices and consequently more energy-efficient circuits, they also present a new opportunity in realizing new system architectures. In this section, we describe how the unique properties of CNTs (specifically, their low-temperature fabrication [23]) enable new three-dimensional (3D) integrated systems with ultradense and fine-grained connectivity between vertical circuit layers. Such systems are key to addressing the major sources of inefficiencies in systems today. As an example: system-level performance today is severely constrained by the growing memory-logic communication bottleneck (referring to the limited connectivity and bandwidth connecting processors to off-chip memory). This is particularly true for future abundant-data applications, which are characterized by their reliance on massive amounts of data with little data locality [1]. Even with an improved transistor or memory cell, this growing communication bottleneck would continue to limit system performance. Rather than rely on improved devices alone, revolutionary digital system architectures—such as three-dimensional integration with fine-grained vertical connectivity—are key for providing massive bandwidth between computation and memory.

Three-dimensional (3D) integration, whereby circuits are layered vertically over one another in a three-dimensional fashion, is typically achieved today through chip-stacking. Using a chip-stacking approach, each vertical layer of the 3D chip is prefabricated on separate two-dimensional (2D) substrates. Afterward, the final 2D substrates are physically stacked and bonded on top of one another (either at the die-level or entire wafer-level). Through-silicon vias (TSVs) are used to connect the different vertical layers of the chip (particularly if multiple layers (>2) are used in the 3D chip). Unfortunately, however, these TSVs occupy a large footprint area (due to the limited aspect ratio processing used to define them, typical TSV dimensions are >5 μm diameter with >20 μm TSV pitch). This large footprint and sparse TSV pitch limits the density of vertical connections between the vertical layers of the 3D chip.

This limit in physical connectivity directly translates into an equally-limited data bandwidth between layers, limiting the potential benefits afforded by conventional 3D chip-stacking techniques for 3D integration.

In contrast, *monolithic* 3D integration enables new 3D system architectures, whereby layers of vertically-layered circuits are fabricated directly over one another, all over the same starting substrate. Therefore, no wafer-stacking or wafer-bonding is necessary, and thus TSVs are not required in order to connect vertical layers of the monolithic 3D chip. Rather, conventional back-end-of-line (BEOL) dense interlayer vias (ILVs) can be used to connect vertical layers of the chip, similar to how ILVs are used to connect multiple layers of metal wiring in the BEOL. These ILVs are fabricated with a traditional damascene process (similar to the global metal wiring in chips today), or can leverage advanced interconnect technologies (e.g., emerging nanotechnologies, such as vertically-oriented CNTs, have been proposed as next-generation ILVs). Importantly, these ILVs have the same pitch and dimensions as tight-pitched metal layer vias used for routing in the BEOL, and are therefore orders of magnitudes denser than TSVs. For instance, given the ratio between state-of-the-art TSV and ILV pitch, monolithic 3D integration enables over $>1000\times$ denser vertical connections compared to 3D chip-stacking today. This massive increase in vertical connectivity translates into an equally large increase in the data bandwidth between vertical layers of a chip. When monolithic 3D integration is used to interleave layers of computation, memory access circuitry, and data storage, such massive vertical connectivity results in a massive increase in the logic-memory data bandwidth. This results in significant performance and energy efficiency benefits, due to the true immersion of computation and memory in a fine-grained manner. In particular, monolithic 3D integration systems offer dramatic benefits for a wide range of next-generation abundant-data applications, that is, applications that access and process massive amounts of loosely structured data, and which thus expose the communication bottleneck between computing engines and memory: the *memory wall* [24]. For these abundant-data applications, projections suggest that monolithic 3D systems can result in $\sim 1000\times$ application-level energy efficiency benefits (quantified by the product of application execution time and energy consumption) compared to 2D silicon-based chips [1] (a case study comparing an example monolithic 3D system vs. a 2D baseline is discussed below).

Despite the promise of monolithic 3D integration, it is extremely challenging to realize with today's silicon-based technologies. With chip-stacking, the fabrication of separate 2D substrates is decoupled, as they are fabricated independently of one another. In contrast, for monolithic 3D integration, the bottom layer of the monolithic 3D chip is exposed to the same processing conditions as the upper layers (since those upper layers are fabricated directly over the circuits on the bottom layers). This imposes stringent limitations on the allowable processing for the upper-layer circuits for monolithic 3D integrated circuits, as the processing on the upper layers cannot impact the devices on the bottom-layer circuits. Specifically, all of the fabrication of the upper-layer circuits must be low-temperature (e.g., $<400\,^{\circ}\mathrm{C}$), so as to not damage or destroy the bottom-layer transistors (e.g., dopant profiles, metal–semiconductor junctions at the MOSFET contacts, or high-k gate dielectrics) and metal interconnects (which can be destroyed or diffused at high processing temperatures). Silicon CMOS circuit fabrication can require temperatures >1000 °C, for steps such as dopant activation annealing after implantation for doping. While techniques for fabricating silicon CMOS below 400 °C have been pursued, they suffer from severe inherent limitations. For instance, they can result in transistors with degraded performance, they have only been demonstrated for a maximum of two vertically-stacked layers, or they constrain the BEOL metal to higher-temperature metals (such as Tungsten) which increase BEOL metal resistances compared to today's copper metal wires (or aluminum, for relaxed technology nodes).

In contrast, many emerging nanotechnologies can be fabricated at low processing temperatures <400 °C, well within the thermal budget for monolithic 3D integration [23]. While the CNT growth process requires high temperature (e.g., >800 °C), the CNTs can be transferred onto a target substrate (e.g., monolithic 3D integrated circuit (IC)) through low-temperature CNT transfer processes (described above and shown in Fig. 1.2a) or low-temperature CNT deposition techniques (e.g., solution-based processing [4]). Importantly, these low-temperature processes decouple the high-temperature CNT growth from the final wafer used for circuit fabrication of the monolithic 3D IC. Therefore, the low-temperature processing of CNFETs naturally enables monolithic 3D integrated circuits (alternative transistor options, e.g., with channels built using 2D materials such as $MoS_2$ or black phosphorus [25], can also be used for monolithic 3D ICs, provided that they can be fabricated at low temperatures, although their energy efficiency benefits may not be as significant as CNFETs). Moreover, all of the circuit design techniques to overcome CNT obstacles described previously can be implemented BEOL on upper layers of a monolithic 3D IC (fabrication flowchart shown in Fig. 1.7).

In addition to fabricating the upper layers of computation (or memory access circuitry) at low temperatures, upper layers of memory must also be fabricated within the thermal budget for monolithic 3D computing systems. Conventional trench or stacked-capacitor DRAM and FLASH are therefore not suitable (moreover, the physical height of the device layers must be small enough to enable dense vias,

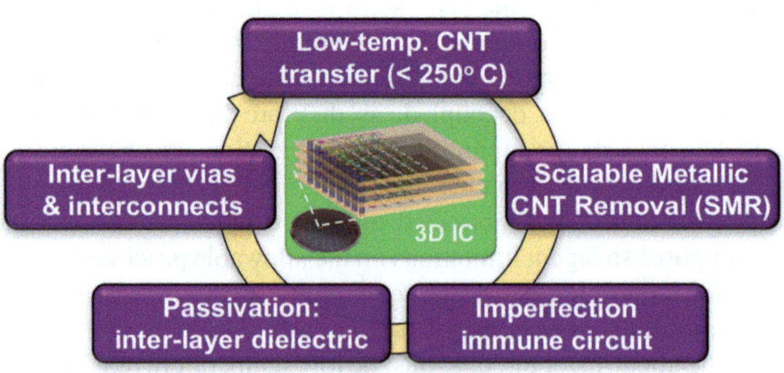


**Fig. 1.7** Monolithic 3D fabrication flowchart (details in [23])

as the aspect ratio of vertical interconnect wires is finite; stacked-capacitor DRAM and stacked control gate FLASH are not suitable for monolithic 3D integration due to this limitation as well). Therefore, emerging memory technologies, such as spin-transfer torque magnetic RAM (STT-MRAM), resistive RAM (RRAM), and conductive-bridging RAM (CB-RAM), are promising options to be integrated as the upper layers of memory [26].

By capitalizing on emerging logic and memory technologies to realize monolithic 3D integration, architectural-level benefits are supplemented by benefits gained at the device level. Therefore, such an approach realizes greater gains than by focusing on improving devices or architectures alone, to realize transformative *nanosystems* that combine advances from across the computing stack: (a) nanomaterials such as carbon nanotubes for high-performance and energy-efficient transistors, (b) high-density on-chip nonvolatile memories, (c) fine-grained 3D integration of logic and memory with ultradense connectivity, (d) new 3D architectures for computation immersed in memory, and e) integration of new materials technologies for efficient heat removal solutions. Figure 1.8 shows an example 3D nanosystem enabled by the logic and memory device technologies mentioned above. The computing elements and memory access circuitry are built

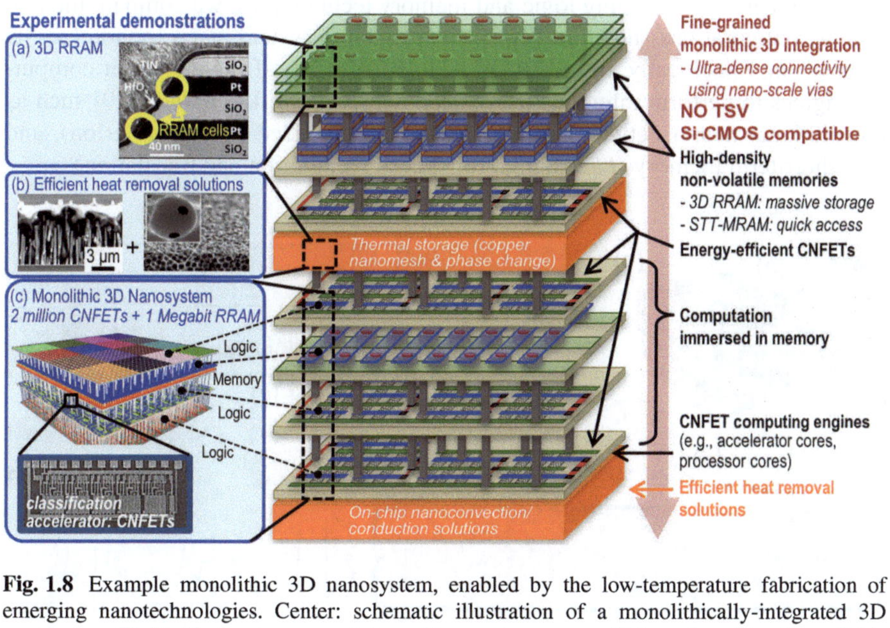


**Fig. 1.8** Example monolithic 3D nanosystem, enabled by the low-temperature fabrication of emerging nanotechnologies. Center: schematic illustration of a monolithically-integrated 3D nanosystem. Right side: key components to enable massive energy efficiency benefits of monolithic 3D nanosystems [1]. Left side: transmission electron microscopy (TEM) and scanning electron microscopy (SEM) images of experimental technology demonstrations; (**a**) TEM of a 3D RRAM for massive storage [26], (**b**) SEMs of nanostructured materials for efficient heat removal: (left) microscale capillary advection and (right) copper nanomesh with phase change thermal storage [27], and (**c**) SEM of a monolithic 3D chip integrating two million CNFETs and 1 Megabit RRAM over a starting silicon substrate [28]

using layers of high-performance and energy-efficient CNFET logic. The memory layers are chosen to best match the properties of the memory technology to the function of the memory subsystem. For instance, STT-MRAM can be used for caches (e.g., L2 cache) to utilize its fast access time, energy retention, and endurance characteristics. RRAM (specifically 3D RRAM [26]) can be used for massive on-chip storage to minimize off-chip communication. The various layers of the 3D nanosystem are connected with conventional fine-grained and dense ILVs, permitting massive connectivity between the vertical layers. Additionally, appropriate interlayer cooling techniques must also be integrated (details below and in [1]).

Recent experimental demonstrations have illustrated the feasibility of this approach. Most recently, a monolithic 3D system, integrating greater than two million CNFETs, >1 Mbit of RRAM, all fabricated over a silicon CMOS substrate, has been experimentally demonstrated (shown in Fig. 1.8c) [28]. While this demonstration highlights the proof-of-concept of 3D nanosystems, these fast maturing demonstrations highlight the promise of exploiting beyond-silicon emerging nanotechnologies to realize improved system architectures (additional experimental demonstrations are shown in Fig. 1.9).

As a case study for quantifying the energy efficiency benefits of 3D nanosystems densely integrating emerging logic and memory technologies, we compare the two system configurations shown in Fig. 1.10: a baseline system and a monolithic 3D nanosystem. Specifically, these systems implement state-of-the-art 16-bit computing engines to perform inference using deep neural networks (*DNNs*) [29] such as convolutional neural networks (*CNNs*, e.g., for embedded computer vision), and long short-term memory (*LSTM*, e.g., for speech recognition and translation).

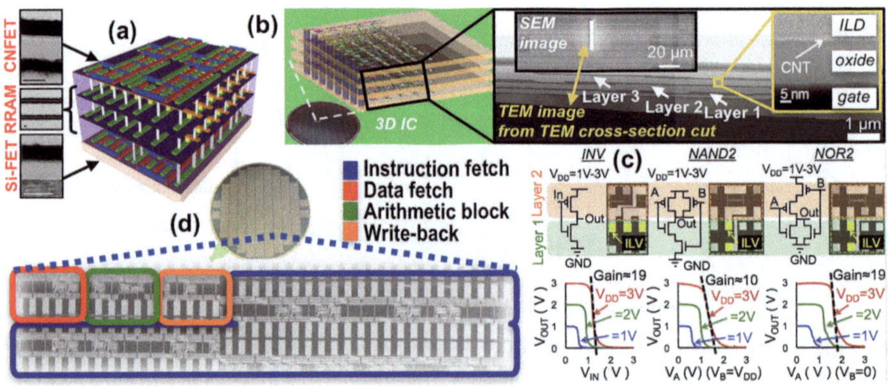


**Fig. 1.9** Larger-scale experimental CNFET circuit and 3D nanosystem demonstrations. (**a**) Four-layer monolithic 3D nanosystem with CNFET + RRAM layers over the bottom layer of silicon FETs [7]; (**b**) three-layer IC implementing static complementary CNFET logic gates (i.e., with both PMOS and NMOS CNFETs), with circuit schematics and voltage transfer curves shown in (**c**) [11]; (**d**) complete microprocessor built entirely out of CNFETs [14]

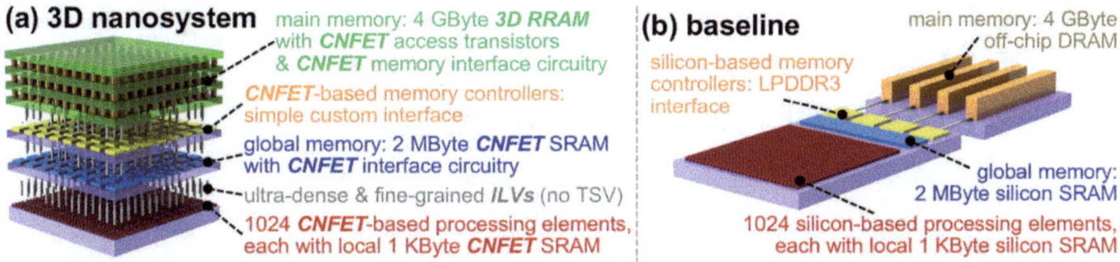


(c)

| sub-system | 3D nanosystem | baseline |
|---|---|---|
| main memory | •4 GBytes on-chip 3D RRAM with CNFET access circuits<br>•64 channels (512 bits/channel) simple custom interface<br>•2.3 ns read, 16 ns write<br>•0.15 pJ/bit read/write | •4 GBytes off-chip DRAM<br>•4 channels (32 bits/channel) LPDDR3<br>•75 ns read/write<br>•15 pJ/bit read/write |
| global memory | •2 MByte CNFET SRAM<br>•1.7 ns read/write<br>•0.15 pJ/bit read/write | •2 MByte silicon SRAM<br>•4 ns read/write<br>•0.32 pJ/bit read/write |
| local memory (for each processing element) | •1 KByte CNFET SRAM per processing element<br>•0.7 ns read/write<br>•0.076 pJ/bit read/write | •1 KByte silicon SRAM per processing element<br>•2 ns read/write<br>•0.23 pJ/bit read/write |
| compute | •1024 16-bit processing elements<br>•1.5 GHz clock speed<br>•0.6 pJ/operation | •1024 16-bit processing elements<br>•0.5 GHz clock speed<br>•1.9 pJ/operation |
| technology | •28 nm CNFET CMOS | •28 nm silicon CMOS |

**Fig. 1.10** System configurations used to quantify EDP benefits of monolithic 3D nanosystems. (**a**) Monolithic 3D nanosystem, (**b**) baseline, and (**c**) summary of architecture parameters and performance metrics for each subsystem

Both systems are designed using the same values for the architectural parameters shown in Fig. 1.10c. In particular, each system contains an array of 1024 processing elements (*PEs*, organized as $2 \times 2$ clusters, with $16 \times 16$ PEs per cluster); each PE comprises a 16-bit multiply and accumulate (*MAC*) unit to perform compute operations, and a local 1-kB SRAM (to store temporary variables). A 2-MB global memory is shared by all PEs, and a 4-GB main memory is used to store the *DNN* model and input data used during inference. The difference between the two systems is in the physical design, including the FET technologies, memories, memory access circuits, and the interfaces between processing elements and memory; these directly affect system-level performance metrics such as read/write access energy/latency, energy per operation, and clock frequency, which are also provided in Fig. 1.10c. These metrics are extracted from physical designs (following place-and-route and parasitic extraction) using 28 nm node process design kits (*PDKs*) for both Si- and CNFET-based technologies (CNFET PDKs are developed using the tools in [30]). We use ZSim for architectural-level simulations [12], and the trace-based simulation framework in [29] to analyze accelerator cores. We perform thermal simulations using 3D-ICE [31].

Our analysis shows that for abundant-data applications running on accelerator cores, 3D nanosystems offer EDP benefits in the range of $1000\times$ compared to computing systems today (Fig. 1.11a). These results are consistent with EDP benefits for general-purpose computing engines analyzed in [1]. Note that, the example 3D nanosystem configuration (Fig. 1.10a) and applications analyzed here are demonstrations of 3D nanosystem energy efficiency benefits, although a wide

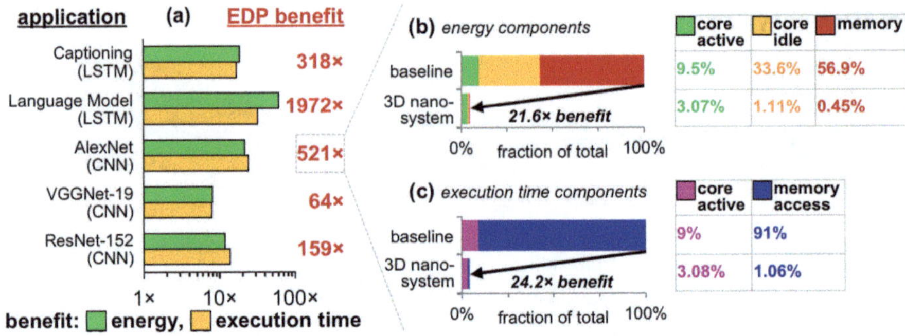


**Fig. 1.11** (**a**) 3D nanosystem energy efficiency benefits for convolutional neural networks (*CNNs*) and long short-term memory (*LSTM*) abundant-data applications (corresponding to the configurations in Fig. 1.10). EDP benefits are typically more significant for applications whose execution time and energy consumption are more highly constrained by memory accesses (e.g., Language Model). Relative energy consumption (**b**) and execution time (**c**) for 3D nanosystem vs. baseline, for a representative abundant-data application (AlexNet)

range of 3D nanosystem configurations are generally applicable to alternative architectures, application domains, and workloads.

Fig. 1.11b, c provides insight into the sources of such massive EDP benefits, shown for the AlexNet application (a representative abundant-data application for CNN workloads, with multiple neural network layers comprising convolutional, pooling, and fully-connected layers for inference). The limited connectivity to off-chip DRAM increases the execution time significantly in the 2D baseline, with 91% of the total time spent in memory accesses (for typical CNN workloads). As a result, the accelerator cores waste considerable energy (33.6% of the total energy consumption) due to leakage power dissipated while stalling for memory accesses (i.e., *core idle* energy in Fig. 1.11). In contrast, the 3D nanosystem configuration leverages wide data buses (i.e., with many bits in parallel, enabled by ultradense and fine-grained monolithic 3D integration), together with quick access to on-chip 3D RRAM, reducing the cumulative time spent accessing memory by 85.8× (due to enhanced memory bandwidth and access latency). Not only does the reduced memory access time contribute to 24.2× application execution time speedup, but also it reduces core idle energy (by 30.3×) since less time is spent stalling during memory accesses. Additional energy benefits include 126× reduced dynamic energy of memory accesses (for on-chip 3D RRAM vs. off-chip DRAM), and 2.9× reduced dynamic energy of accelerator cores executing compute operations (using energy-efficient CNFETs). In total, the application energy is reduced by 21.6× (with simultaneous 24.2× execution time speedup) resulting in 521× EDP benefit.

Furthermore, 3D nanosystems achieve these significant benefits while maintaining similar average power density (~10 W/cm$^2$ of footprint) and peak operating temperature (~35 °C) as the baseline system; as shown in Fig. 1.8, the computing engines, which account for most of the power consumption, are implemented only on the bottom layer, whereas the upper layers consist of memory access circuits and memories (relatively lower power). Thus, the average power density for the 3D nanosystem is 9.5 W/cm$^2$ and the peak operating temperature is 35 °C (vs. 10.4 W/cm$^2$ and 36 °C for the baseline), even without integrating advanced heat removal solutions (e.g., on upper circuit layers as shown in Fig. 1.8). Potential heat removal solutions include (but are not limited to) 2D materials with improved heat conduction [32], and advanced convective structures such as copper nanowire arrays and copper-based nanostructures [27], which not only can manage heat flux densities from 10 to 5000 W/cm$^2$ [33] but also can encapsulate phase change materials (e.g., paraffin) to suppress thermal transients. The capability to meet system-level temperature constraints despite higher power densities (e.g., for computing engines integrated on multiple layers) can enable additional EDP benefits. Moving forward, opportunities for even larger benefits exist when making additional modifications across the computing stack, for example, through rethinking algorithms, co-design of hardware and software, brain-inspired architectures, domain-specific languages, compilers targeting computation immersed in memory, and new computing paradigms.

## 1.5 Outlook

It should be clear to the reader that emerging nanotechnologies promise to revolutionize computing by enabling significant gains in EDP. Yet it should also be clear that to do so, circuit-level and architectural-level considerations of these emerging nanotechnologies must be taken into account. On the one hand, doing so is key for overcoming their inherent imperfections and variations in order to realize working systems. On the other hand, leveraging the unique properties of these emerging nanotechnologies to realize novel system architectures allows device-level benefits to be combined with architectural-level benefits, realizing EDP gains that far exceed the sum of their individual parts. Using CNTs as a case study, circuit-level considerations allow one to design circuits that are immune to the major challenges facing a VLSI CNFET technology, such as mis-positioned and metallic CNTs. Moreover, exploiting low-temperature fabrication of CNFETs (as well as the low-temperature fabrication of several beyond-silicon emerging memory technologies) enables monolithic 3D chips, with fine-grained interleaved layers of computing, memory access circuitry, and data storage. Such new architectures, which are enabled by using these emerging nanotechnologies, are key to enabling the new generation of impactful abundant-data applications. While challenges still exist, this vision is quickly morphing from ideas to reality.

## References


1. M.M.S. Aly et al., Energy-efficient abundant-data computing: The N3XT 1,000. Computer **48**(12), 24–33 (2015)
2. J. Zhang et al., Robust digital VLSI using carbon nanotubes. IEEE Trans. Comput. Aided Des. Integr. Circuits Syst. **31**(4), 453–471 (2012)
3. C. Qiu et al., Scaling carbon nanotube complementary transistors to 5-nm gate lengths. Science **355**(6322), 271–276 (2017)
4. G.J. Brady et al., Quasi-ballistic carbon nanotube array transistors with current density exceeding Si and GaAs. Sci. Adv. **2**(9), e1601240 (2016)
5. A.D. Franklin et al., Sub-10 nm carbon nanotube transistor. Nano Lett. **12**(2), 758–762 (2012)
6. M.M. Shulaker et al., Sensor-to-digital interface built entirely with carbon nanotube fets. IEEE J. Solid State Circuits **49**(1), 190–201 (2014)
7. M.M. Shulaker et al., Monolithic 3D integration of logic and memory: Carbon nanotube FETs, resistive RAM, and silicon FETs, in *2015 IEEE International Electron Devices Meeting (IEDM)* (IEEE, 2015), pp. 27.4.1–27.4.4
8. M.M. Shulaker et al., High-performance carbon nanotube field-effect transistors, in *2014 IEEE International Electron Devices Meeting (IEDM)* (IEEE, 2014)
9. R. Park et al., Hysteresis-free carbon nanotube field-effect transistors. ACS Nano **11**(5), 4785–4791 (2017)
10. T. Srimani, G. Hills, M.D. Bishop, U. Radhakrishna, A. Zubair, R.S. Park, Y. Stein, T. Palacios, D. Antoniadis, M.M. Shulaker, Negative capacitance carbon nanotube FETs. IEEE Electron Device Lett **39**(2), 304–307 (2017). https://doi.org/10.1109/LED.2017.2781901
11. H. Wei et al., Monolithic three-dimensional integration of carbon nanotube FET complementary logic circuits, in *2013 IEEE International Electron Devices Meeting (IEDM)*, (IEEE, 2013), pp. 511–514
12. D. Sanchez et al., ZSim: fast and accurate microarchitectural simulation of thousand-core systems, in *ISCA '13*, (ACM, New York, 2013)
13. M. Shulaker et al., Experimental demonstration of a fully digital capacitive sensor interface built entirely using carbon-nanotube FETs. IEEE Int. Solid State Circuits Conf. **56**, 112–113 (2013)
14. M.M. Shulaker et al., Carbon nanotube computer. Nature **501**(7468), 526–530 (2013)
15. N. Patil et al., Design methods for misaligned and mispositioned carbon-nanotube immune circuits. IEEE Trans. Comput. Aided Des. Integr. Circuits Syst. **27**(10), 1725–1736 (2008)
16. N. Patil et al., VMR: VLSI-compatible metallic carbon nanotube removal for imperfection-immune cascaded multi-stage digital logic circuits using carbon nanotube FETs, in *2009 IEEE International Electron Devices Meeting (IEDM)* (IEEE, 2009)
17. G. Hills et al., Rapid co-optimization of processing and circuit design to overcome carbon nanotube variations. IEEE Trans. Comput. Aided Des. **34**(7), 1082–1095 (2015)
18. N. Patil et al., Wafer-scale growth and transfer of aligned single-walled carbon nanotubes. IEEE Trans. Nanotechnol. **8**(4), 498–504 (2009)
19. J. Zhang et al., Probabilistic analysis and design of metallic-carbon-nanotube-tolerant digital logic circuits. IEEE Trans. Comput. Aided Des. Integr. Circuits Syst. **28**(9), 1307–1320 (2009)
20. M.M. Shulaker et al., Efficient metallic carbon nanotube removal for highly-scaled technologies, in *2015 IEEE International Electron Devices Meeting (IEDM)* (IEEE, 2015)
21. J. Zhang et al., Carbon nanotube circuits in the presence of carbon nanotube density variations, in *46th Annual Design Automation Conference (DAC)*, (IEEE, 2009) pp. 71–76
22. J. Zhang et al., Carbon nanotube correlation: promising opportunity for CNFET circuit yield enhancement, in *47th Annual Design Automation Conference (DAC)* (IEEE, 2010), pp. 889–892
23. H. Wei et al., Monolithic three-dimensional integrated circuits using carbon nanotube FETs and interconnects, in *2009 IEEE International Electron Devices Meeting (IEDM)* (IEEE, 2009), pp. 577–580

24. P. Stanley-Marble et al., Pinned to the walls – Impact of packaging and application properties on the memory and power walls, in *ISLPED 2011* (IEEE, 2011)
25. G. Fiori et al., Electronics based on two-dimensional materials. Nat. Nanotechnol. **9**(10), 768–779 (2014)
26. H.Y. Chen et al., HfOx based vertical resistive random access memory for cost-effective 3D cross-point architecture without cell selector, in *2012 IEEE International Electron Devices Meeting (IEDM)* (IEEE, 2012)
27. M.T. Barako et al., Thermal conduction in vertically aligned copper nanowire arrays and composites. ACS Appl. Mater. Interfaces **7**(34), 19251–19259 (2015)
28. M.M. Shulaker et al., Three-dimensional integration of nanotechnologies for computing and data storage on a single chip. Nature **547**(7661), 74–78 (2017)
29. M. Gao et al., TETRIS: scalable and efficient neural network accelerator with 3D memory, in *ASPLOS*, (ACM, New York, 2017)
30. G. Hills, Variation-aware nanosystem design kit, https://nanohub.org/resources/22582
31. A. Sridhar et al., 3D-ICE: A compact thermal model for early-stage design of liquid-cooled ICs. IEEE Trans. Comput. **63**(10), 2576–2589 (2014)
32. E. Pop et al., Thermal properties of graphene: fundamentals and applications. MRS Bull. **37**(12), 1273–1281 (2012)
33. M. Fuensanta et al., Thermal properties of a novel nanoencapsulated phase change material for thermal energy storage. Thermochim. Acta **565**, 95–101 (2013)

# Chapter 2
# Functionality-Enhanced Devices: From Transistors to Circuit-Level Opportunities

**Giovanni V. Resta, Pierre-Emmanuel Gaillardon, and Giovanni De Micheli**

## 2.1 Introduction

Since the invention of the complementary metal-oxide-semiconductor field-effect transistor (CMOS-FET), the main drive of the semiconductor industry has been the downscaling of the devices, exemplified by Moore's law, which allowed to greatly reduce the cost per transistor, by increasing the number of transistors per unit area. Conventional CMOS logic circuits are based on doped, *n* or *p*, unipolar devices. The doping is introduced by ion-implantation: Boron atoms lead to *p*-type FET, while Arsenic is used to realize *n*-type devices. The doping process irreversibly sets the transistor polarity by providing an excess of majority carriers, electrons for *n*-doping, and holes for *p*-doping, and moreover, allows the creation of Ohmic contacts at source and drain. With physical gate length as small as 14 nm in modern devices, doping processes have become increasingly complicated to control. Very abrupt doping profiles are needed, and due to random fluctuations in the number of dopants in the channel, that cause an undesired shift in the threshold voltage of the FET, device variability has been increasing. Moreover, short channel effects have already forced the transition to a 3D geometry (Fin-FETs) in order to improve the gate control over the transistor channel. As further downscaling has become increasingly

G. V. Resta (✉) · G. De Micheli
Integrated System Laboratory (LSI), École Polytechnique Fédérale de Lausanne (EPFL), Lausanne, Switzerland
e-mail: giovanni.resta@epfl.ch; giovanni.demicheli@epfl.ch

P.-E. Gaillardon
Laboratory of NanoIntegrated System (LNIS), University of Utah, Salt-Lake City, UT, USA
e-mail: pierre-emmanuel.gaillardon@utah.edu



R. O. Topaloglu, H.-S. P. Wong (eds.), *Beyond-CMOS Technologies for Next Generation Computer Design*, https://doi.org/10.1007/978-3-319-90385-9_2

more expensive in terms of fabrication and facility costs,[1] an alternative path to Moore's Law has been proposed, that, instead of focusing on further decreasing the transistor dimensions, aims at increasing their functionality per unit area. Using the words of Shekhar Borkar, former head of Intel's microprocessor technology research: "Moore's law simply states that user value doubles every 2 years", and in this form, the law will continue as long as the industry will be able to keep increasing the device functionality [1].

This alternative scaling approach is based on the concept of multiple-independent-gate (MIG) FETs which introduce novel functionalities at the device scale and innovative circuit-level design opportunities. MIG-FETs are a novel class of devices with multiple gate regions that independently control the switching properties of the device. The key enabler of such concept is the exploitation and control of the inherently ambipolar behavior, also known as ambipolarity, of Schottky-barrier transistors (SB-FETs). Here we will only focus on SB-FETs as the building block of MIG-FETs, and for a more general coverage of Schottky-barrier physics and applications, the interested reader can refer to [2]. Ambipolarity arises in SB-FETs since the conduction property is determined by the bands alignment at the source and drain contacts, and by the gate-induced modulation of the Schottky barriers. Both electrons and holes can be injected in the intrinsic device channel depending on the voltage applied to the gate. Ambipolarity is usually considered a drawback in standard CMOS devices, since it allows the conduction of both charge carriers in the same device, deteriorating the OFF-state of the transistor. As a result, ambipolarity is suppressed thanks to the doping process that creates strictly unipolar devices. In MIG-FETs instead, the device polarity is not set during the fabrication process, and can be dynamically changed thanks to the additional gate electrodes, which modulate the SB at source and drain and therefore enable to select the carrier type injected in the device. In principle, no dopant implantation is required in the fabrication process of the device, thus there is no need for the separate development of *n*- and *p*-type devices, to the benefit of fabrication simplicity and device regularity. The gate-induced modulation of the SB enables dynamic control of the polarity and of the threshold voltage of the device at run-time. Moreover, with the peculiar gates configuration, the subthreshold slope (SS) can be controlled when increasing the $V_{DS}$ applied to the device. In particular, a dynamic control of the transistor polarity enables the realization of compact binate operators, such as 4-transistor XOR operator, that can be used as the building block to realize circuits with higher computational density with respect to CMOS.

The chapter is organized as follows. In Sect. 2.2, MIG devices realized with silicon nanowires and silicon Fin-FETs, which are appealing for near-term scaling, are presented. Particular focus is given to the explanation of the main operation principle and to the different operation modes of such MIG-FETs. Section 2.3 is focused on long-term scaling opportunities for beyond-CMOS electronics and different promising materials for the realization of the next-generation MIG-FETs

[1]For example, a state-of-the-art research clean room, such as the one of IMEC research center in Belgium, calls for more than €1 billion investment.

are presented. Finally, Sect. 2.4 illustrates the circuit design opportunities allowed by the use of MIG-FETs, such as compact arithmetic logic gates and novel design methodology. The chapter is concluded in Sect. 2.5 with a brief summary.

## 2.2 Multiple-Independent-Gate Silicon Nanowires Transistors

As introduced in Sect. 2.1, MIG FETs are devices whose conduction properties can be dynamically controlled via additional gate terminals. These additional gates, usually referred to as polarity (or program) gates (PG), act on the Schottky barriers present at the drain and source contact and allow to exploit different functionality and selecting different operation modes. Here we focus on double-independent-gate (DIG) devices, with different gates configurations, for polarity and subthreshold swing control mode, Sects. 2.2.1 and 2.2.2, and then, as a natural evolution, we highlight three-independent-gate (TIG) transistors, which additionally allow the control of the threshold voltage of the device, Sect. 2.2.3.

### *2.2.1 Polarity Control*

The first experimental reports on silicon-nanowires double-independent-gate (DIG) devices were presented in [3, 4] and both adopted a double-gate geometry with a top gate acting as control gate in the central region of the channel and the wafer substrate acting as program gate at the source and drain contacts. These first reports paved the way for the realization of more advanced design with Ω-gates for both control and program terminals first realized in [5] and optimized in [6], as shown in Fig. 2.1a, b. The devices were fabricated using a bottom-up approach, with a single silicon nanowire grown and transferred on a final substrate where two Ω-gates were then patterned. In this reconfigurable device (RFET), one Schottky junction is controlled to block the undesired charge carrier type, while the other junction controls the injection of the desired carriers into the channel, which is ungated in the central region. In the *p*-type configuration, shown in Fig. 2.1c, the program gate (PG) is set to a negative value and blocks the injection of electrons from the drain contact. The ON/OFF status of the device is then determined by the voltage applied to the control gate (CG) at source. In a similar fashion, when PG is kept at a positive voltage, it blocks hole injection from the drain, while the CG acting at source controls the injection of electrons, Fig. 2.1c. It should also be noted that with this gate configuration, in order to switch the device from *p*-type to *n*-type behavior, both the polarization of the PG at drain contact and $V_{DS}$ and have to be reversed. The experimental transfer characteristics for both *p*- and *n*-type operation of the RFET are reported in Fig. 2.1d and show extremely low leakage current and negligible hysteresis.

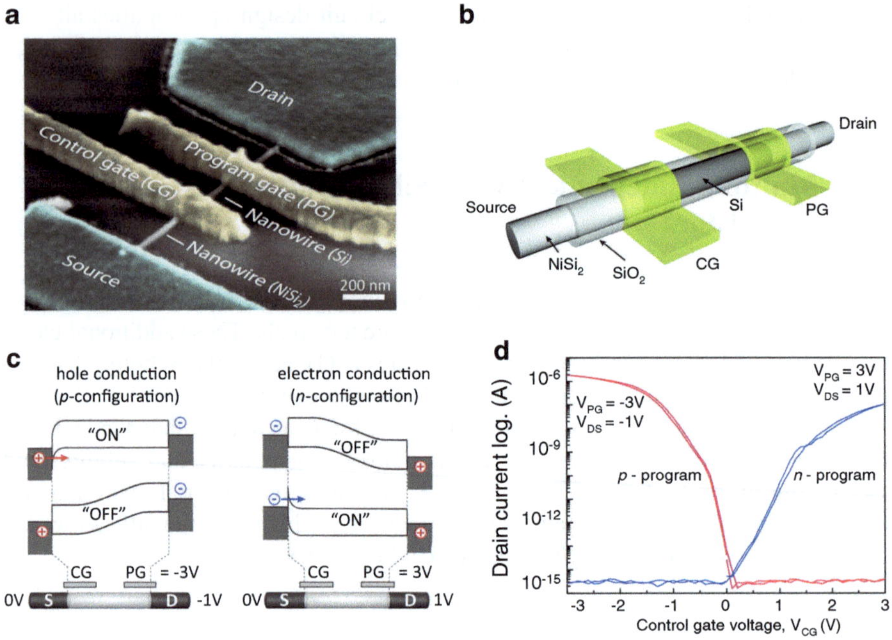


**Fig. 2.1** The reconfigurable silicon nanowire FET with independent gates. (**a**) Tilted SEM view of a fabricated device. (**b**) Schematic description of the device structure, highlighting the different materials and terminals. (**c**) Schematic band diagram of the different operation state of the reconfigurable nanowire FET. Arrows indicate electron (*n*-type) and hole (*p*-type) injection from contacts to the channel. The voltage values of all terminal are reported. (**d**) Measured transfer characteristics of the reconfigurable nanowire FET. The characteristics are plotted in both forward and backward sweeping and show insignificant hysteresis. Adapted with permission from Heinzig et al. [5, 6]. Copyright (2012) and (2013), American Chemical Society

Although the devices reported in [3–6] show the great potential of reconfigurable transistors, in order to realize a viable alternative to standard CMOS technology, a scalable top-down fabrication process that doesn't require using bottom-gate electrodes or complex transfer procedure of pre-grown nanowires is necessary. The first experimental demonstration of a top-down fabrication method for silicon-nanowires polarity-controllable devices was reported by De Marchi et al. [7] using vertically stacked nanowires, which represent a natural evolution of current Fin-FET technology, and provide greater electrostatic control on the channel, thanks to the gate-all-around (GAA) structure. The fabrication process starts from a lightly p-doped ($10^{15}$ $cm^{-3}$) silicon-on-insulator (SOI) wafer, where the vertically-stacked silicon nanowires are defined using a Bosch process based on deep reactive ion etching (DRIE) [7, 8]. The nominal length of the defined nanowires is 350 nm with a diameter of 50 nm, while the typical vertical spacing between them is 40 nm. 15 nm of $SiO_2$ is then formed via thermal oxidation of the silicon nanowires to act as gate oxide. The polarity gates are then patterned on conformal deposited polycrystalline silicon. A second thermal oxidation is performed in order to ensure

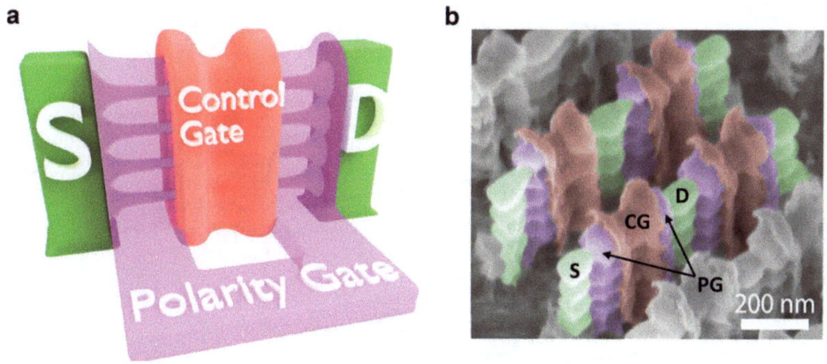


**Fig. 2.2** Double-independent-gates silicon nanowires. (**a**) 3D conceptual view of the vertically-stacked silicon nanowires FETs. (**b**) Tilted SEM micrograph of the fabricated devices. The SEM view shows several devices fabricated with regular arrangement. For a single device the main terminals are indicated, and the same terminals can also be visually identified on the other transistors. Adapted with permission from De Marchi et al. [7, 8]. Copyright (2012) and (2014) IEEE

the separation between the polarity gates and the control gate, which is patterned on polycrystalline silicon in a self-aligned way with respect to the polarity gates. At the end of the process, considering the silicon consumed during the oxidation process, the diameter of the nanowires has been reduced to around 30 nm. After the definition of the nanowires and the gates, a nickel layer is sputtered and then annealed to form nickel silicide contacts at source and drain. The annealing temperature and duration are controlled in order to ensure the formation of the proper $Ni_1Si_1$ phase which provides a near mid-gap work function (∼4.8 eV) and low resistivity [9, 10]. The process can be further optimized to replace the $SiO_2$ with a high-*k* dielectric and to more aggressively scale both the oxide thickness and the channel length. A 3D schematic view and a SEM micrograph of the final fabricated device are shown in Fig. 2.2. As can be appreciated in Fig. 2.2, the device geometry is different from the one reported in [5, 6], as now the polarity gate is acting simultaneously on both source and drain Schottky junctions, while the CG is now acting in the central region of the channel. With this particular gate configuration, the device polarity, *n*- or *p*-type, can be dynamically set by only the voltage applied to the PG, without having to invert the applied $V_{DS}$. This new gate configuration will enable tremendous advantages at the circuit level, as it will be further elucidated in Sect. 2.4. The device has four regions of operation, corresponding to the four combinations of high/low bias voltages applied on the two gates, namely CG and PG. In order to clearly illustrate the operation principle and the band structure relative to each operation mode, we refer to Fig. 2.3:

1. ***p-type ON state:*** For low voltage values (logic ‘0’) of the PG, the band bending at source and drain allows for holes conduction in the channel, which are injected through the thin tunneling barrier at drain, while electron conduction is blocked by the thick Schottky barrier at source (see Fig. 2.3a). In this configuration, the CG is kept at a low bias allowing for holes conduction through the channel.

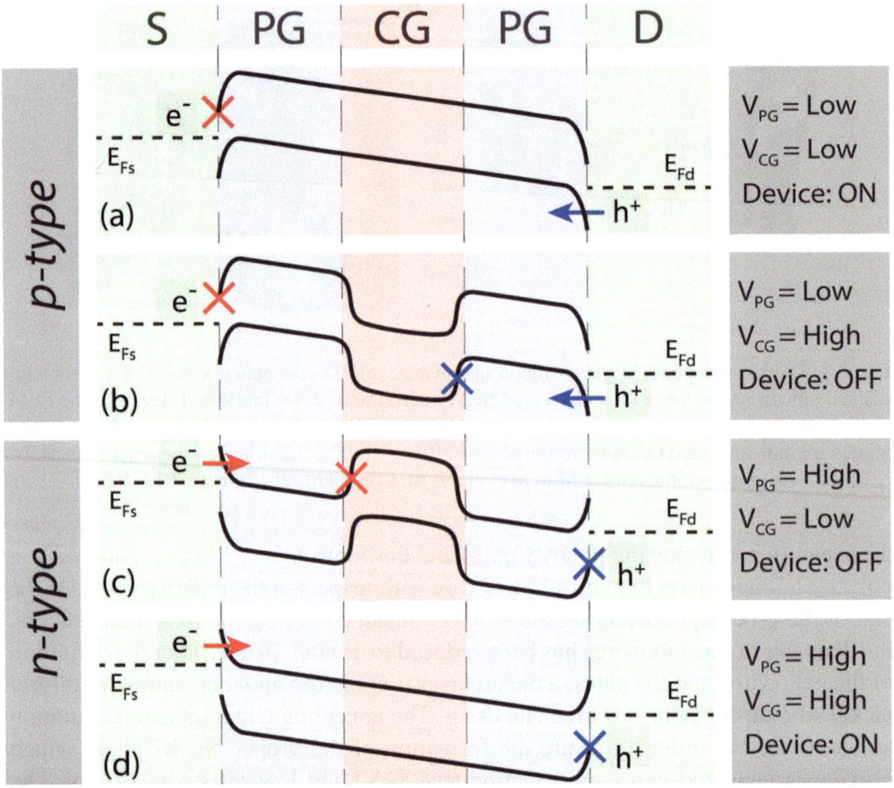


**Fig. 2.3** Conceptual band diagrams for the DIG-FET in the different operation modes. Adapted with permission from De Marchi et al. [7]. Copyright (2012) IEEE

2. ***p-type OFF state:*** To switch off hole conduction, the voltage applied to the CG is inverted to high values (logic '1'). The potential barrier created in the central region of the channel does not allow for hole conduction, while electron conduction is still blocked by the Schottky barrier at source (see Fig. 2.3b).
3. ***n-type OFF state:*** For high voltage values of the PG, the band bending at source and drain allows for injection of electrons in the channel through the thin tunneling barrier at the source contact, while hole conduction is blocked by the thick Schottky barrier at drain. Similarly to the *p*-type OFF state, electron conduction is blocked by the potential barrier created by the CG, which is now kept at logic '0' (see Fig. 2.3c).
4. ***n-type ON state:*** The bias on the PG gate is not changed with respect to the *n*-type OFF state, and conduction of electrons is enabled by applying a high voltage value to CG. In this bias configuration, no potential barrier is created in the CG region, and electrons are able to flow from source to drain (see Fig. 2.3d).

The device transfer characteristics are presented in Fig. 2.4 and show the polarity-controllable behavior of the fabricated DIG-SiNWFET. The device showed

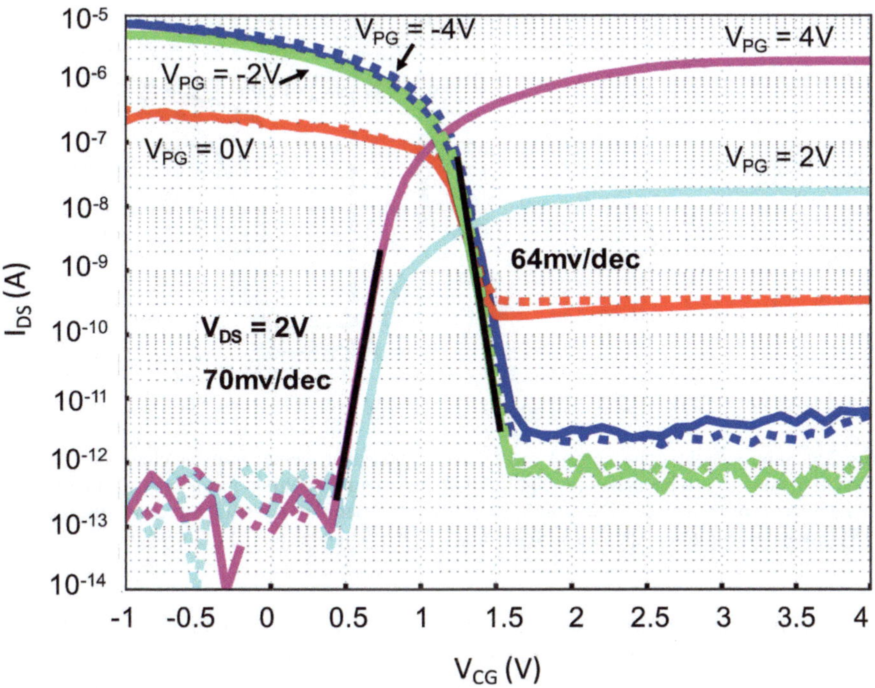


**Fig. 2.4** DIG-FET transfer characteristics obtained at different bias voltage of the PG gate, sweeping the CG voltage. The device shows controllable unipolar behavior with subthreshold slopes for both *n*- and *p*-type conduction branches below 70 mV/dec. $I_{ON}/I_{OFF}$ ratios for both polarities are above $10^6$. Adapted with permission from De Marchi et al. [7]. Copyright (2012) IEEE

subthreshold slope (SS) of 64 mV/dec with $I_{ON}/I_{OFF}$ ratio of $10^6$ for *p*-type conduction and SS of 70 mV/dec with $I_{ON}/I_{OFF}$ ratio of $10^7$ for *n*-type conduction in the same device.

## 2.2.2 *Subthreshold Slope Control*

Using the same gating configuration described in [7], we can also exploit the possibility of controlling the subthreshold slope (SS) of the device and operate with sub-60 mV/dec subthreshold swings [11]. This feature can be achieved by increasing the $V_{DS}$ voltage in order to create enough electric field in the channel to trigger weak-impact ionization [12], and thanks to a positive-feedback mechanism provided by the potential well created by the CG region. This operation regime was first demonstrated on DIG-FETs in [11] using a DIG-FinFET device, but the same working principle is applicable to silicon-nanowires FETs.

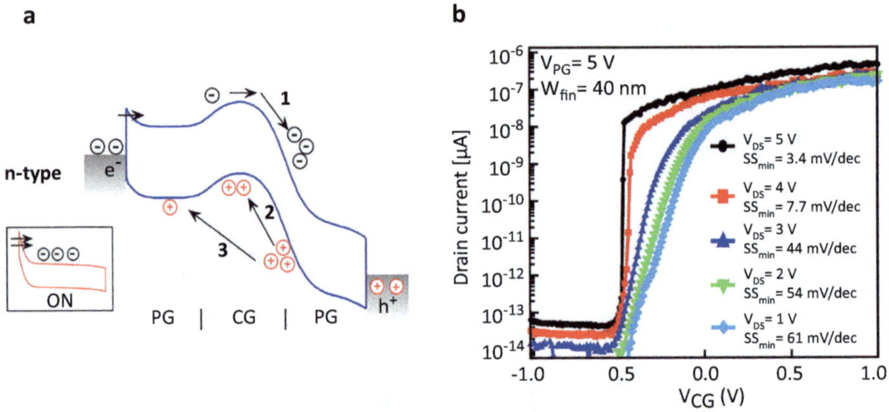


**Fig. 2.5** Subthreshold-slope control mechanisms and experimental transfer characteristics. (**a**) Band diagram of *n*-type behavior highlighting the main switching mechanisms. (**b**) Transfer characteristics of the device in the subthreshold control operation mode measured at different $V_{DS}$ and at room temperature. Adapted with permission from Zhang et al. [11]. Copyright (2014) IEEE

The operation principle of the *n*-type device is depicted in Fig. 2.5a, with a schematic band diagram of the device structure, and the same operation principle applies in the case of *p*-type operation. For *n*-type behavior, corresponding to a positive voltage ('1') applied to the PG, electrons are injected in the channel from the source contact. When sweeping $V_{CG}$ from logic '0' to logic '1', the full transition between OFF and ON states occurs when the threshold value $V_{TH}$ is reached. At this point, electrons flowing in the channel gain enough energy to trigger weak-impact ionization, generating a greater number of electron–hole pairs, see step 1 in Fig. 2.5a. The generated electrons drift to the drain, thanks to the high electric field in the channel, while holes accumulate in the potential well induced by the CG in the central region of the channel (Fig. 2.5 step 2). A net positive charge is thus created in this region, which lowers the potential barrier in the channel, providing more electrons for the impact ionization process. A positive feedback mechanism is thus created: The generation of more electron–hole pairs leads to a greater amount of holes accumulating under the potential well which continue to lower the potential barrier, providing even more electrons injected in the channel [13]. Parts of the generated holes are swept toward the source, increasing the hole density in the PG region at source and thinning the Schottky barrier at source even further. In the meantime, the potential well under the CG gate is kept until the final ON state (Fig. 2.5 step 3). The positive feedback mechanism described ultimately enables the steep device turn-on as it is able to provide a faster modulation of the Schottky barriers at source and drain. The operation for *p*-type behavior is similar but for $V_{PG}$ set to logic '0'. The positive feedback mechanism allows to break the theoretical limit of 60 mV/dec subthreshold swing and, as shown in Fig. 2.5b, for $V_{DS} = 5$ V the minimum subthreshold slope measured is 3.4 mV/dec and remains below 10 mV/dec over five decades of current. However, further research on the operation principle and scaling of the device dimensions would be needed to reduce the $V_{DS}$ necessary to achieve steep subthreshold slope operation.

### 2.2.3 Threshold Voltage Control

Control over the threshold voltage ($V_{TH}$) of the device can be achieved thanks to the separate modulation of the two Schottky barriers at drain and source. To do so, each device has now three-independent-gates (TIG), namely polarity gate at source ($PG_S$), control gate (CG), and polarity gate at drain ($PG_D$) [14, 15], as depicted in Fig. 2.6. The experimental demonstration of dual-$V_{TH}$ operation was done on TIG vertically-stacked SiNWFETs built with the same top-down process described in Sect. 2.2.1, with the only key difference that the two program gates were not connected together. It is straightforward to see that this device concept embeds the polarity-control function described in Sect. 2.2.1, which is achieved when the same voltage is applied on $PG_S$ and $PG_D$. A total of eight operation modes are possible by independently biasing the three gates to either '0' (GND) or '1' ($V_{DD}$). We can identify two ON states, *n*- and *p*-type, two low-$V_{TH}$ OFF states, two high-$V_{TH}$ OFF states, and two uncertain states which are not going to be used. Referring to the band diagrams reported in Fig. 2.7, where all the relevant operation modes are depicted, we have

1. ***ON states:*** As shown in Fig. 2.7a, b, when $PG_S = PG_D = CG$, one of the Schottky barriers is thin enough to allow injections of holes from the drain (*p*-type) or of electrons from the source (*n*-type), and there is no potential barrier created by the CG. Remarkably, and in contrast to multi-threshold CMOS devices, where the shift in threshold voltage is achieved by changing the channel doping, the ON state remains the same for both low- and high-$V_{TH}$ operation modes.

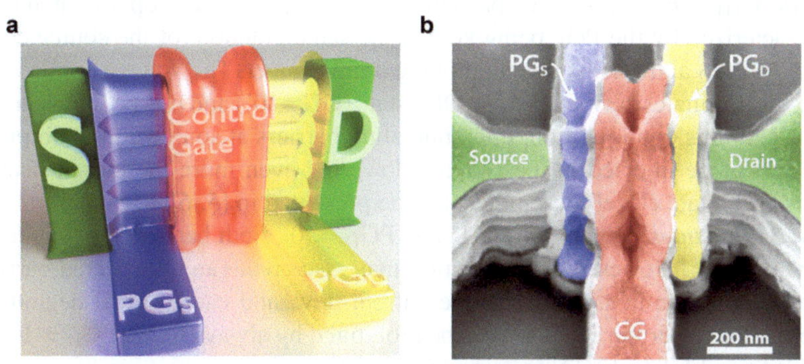


**Fig. 2.6** Three-independent-gate FET. (**a**) Schematic structure of the device. (**b**) Tilted SEM view of the fabricated device, with gates and contacts marked. Adapted with permission from Zhang et al. [15]. Copyright (2014) IEEE

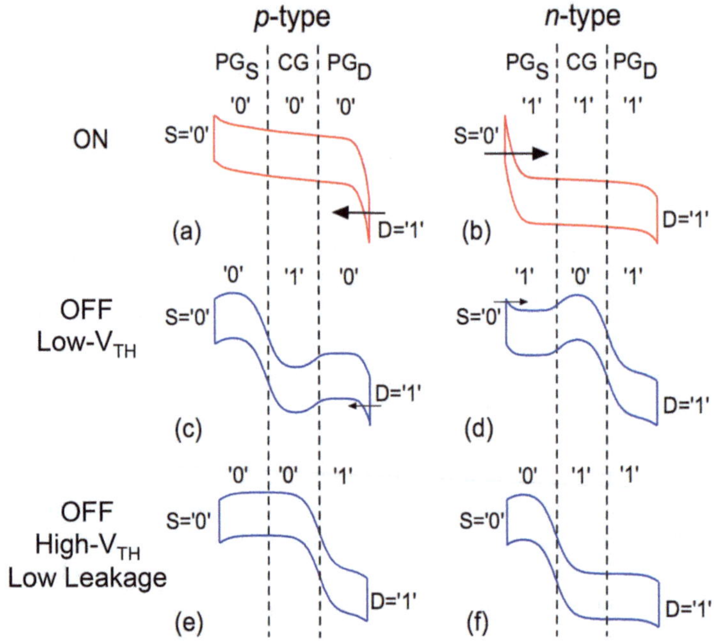


**Fig. 2.7** Band diagrams relative to the six allowed operation modes. Adapted with permission from Zhang et al. [15]. Copyright (2014) IEEE

2. ***Low-$V_{TH}$ OFF states:*** Current flow is blocked by the potential barrier created by the opposite biasing of the control gate with respect to the polarity gates (Fig. 2.7c, d). However, the tunneling barrier at drain (*p*-type) or at source (*n*-type) is still thin enough to allow tunneling of carriers in the channel, and few of them can still be transmitted through the channel, thanks to thermionic emission over the potential barrier created by the CG. This OFF-state is identical to the one presented in Sect. 2.2.1 for the DIG-SiNWFET [7].
3. ***High-$V_{TH}$ OFF states:*** As presented in Fig. 2.7e, f, this operation mode is characterized by the $PG_S$ being kept at the same potential of the source contact and the $PG_D$ at the same potential of the drain contact. The voltage applied to the CG discriminates between *p*-type OFF state (CG = '0') and *n*-type OFF state (CG = '1'). In this configuration, thick tunneling barriers at source and drain prevent carriers to be injected in the channel, lowering even more the current leakage. This OFF-state closely resembles the one presented in [5, 6].
4. ***Uncertain states:*** When $PG_S$ = '1' and $PG_D$ = '0', both barriers are thin enough for tunneling. However, this condition may also create an unexpected barrier in the inner region that will block the current flow, and cause signal degradation. Hence, the uncertain states should be prohibited by always fixing $PG_D$ ='1' ($PG_S$ ='0') for *n*-FET (*p*-FET), or using $PG_D = PG_S$.

The experimental transfer characteristics are shown in Fig. 2.8. Both *p*- and *n*-type behaviors with different threshold voltages (low-$V_{TH}$ and high-$V_{TH}$) were

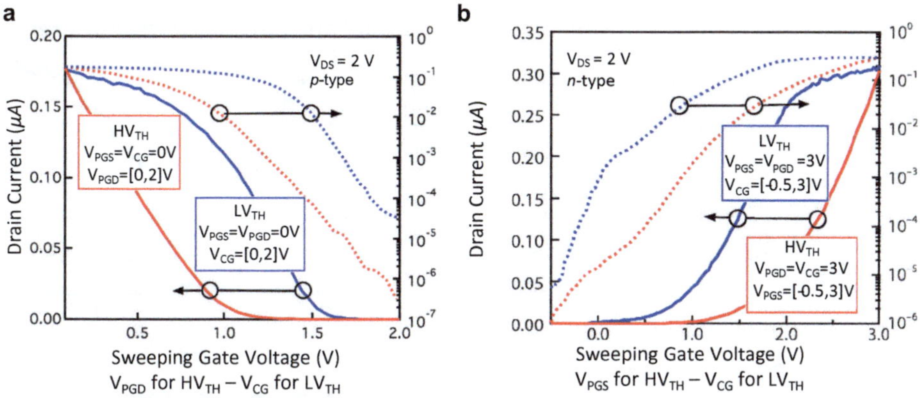


**Fig. 2.8** Experimental transfer characteristics of the three-independent-gate device, showing the threshold-voltage control mode. (**a**) *p*-Type transfer characteristics. (**b**) *n*-Type transfer characteristics. The reader can appreciate how, for both *n*- and *p*-type operation, there is no current degradation in the device ON-state between the low- and high-$V_{TH}$ modes. Adapted with permission from Zhang et al. [15]. Copyright (2014) IEEE

observed in the same device. By extracting the threshold voltages at 1 nA drain current the threshold difference in *p*-FET configuration is 0.48 and 0.86 V in *n*-FET configuration. As mentioned previously, the device ON-state is unchanged when switching from low- and high-$V_{TH}$ and there is no degradation in the device ON-current, as can be appreciated in Fig. 2.8. This represents a competitive advantage for this technology, as the transistor is able to maintain the same current drive in both configurations.

## 2.3 Novel Materials for Polarity-Controllable Devices

Scaling of conventional silicon-based electronics is reaching its ultimate limit and the quest for a new material, with the potential to outperform silicon, is now open. Here, we focus on materials that have been proven to be adaptable for beyond-CMOS polarity-controllable electronics and show the most recent experimental results achieved by worldwide research groups.

### 2.3.1 *Carbon Nanotubes*

Carbon nanotubes FETs (CNFETs) with Schottky metal contacts have been frequently reported in literature and their ambipolar switching behavior has been studied extensively (see [16–18] for a in-depth review). Electrostatic doping was

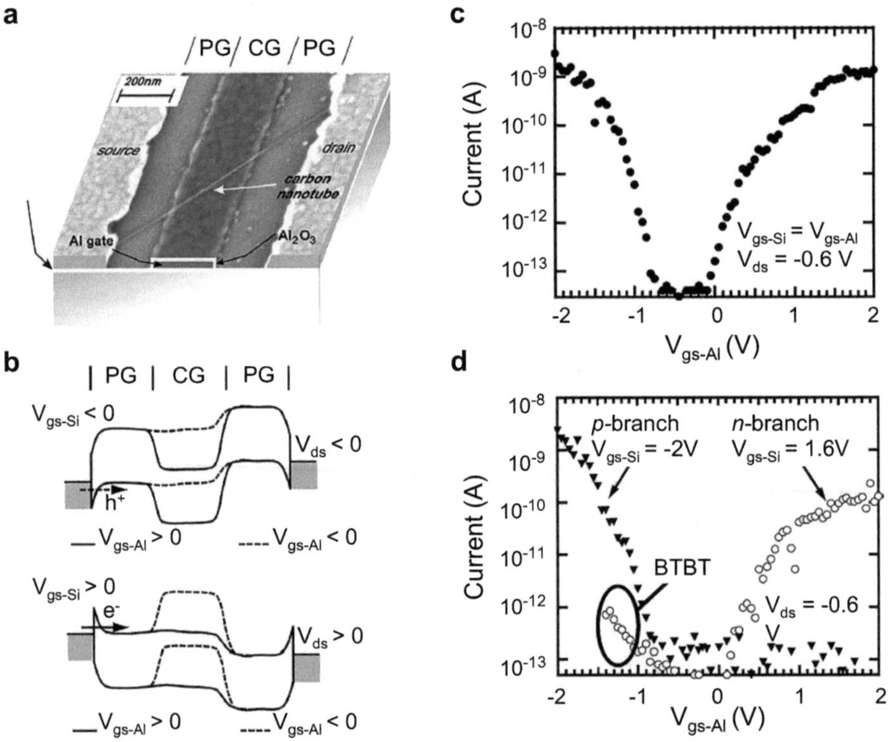


**Fig. 2.9** Polarity control in carbon nanotubes FETs. (**a**) SEM view of the CNFET fabricated. (**b**) Schematic band diagram of the different operation modes. (**c**) Ambipolar transfer characteristic measured without the use of the program gate. (**d**) Transfer characteristics of the same dual-gate CNFET exploiting the polarity-control mechanisms and showing clear *p*- and *n*-type unipolar behavior. Band-to-band tunneling (BTBT) can be observed in *n*-type operation mode for $V_{gs-Al} < -1$ V. Adapted with permission from Lin et al. [20]. Copyright (2005) IEEE

first used in CNTs to demonstrate tunable *p-n* junction diodes [19]. Researchers at IBM then exploited electrostatic doping for the realization of field-effect transistors with tunable polarity, using a double-independent-gate (DIG) CNFET [20]. In the proposed DIG device structure, an aluminum back-gate is fabricated on a $Si/SiO_2$ substrate to act as the control gate and a single CNT is transferred on the substrate and aligned with respect to the pre-patterned gate. The silicon substrate acts as a program gate in the contact region, creating a gate configuration similar to the one presented in [7], where the program gate is acting simultaneously on the source and drain Schottky barrier. The control-gate, placed in the central region of the channel, controls the ON/OFF state of the device, as shown in Fig. 2.9a, b. As previously discussed in Sect. 2.1, Schottky-barrier undoped FETs are intrinsically ambipolar, as they permit to have conduction of both charge carriers. The additional PG allows to selectively choose the charge carriers that are injected in the channel. This effect is clearly shown by the comparison between the experimental transfer characteristics shown in Fig. 2.9c, d. When the PG bias ($V_{gs-Si}$) is set to be equal to the CG ($V_{gs-Al}$), the polarity control mechanism is not used, and the device shows its ambipolar behavior, see Fig. 2.9c. Instead when using PG as a second independent gate, the selection of the charge carriers can be used to create two separate unipolar behaviors on the same device, as shown in Fig. 2.9d. The device transfer characteristics obtained show low ON-current values, 3 nA for *p*-type and 0.1 nA for *n*-type behavior, with low current leakage (0.1 pA) for both polarities. The low ON-currents are a consequence of the reduced dimensionality of the CNT, and could be improved by placing multiple CNTs between the source and drain contacts.

### 2.3.2 Graphene

Graphene is a two-dimensional allotrope of carbon first discovered in 2005 by Geim and Novoselov at Manchester University [21]. Graphene has been used to realize electronics devices, but due to the absence of a semiconducting bandgap, it has been difficult to achieve low OFF-current and subsequently high ON/OFF current ratios. An improvement in the performances of graphene devices can come from managing to open a transport bandgap in graphene. In [22, 23], a defect-induced bandgap was created in a graphene flake by helium ion-beam irradiation [24]. By using a structure with two independent top gates, similar to what already described for silicon nanowires in Fig. 2.1a, b, polarity-controllable behavior on the graphene FET was demonstrated (Fig. 2.10c, d). For both polarities the ON-currents are lower than 0.1 nA, indicating how the creation of the defect-induced bandgap causes an increased scattering rate in the channel, and destroys the conventional high mobility of graphene. Moreover the characteristics were measured at 200 K, indicating that behavior at room temperature might be even more degraded.

### 2.3.3 Two-Dimensional Transition Metal Dichalcogenides

Two-dimensional graphene-like monolayers and few-layers semiconducting transition metal dichalcogenides (TMDCs) have recently drawn considerable attention as viable candidates for flexible and beyond-CMOS electronics and have shown the potential for the realization of polarity controllable devices. The most studied material among TMDCs, molybdenum disulphide ($MoS_2$), suffers from Fermi-level pinning to the conduction band at the metal-semiconductor interface which makes it challenging to achieve an ambipolar behavior, necessary for the realization of polarity-controllable devices. Thus, researchers have focused on different 2D-TMDCs, such as tungsten diselenide ($WSe_2$) and molybdenum telluride ($MoTe_2$), that have shown the ability to efficiently conduct both types of charge carriers. Electrostatically-reversible polarity transistors have been realized with multilayer

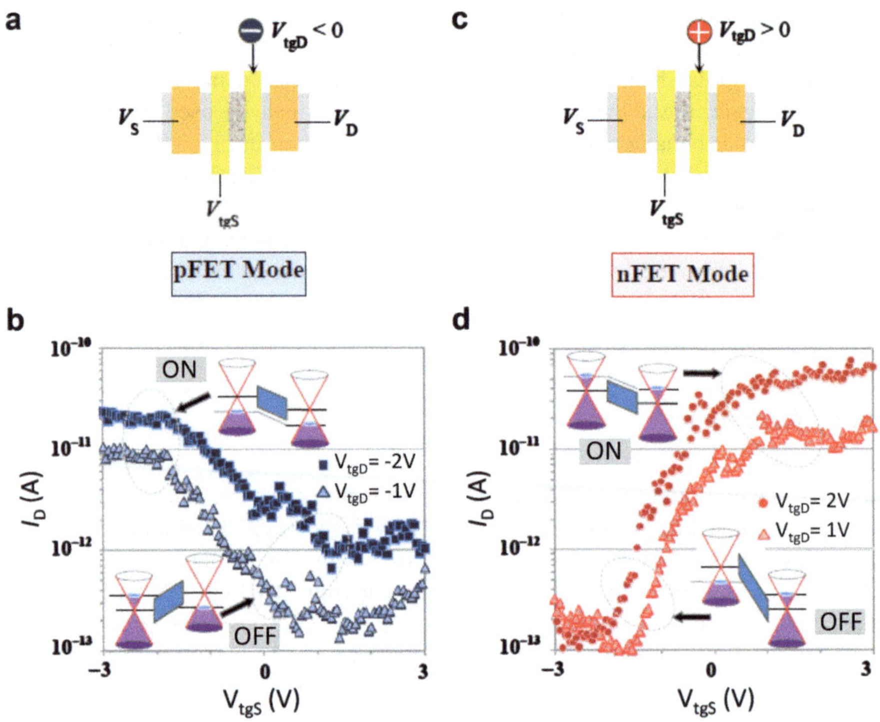


**Fig. 2.10** Polarity-control in graphene FETs. (**a**) Schematic representation of the device and of the voltages applied at the terminals for *p*-type operation. (**b**) Transfer characteristics of the *p*-type operation mode with representation of the distribution of the charge carriers at the contacts. The measurements were taken at 200 K and the applied $V_{DS}$ was 200 mV. (**c**) Schematic depiction of the device and of the voltages applied at the terminals for *n*-type operation. (**d**) Transfer characteristics of the *n*-type operation mode with representation of the distribution of the charge carriers at the contacts. The measurements were taken at 200 K and the applied $V_{DS}$ was 200 mV. Adapted with permission from Nakaharai et al. [22]. Copyright (2012) IEEE

$MoTe_2$ [25, 26], but with $I_{ON}/I_{OFF}$ ratios of the order of $10^2$ for hole conduction and $10^3$ for electron conduction. $WSe_2$ has been explored for the realization of both CMOS-like devices [27] and Schottky-barrier ambipolar FETs [28], and has shown excellent electrical properties for both *p*- and *n*-type conduction. Recently, the ambipolar behavior of $WSe_2$ has been exploited to realize double-independent back-gated devices and polarity-controllable behavior has been demonstrated with ON/OFF current ratios $> 10^6$ for both polarities, on the same device [29]. The device was realized on multilayer $WSe_2$ flake (7.5 nm thick), which was transferred and aligned on a substrate where buried metal lines were used as PG and the silicon substrate as CG (Fig. 2.11a). The metal contacts were realized using evaporated Titanium (Ti)/Palladium (Pd), which provide a band-alignment suitable for the injection of both charge carriers (near mid-gap contacts). The ambipolar behavior of the device can be seen in Fig. 2.11b, where the PG and CG gates were kept at the

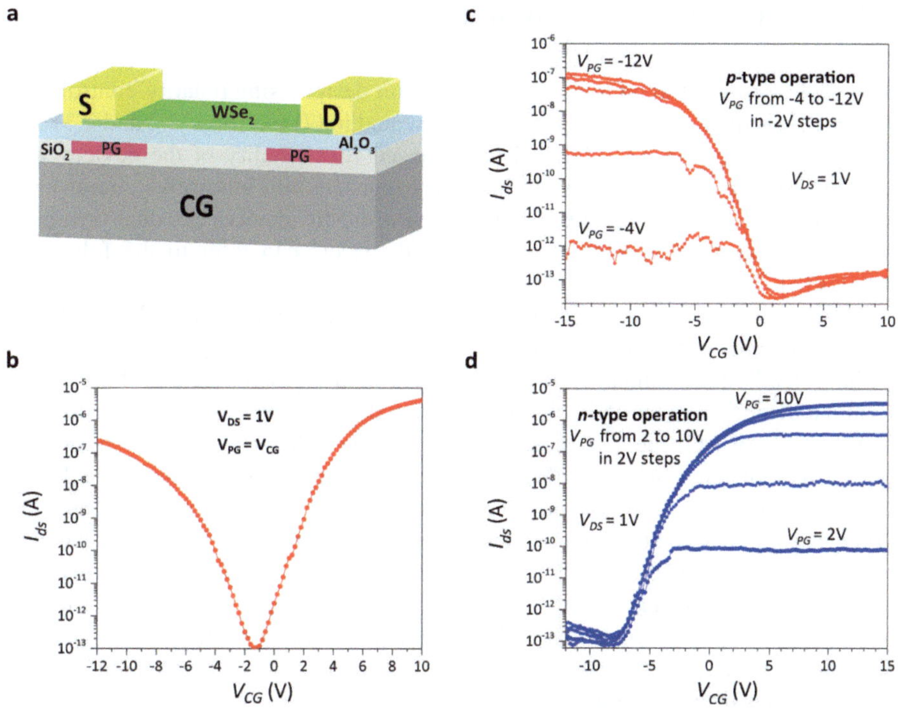


**Fig. 2.11** Polarity-control in 2D-$WSe_2$ DIG-FETs. (**a**) 3D-schematic view of the device. The silicon wafer acts as CG, while the PG have been patterned before the flake transfer. (**b**) Experimental transfer characteristic of the device measured with the same bias applied to CG and PG. The gate leakage current is also plot. (**c**) *p*-Type transfer characteristics obtained for multiple negative biases of the PG gate and sweeping the CG. The leakage currents of both gates are also plotted. (**d**) *n*-Type transfer characteristics obtained for multiple positive biases of the PG gate and sweeping the CG. The leakage currents of both gates are also plotted. Adapted with permission from Resta et al. [29]. Copyright (2016) Nature publishing group

same potential during the voltage sweep. When using the two gates independently, the transistor polarity could be dynamically changed by the PG, while the CG controlled the ON/OFF status of the device (Fig. 2.11c, d). The experimental transfer characteristics measured showed a *p*-type behavior for $V_{PG} < -6\,V$, Fig. 2.11c, while *n*-type conduction properties are shown for $V_{PG} > 4\,V$, Fig. 2.11d, on the same device.

The proposed approach of controlling the polarity of an undoped SB-FET through an additional gate is relatively simple to implement, as Schottky barriers are much easier to create than Ohmic contacts in low-dimensionality materials, and is adaptable to any 2D-semiconductor. For example, promising work on 2D-phosphorene (black phosphorus) has shown ambipolar conduction [30], which, as explained when discussing Figs. 2.9c and 2.11b, is a key step toward the demonstration of controllable-polarity.

## 2.4 Circuit-Level Opportunities

This section is focused on logic gates and circuit design using polarity-controllable DIG-FETs, with the particular geometry presented in [7] (see Figs. 2.2 and 2.3) and described in depth in Sect. 2.2.1. The enhanced functionality of the devices will be addressed, and it will be shown how they translate into innovative circuit-level opportunities. For further references on design with MIG devices and dual-threshold operation, presented in Sect. 2.2.3, interested readers can refer to the following articles [31–33].

Digital circuits based on polarity-controllable DIG-FETs can exploit both PG and CG as inputs, thereby enabling more expressive switching properties. Indeed, while a standard 3-terminal device behaves as a binary switch, the DIG-FET is a 4-terminal device, with the PG being the additional input (see Fig. 2.12a). According to the value of PG, the device abstraction can be either a *p*-MOS or a *n*-MOS device, as shown in Fig. 2.12b. The general switching properties of the single device can be regarded as a comparison-driven switch, that is, the DIG-FET compares the voltages applied at the two independent gates [7, 32], and when loaded implements an exclusive OR function (XOR), see Fig. 2.13. Indeed, when the transistor is not conducting, cases B and C in Fig. 2.13 corresponding to opposite logic values of CG and PG, the output is kept at logic '1'. When the voltages on CG and PG have the same logic value, cases A and D in Fig. 2.13, the transistor is conducting and the output voltage drops to logic '0'.

The unique switching properties of the device are the key for the realization of fully-complementary compact logic gates that can be used for the realization of digital circuits. Adopting a pass-transistor configuration we can realize both unate (NAND) and binate functions (XOR), together with highly compact majority-gates, see Fig. 2.14. As it can be appreciated in Fig. 2.14a, there is no real advantage in using DIG-FETs to realize unate functions such as NAND gates. In this case the polarity of the devices is set, polarity gates are not connected to any logic-input, and the number of transistors used, 4, is the same as in the standard CMOS realization of NAND gates. As previously mentioned, the real advantage of DIG-FETs can be appreciated in the implementation of binate functions, such as XOR, where

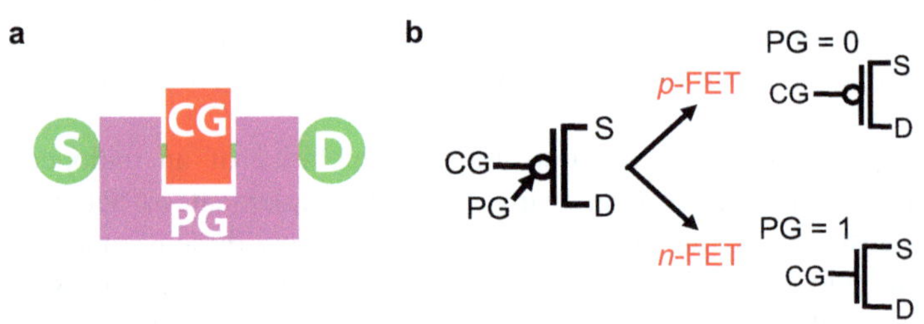


**Fig. 2.12** Device abstraction. (**a**) Stick diagram of the DIG-FETs showing the four terminals. (**b**) Circuit symbol of the DIG-FETs and effect of PG gate on device behavior

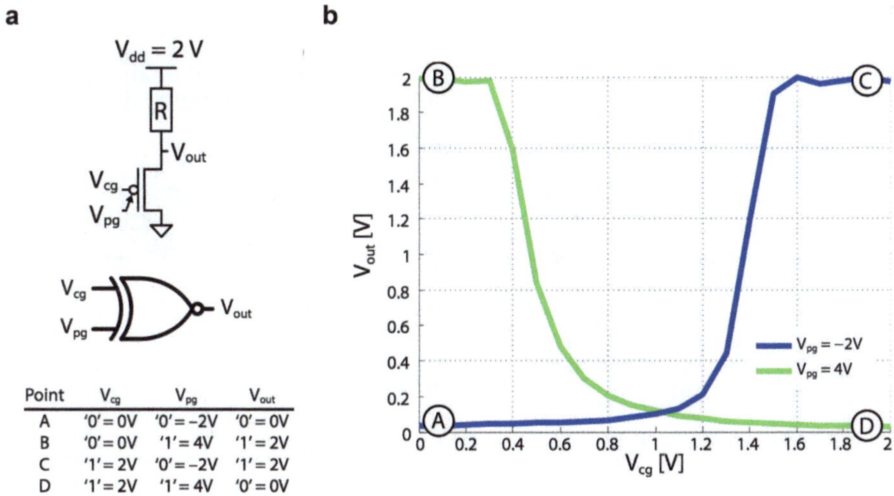


| Point | $V_{cg}$ | $V_{pg}$ | $V_{out}$ |
|---|---|---|---|
| A | '0' = 0V | '0' = −2V | '0' = 0V |
| B | '0' = 0V | '1' = 4V | '1' = 2V |
| C | '1' = 2V | '0' = −2V | '1' = 2V |
| D | '1' = 2V | '1' = 4V | '0' = 0V |

**Fig. 2.13** XOR behavior of the DIG-FET when loaded with resistor. (**a**) Circuit schematic of the loaded device, with logic-level abstraction and summary of the different bias points. (**b**) Experimental characteristic showing the XOR-behavior. Adapted with permission from De Marchi et al. [8]. Copyright (2014) IEEE

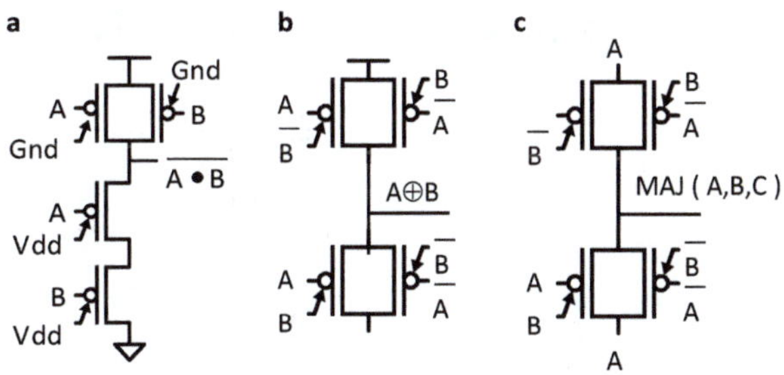


**Fig. 2.14** Fully complementary logic gates. (**a**) Schematic for a NAND gate realized using DIG-FETs. (**b**) Schematic of a fully-complementary 4-transistor XOR gate realized with DIG-FETs. (**c**) Highly compact implementation of a 3-input majority gate with only four DIG-FETs

both transistor gates can be used as logic inputs. Figure 2.14b shows the efficient implementation of a XOR logic gate with only four DIG-FETs (in regular CMOS, we would need eight transistors) that will be used as the building block for the implementation of XOR-rich circuits. Experimental demonstration of NAND and XOR logic gates realized using DIG-SiNWFETs [34] can be found in Fig. 2.15. It should be noted that in order to obtain fully cascadable logic gates only positive gate voltages would be required to be applied to both CG and PG. To achieve this tuning of the process, parameters would be needed to obtain the desired PG and CG thresholds. This problem could be addressed by applying strain to the nanowires or tuning the work function of the metal gate.

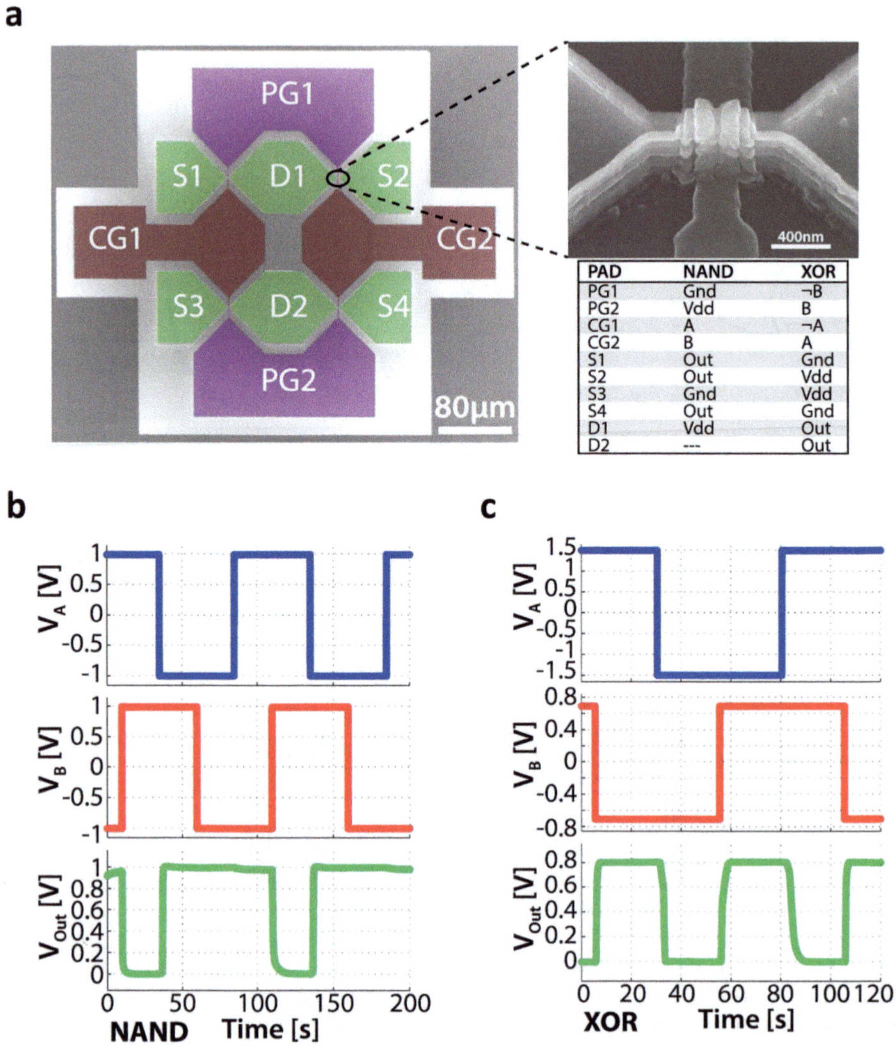


| PAD | NAND | XOR |
|---|---|---|
| PG1 | Gnd | ¬B |
| PG2 | Vdd | B |
| CG1 | A | ¬A |
| CG2 | B | A |
| S1 | Out | Gnd |
| S2 | Out | Vdd |
| S3 | Gnd | Vdd |
| S4 | Out | Gnd |
| D1 | Vdd | Out |
| D2 | --- | Out |

**Fig. 2.15** Experimental demonstartion of NAND and XOR logic gates with DIG-FETs. (**a**) SEM micrograph of the fabricated devices with PAD names and zoomed view on the gated region of the transistor. The voltages applied to each PAD in both NAND and XOR operation are also listed. (**b**) Experimental characteristics showing NAND behavior. (**c**) Experimental behavior of XOR logic gate. Reprinted with permission from De Marchi et al. [34]. Copyright (2014) IEEE

The impact of this device concept on circuit and logic gates design does not only come from its peculiar switching properties, but also from the doping-free process that can be used for the realization of DIG-FETs. The great advantage in the realization of reconfigurable devices is that they eliminate the need to separate *p*- and *n*-type devices, that is, in standard CMOS technology, *p*-type devices need to be realized on a $n$-doped region ($n$-well), opening alternative ways to place devices and allowing to achieve a much higher degree of regularity in the design of digital circuits [35, 36].

The observant reader might, at this point, have noticed that from a device-level perspective the additional polarity gate introduces larger parasitic capacitance and area consumption (a DIG-FET is intrinsically larger than a conventional single-gate device). Thus, to unlock the full potential of DIG-FETs and ultimately achieve a higher computational density than CMOS technology, not only logic functions need to be redesigned, but also novel circuit synthesis techniques have to be developed [37]. Current logic-synthesis techniques derive from the abilities of CMOS technology, that is, compact and efficient realization of NAND, NOR, and, in general, unate inverting functions, and tend to be less effective in synthesizing XOR-rich circuits, such as arithmetic operators and data paths. The compact implementations of XOR and MAJ functions with DIG-FETs bear the promise for superior automated design of arithmetic circuits and datapaths. However, conventional logic synthesis tools are not adequate to take full advantage of the possibilities opened by controllable-polarity feature, as they are missing some optimization opportunities. To overcome these limitations, it is necessary to better integrate the efficient primitives of controllable-polarity FETs (XOR and MAJ) in the logic synthesis tools. On the one hand, it is possible to propose innovations in the data representation form. For instance, biconditional binary decision diagrams (BBDDs) [38, 39] are a canonical logic representation form based on the biconditional (XOR) expansion. They provide a one-to-one correspondence between the functionality of a controllable-polarity transistor and its core expansion, thereby enabling an efficient mapping of the devices onto BBDD structures. On the other hand, it is also possible to identify the logic primitives efficiently realized by controllable-polarity FETs in existing data structures. In particular, BDD Decomposition System based on MAJority decomposition (BDS-MAJ) [40] is a logic optimization system driven by binary decision diagrams that support integrated MUX, XOR, AND, OR, and MAJ logic decompositions. Since it provides both XOR and MAJ decompositions, BDS-MAJ is an effective alternative to standard tools to synthesize datapath circuits. In the controllable-polarity transistor context, BDS-MAJ natively and automatically highlights the efficient implementation of arithmetic gates. Finally, very efficient logic optimization can be directly performed on data structures supporting MAJ operator. In [41], a novel data structure, called *Majority-Inverter-Graph* (MIG), exploiting only MAJ and INV operators has been introduced. Such data structure is supported by an expressive Boolean algebra allowing for powerful logic optimization of both standard general logic and arithmetic oriented logic. By applying these logic-synthesis techniques to various industry-standard benchmark circuits, such as adders, multipliers, compressors, and counters, an average improvement in both area (32%) and delay (38%), with respect to conventional CMOS technology, can be achieved using MIG-FETs [31].

## 2.5 Summary

This chapter was dedicated to functionality-enhanced devices in the form of multiple-independent-gate field-effect transistors, which have been referred throughout as MIG-FETs. We aimed at giving a broad overlook of the field, and of the advantages that this technology could bring to future electronic-circuit design. We focused on experimental devices realized with different materials and structures, and showed how the device concept is flexible and adaptable to both silicon and novel emerging semiconductors. Some of the key aspects that have been elucidated in this chapter are

1. Undoped Schottky-barrier (SB) FETs for the conduction of both types of charge carriers (ambipolar behavior).
2. The ambipolar behavior provides an added degree of freedom for the realization of doping-free or lightly-doped devices.
3. Double-independent-gate (DIG) devices allow to control both the polarity and the subthreshold slope of the transistor.
4. Three-independent-gate (TIG) transistors also enable the control of the threshold voltage of the device. The low- and high-$V_{TH}$ operation mode share the same ON-state avoiding any degradation in the drive capability of the device.
5. Novel materials can be adapted to this technology as Schottky-contacts are easily created (realizing Ohmic contacts to 1D and 2D materials is still a great challenge).
6. With particular gate configurations, the device switching properties lead to highly compact logic gates, that is, XOR and MAJ, and create the opportunity to explore novel design styles and tools for logic synthesis.

The unique properties of the proposed technology are routed in the fine-grain reconfigurability of the single device and, when paired with novel semiconducting materials and innovative logic-synthesis techniques, could provide a significant advantage over standard silicon-based CMOS logic circuits.

## References


1. M.M. Waldrop, The chips are down for Moore's law. Nat. News **530**(7589), 144 (2016)
2. B. Sharma, *Metal-Semiconductor Schottky Barrier Junctions and Their Applications* (Springer Science & Business Media, Berlin, 2013)
3. S.-M. Koo, Q. Li, M.D. Edelstein, C.A. Richter, E.M. Vogel, Enhanced channel modulation in dual-gated silicon nanowire transistors. Nano Lett. **5**(12), 2519 (2523)
4. J. Appenzeller, J. Knoch, E. Tutuc, M. Reuter, S. Guha, Dual-gate silicon nanowire transistors with nickel silicide contacts, in *Electron Devices Meeting, 2006. IEDM'06. International* (IEEE, New York, 2006), pp. 1–4
5. A. Heinzig, S. Slesazeck, F. Kreupl, T. Mikolajick, W.M. Weber, Reconfigurable silicon nanowire transistors. Nano Lett. **12**(1), 119–124 (2011)

6. A. Heinzig, T. Mikolajick, J. Trommer, D. Grimm, W.M. Weber, Dually active silicon nanowire transistors and circuits with equal electron and hole transport. Nano Lett. **13**(9), 4176–4181 (2013)
7. M. De Marchi, D. Sacchetto, S. Frache, J. Zhang, P.-E. Gaillardon, Y. Leblebici, G. De Micheli, Polarity control in double-gate, gate-all-around vertically stacked silicon nanowire FETs, in *2012 IEEE International Electron Devices Meeting (IEDM)* (IEEE, New York, 2012), pp. 8–4
8. M. De Marchi, D. Sacchetto, J. Zhang, S. Frache, P.-E. Gaillardon, Y. Leblebici, G. De Micheli, Top-down fabrication of gate-all-around vertically stacked silicon nanowire fets with controllable polarity. IEEE Trans. Nanotechnol. **13**(6), 1029 (1038)
9. Y.-J. Chang, J. Erskine, Diffusion layers and the schottky-barrier height in nickel silicide–silicon interfaces. Phys. Rev. B **28**(10), 5766 (1983)
10. Q. Zhao, U. Breuer, E. Rije, S. Lenk, S. Mantl, Tuning of NiSi/Si Schottky barrier heights by sulfur segregation during Ni silicidation. Appl. Phys. Lett. **86**(6), 62108 (62108)
11. J. Zhang, M. De Marchi, P.-E. Gaillardon, G. De Micheli, A Schottky-barrier silicon FinFet with 6.0 mv/dec subthreshold slope over 5 decades of current, in *Proceedings of the International Electron Devices Meeting (IEDM'14)*, no. EPFL-CONF-201905 (2014)
12. S.M. Sze, K.K. Ng, *Physics of Semiconductor Devices* (Wiley, New York, 2006)
13. Z. Lu, N. Collaert, M. Aoulaiche, B. De Wachter, A. De Keersgieter, J. Fossum, L. Altimime, M. Jurczak, Realizing super-steep subthreshold slope with conventional fdsoi cmos at low-bias voltages, in *2010 IEEE International Electron Devices Meeting (IEDM)* (IEEE, New York, 2010), pp. 16–6
14. J. Zhang, P.-E. Gaillardon, G. De Micheli, Dual-threshold-voltage configurable circuits with three-independent-gate silicon nanowire FETs, in *2013 IEEE International Symposium on Circuits and Systems (ISCAS)* (IEEE, New York, 2013), pp. 2111–2114
15. J. Zhang, M. De Marchi, D. Sacchetto, P.-E. Gaillardon, Y. Leblebici, G. De Micheli, Polarity-controllable silicon nanowire transistors with dual threshold voltages. IEEE Trans. Electron Devices **61**(11), 3654–3660 (2014)
16. S. Heinze, J. Tersoff, R. Martel, V. Derycke, J. Appenzeller, P. Avouris, Carbon nanotubes as schottky barrier transistors. Phys. Rev. Lett. **89**(10), 106801 (2002)
17. R. Martel, V. Derycke, C. Lavoie, J. Appenzeller, K. Chan, J. Tersoff, P. Avouris, Ambipolar electrical transport in semiconducting single-wall carbon nanotubes. Phys. Rev. Lett. **87**(25), 256805 (2001)
18. P. Avouris, Z. Chen, V. Perebeinos, Carbon-based electronics. Nat. Nanotechnol. **2**(10), 605–615 (2007)
19. J.U. Lee, P. Gipp, C. Heller, Carbon nanotube p-n junction diodes. Appl. Phys. Lett. **85**(1), 145–147 (2004)
20. Y.-M. Lin, J. Appenzeller, J. Knoch, P. Avouris, High-performance carbon nanotube field-effect transistor with tunable polarities. IEEE Trans. Nanotechnol. **4**(5), 481–489 (2005)
21. K. Novoselov, A.K. Geim, S. Morozov, D. Jiang, M. Katsnelson, I. Grigorieva, S. Dubonos, A. Firsov, Two-dimensional gas of massless dirac fermions in graphene. Nature **438**(7065), 197–200 (2005)
22. S. Nakaharai, T. Iijima, S. Ogawa, S. Suzuki, K. Tsukagoshi, S. Sato, N. Yokoyama, Electrostatically-reversible polarity of dual-gated graphene transistors with He ion irradiated channel: toward reconfigurable CMOS applications, in *2012 IEEE International Electron Devices Meeting (IEDM)* (IEEE, New York, 2012), pp. 4–2
23. S. Nakaharai, T. Iijima, S. Ogawa, S.-L. Li, K. Tsukagoshi, S. Sato, N. Yokoyama, Electrostatically reversible polarity of dual-gated graphene transistors. IEEE Trans. Nanotechnol. **13**(6), 1039–1043 (2014)
24. S. Nakaharai, T. Iijima, S. Ogawa, S. Suzuki, S.-L. Li, K. Tsukagoshi, S. Sato, N. Yokoyama, Conduction tuning of graphene based on defect-induced localization. ACS Nano **7**(7), 5694–5700 (2013)
25. Y.-F. Lin, Y. Xu, S.-T. Wang, S.-L. Li, M. Yamamoto, A. Aparecido-Ferreira, W. Li, H. Sun, S. Nakaharai, W.-B. Jian et al., Ambipolar mote2 transistors and their applications in logic circuits. Adv. Mater. **26**(20), 3263–3269 (2014)

26. S. Nakaharai, M. Yamamoto, K. Ueno, Y.-F. Lin, S.-L. Li, K. Tsukagoshi, Electrostatically reversible polarity of ambipolar $\alpha$-mote2 transistors. ACS Nano **9**(6), 5976–5983 (2015)
27. L. Yu, A. Zubair, E.J. Santos, X. Zhang, Y. Lin, Y. Zhang, T. Palacios, High-performance wse2 complementary metal oxide semiconductor technology and integrated circuits. Nano Lett. **15**(8), 4928–4934 (2015)
28. S. Das, J. Appenzeller, Wse2 field effect transistors with enhanced ambipolar characteristics. Appl. Phys. Lett. **103**(10), 103501 (2013)
29. G.V. Resta, S. Sutar, Y. Blaji, D. Lin, P. Raghavan, I. Radu, F. Catthoor, A. Thean, P.-E. Gaillardon, G. De Micheli, Polarity control in $wse_2$ double-gate transistors. Sci. Rep. **6**, 29448 (2016)
30. S. Das, M. Demarteau, A. Roelofs, Ambipolar phosphorene field effect transistor. ACS Nano **8**(11), 11730–11738 (2014)
31. P.-E. Gaillardon, L. Amaru, J. Zhang, G. De Micheli, Advanced system on a chip design based on controllable-polarity FETs, in *Proceedings of the Conference on Design, Automation & Test in Europe* (European Design and Automation Association, Leuven, 2014), p. 235
32. P.-E. Gaillardon, L.G. Amarù, S. Bobba, M. De Marchi, D. Sacchetto, G. De Micheli, Nanowire systems: technology and design. Philos. Trans. R. Soc. Lond. A Math. Phys. Eng. Sci. **372**(2012), 20130102 (2014)
33. J. Zhang, X. Tang, P.-E. Gaillardon, G. De Micheli, Configurable circuits featuring dual-threshold-voltage design with three-independent-gate silicon nanowire FETs. IEEE Trans. Circuits Syst. Regul. Pap. **61**(10), 2851–2861 (2014)
34. M. De Marchi, J. Zhang, S. Frache, D. Sacchetto, P.-E. Gaillardon, Y. Leblebici, G. De Micheli, Configurable logic gates using polarity-controlled silicon nanowire gate-all-around FETs. IEEE Electron Device Lett. **35**(8), 880–882 (2014)
35. S. Bobba, P.-E. Gaillardon, J. Zhang, M. De Marchi, D. Sacchetto, Y. Leblebici, G. De Micheli, Process/design co-optimization of regular logic tiles for double-gate silicon nanowire transistors, in *2012 IEEE/ACM International Symposium on Nanoscale Architectures (NANOARCH)*. (IEEE, New York, 2012), pp. 55–60
36. O. Zografos, P.-E. Gaillardon, G. De Micheli, Novel grid-based power routing scheme for regular controllable-polarity fet arrangements, in *2014 IEEE International Symposium on Circuits and Systems (ISCAS)* (IEEE, New York, 2014), pp. 1416–1419
37. L. Amarú, P.-E. Gaillardon, S. Mitra, G. De Micheli, New logic synthesis as nanotechnology enabler. Proc. IEEE **103**(11), 2168–2195 (2015)
38. L. Amarú, P.-E. Gaillardon, G. De Micheli, Biconditional BDD: a novel canonical BDD for logic synthesis targeting XOR-rich circuits, in *Proceedings of the Conference on Design, Automation and Test in Europe* (EDA Consortium, 2013), pp. 1014–1017
39. L. Amarú, P.-E. Gaillardon, G. De Micheli, An efficient manipulation package for biconditional binary decision diagrams, in *Proceedings of the conference on Design, Automation & Test in Europe* (European Design and Automation Association, Leuven, 2014), p. 296
40. L. Amarú, P.-E. Gaillardon, G. De Micheli, BDS-MAJ: a BDD-based logic synthesis tool exploiting majority logic decomposition, in *Proceedings of the 50th Annual Design Automation Conference* (ACM, New York, 2013), p. 47
41. L. Amarú, P.-E. Gaillardon, G. De Micheli, Majority-inverter graph: a novel data-structure and algorithms for efficient logic optimization, in *Proceedings of the 51st Annual Design Automation Conference* (ACM, New York, 2014), pp. 1–6

# Chapter 3
# Heterogeneous Integration of 2D Materials and Devices on a Si Platform

**Amirhasan Nourbakhsh, Lili Yu, Yuxuan Lin, Marek Hempel, Ren-Jye Shiue, Dirk Englund, and Tomás Palacios**

## 3.1 Introduction

Two-dimensional (2D) materials are atomically thin films originally derived from layered crystals such as graphite, hexagonal boron nitride (h-BN), and the family of transition metal dichalcogenides (TMDs, such as $MoS_2$, $WSe_2$, $MoTe_2$, and others). Atomic planes in such crystals are weakly stacked on each other by van der Waals forces so that they can be easily isolated, leaving no dangling bonds. This is in distinct contrast to their counterpart, quasi-low-dimensional semiconductors, which are produced by thinning down conventional bulk or epitaxial crystals. The lack of dangling bonds at the interfaces and surfaces of 2D materials enables new devices with unprecedented performance.

The merits of 2D materials are not limited to the absence of dangling bonds. They also show a high degree of mechanical stability, as well as unique electronic and optoelectronic properties. This makes 2D materials highly suitable for a wide range of applications, from high performance transistors to extremely sensitive photodetectors and sensors. In addition, the few-atom thickness of many of these novel devices and systems and the low temperatures required during the device fabrication allow their seamless integration with conventional silicon electronics. It is possible to fabricate many of these devices on top of a fully fabricated silicon CMOS wafer without degrading the Si transistors underneath, bringing new functionality to the silicon chip. This integration process can be repeated numerous times to build complex 3D systems.

This chapter provides an overview of the technology and advantages of the heterogeneous integration of various 2D materials-based devices with a standard

A. Nourbakhsh · L. Yu · Y. Lin · M. Hempel · R.-J. Shiue · D. Englund · T. Palacios (✉)
Department of Electrical Engineering and Computer Science, Massachusetts Institute of Technology, Cambridge, MA, USA
e-mail: tpalacios@mit.edu



R. O. Topaloglu, H.-S. P. Wong (eds.), *Beyond-CMOS Technologies for Next Generation Computer Design*, https://doi.org/10.1007/978-3-319-90385-9_3

Si platform. Some of the system-level examples that will be discussed include chemical and infrared sensors, large area electronics, and optical communication systems. Section 3.1 describes the advantages of wide bandgap $MoS_2$ and other TMDs over conventional semiconductors for aggressive scaling of the transistors channel length as well as for ultra-low power applications. Section 3.2 summarizes the research on graphene-based infrared sensors and the methods for building such sensors on top of conventional Si-CMOS readout chips. Section 3.3 focuses on the heterogeneous integration of 2D materials with Si nanophotonics, while Sect. 3.4 discusses different approaches on how 2D materials can be used as chemical or biological sensors.

## 3.2 Scaling and Integration of $MoS_2$ Transistors

### *3.2.1 $MoS_2$ Transistors for Ultimate Scaling and Power Gating*

As the channel length of transistors has shrunk over the years, short-channel effects have become a major limiting factor to further the transistor miniaturization. Current state-of-the-art silicon-based transistors at the 14-nm technology node have channel lengths of around 20 nm, and several technological reasons are compromising further reductions in the channel length. In addition to the inherent difficulties of high-resolution lithography, the direct source-drain tunneling is expected to become a very significant fraction of the off-state current in sub-10-nm silicon transistors, dominating in this way the standby power. Therefore, new transistor structures that reduce the direct source-drain tunneling are needed to achieve further reductions in the transistor channel length. Transistors based on high mobility III–V materials [1, 2], nanowire field-effect transistors (FETs) [3, 4], internal gain FETs [5, 6] (such as negative capacitance devices), and tunnel FETs are among those that have been considered to date. More recently, layered 2D semiconducting crystals of transition metal dichalcogenides (TMDs), such as molybdenum disulfide ($MoS_2$) and tungsten diselenide ($WSe_2$), have also been proposed to enable aggressive miniaturization of FETs [7–10]. In addition to the reduced direct source-drain tunneling current possible in these wide-bandgap materials, the atomically thin body of these novel semiconductor materials is expected to improve the transport properties in the channel thanks to the lack of dangling bonds. Some studies have reported, for example, that single and few layer $MoS_2$ could potentially outperform ultrashort channel and ultrathin body silicon at similar thicknesses [11].

Moreover, the atomically thin body thickness of TMDs also improves the gate modulation efficiency. This can be seen in their characteristic scaling length ($\lambda = \sqrt{\frac{\varepsilon_{semi}}{\varepsilon_{ox}} t_{ox} . t_{semi}}$), which determines important short channel effects such as the drain-induced barrier lowering (DIBL) and the degradation of the subthreshold swing (SS). In particular, $MoS_2$ has low dielectric constant $\varepsilon = 4 - 7$ [12, 13] and an atomically thin body ($t_{semi} = 0.7$ nm $\times$ number of layers) which facilitate the decrease of $\lambda$ while its relatively high bandgap energy (1.85 eV for a monolayer) and high effective mass allow for a high on/off current ratio ($I_{on}/I_{off}$) via reduction of direct source–drain tunneling. These features make $MoS_2$ in particular, and wide-bandgap 2D semiconductors in general, highly desirable for low-power subthreshold electronics.

Ni et al. [14] used first principles quantum transport investigations to predict that monolayer $MoS_2$ FETs would show good performance at sub-10-nm channel lengths and also display small SS values, comparable to the current best sub-10-nm silicon FETs. In addition, its large bandgap makes $MoS_2$ an excellent semiconductor for low power applications, while its ability to form atomically thin films allows excellent electrostatic gate control over the FET channel.

To experimentally demonstrate and benchmark $MoS_2$ transistors with channel lengths below 10 nm, two important challenges need to be overcome. Firstly, a suitable lithography technology is required; secondly, a low-contact resistance is needed for the source and drain to prevent the channel resistance from dominating the device behavior.

Liu et al. [15] demonstrated channel length scaling in $MoS_2$ FETs from 2 μm down to 50 nm in devices built with a 300-nm $SiO_2$ gate dielectric. Despite the thick dielectric oxide layer used in this study, short channel effects were limited for channel lengths as low as 100 nm. However, devices with channels below 100 nm showed a high off current and DIBL (Fig. 3.1).

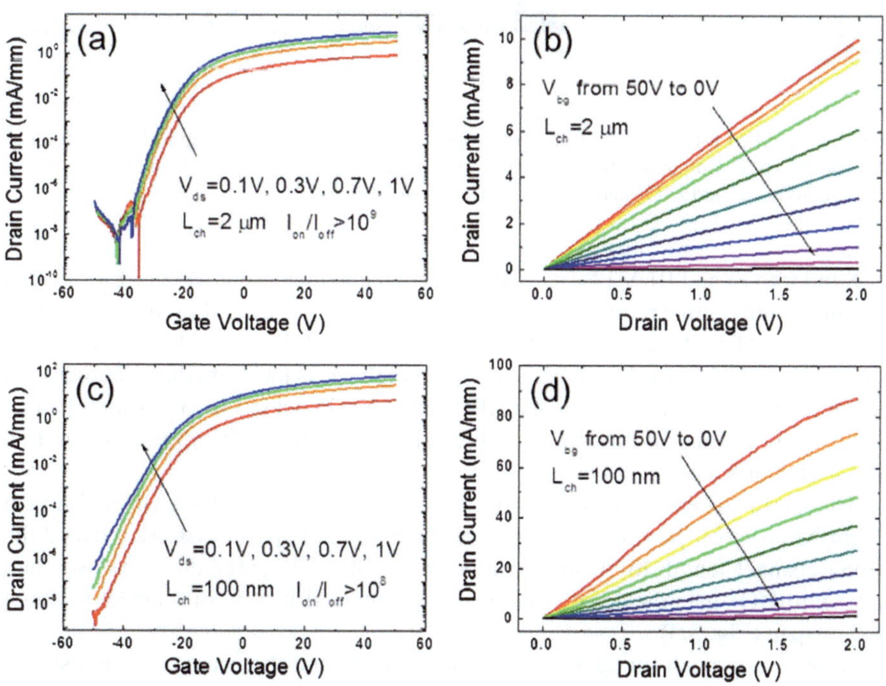


**Fig. 3.1** (**a, b**) Transfer and output characteristics of a 12-nm layer of $MoS_2$ with a channel length of 50 nm. (**c, d**) Channel length dependence of the current on/off ratio and DIBL for $MoS_2$ devices with channel thickness of 5 and 12 nm. Liu et al. [15]

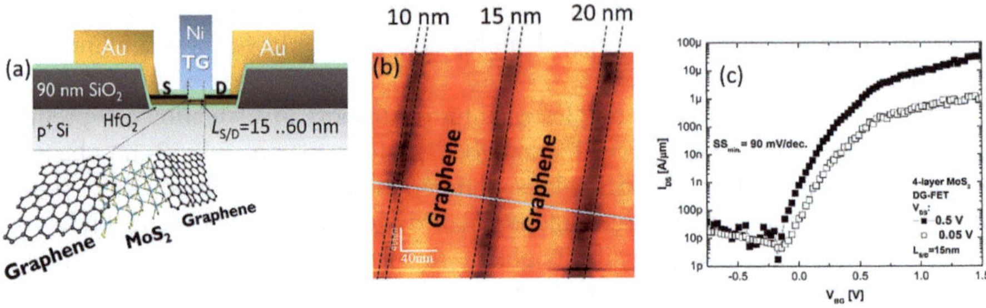


**Fig. 3.2** (**a**) Schematic cross section of a short channel double-gate (DG) $MoS_2$ FET with graphene source/drain contacts. (**b**) AFM images showing 10, 15, 20 nm graphene slits that define the channel length. (**c**) Transfer characteristics ($I_d$–$V_g$) for a 15-nm 4-layer DG- $MoS_2$ FET with $SS_{min}$ = 90 mV/dec and $I_{off}$ < 10 pA. Nourbakhsh et al. [9]

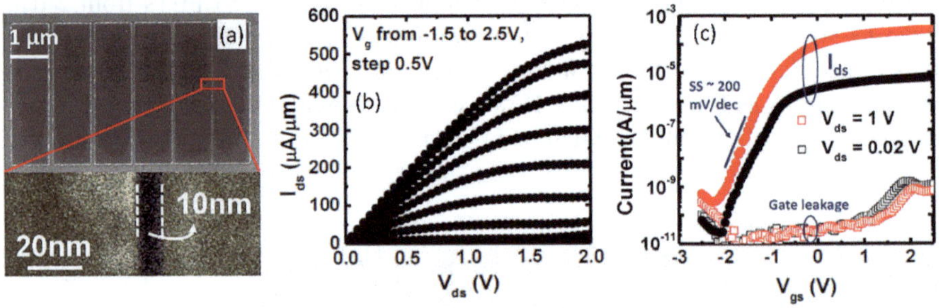


**Fig. 3.3** (**a**) SEM image of $MoS_2$ channel lengths ranging from 10 to 80 nm after deposition of Ni contacts. (**b, c**) Output and transfer characteristics of the 10-nm nominal channel length $MoS_2$ FET built on a 7.5-nm $HfO_2$ gate dielectric. Yang et al. [10]

Similar to Si and III-V FETs, reducing the channel length of $MoS_2$ FETs toward the sub-10 nm regime requires state-of-the-art high-k dielectric thin films to be used in the place of $SiO_2$ dielectrics. Nourbakhsh et al. [9] demonstrated a $MoS_2$ FET with a channel length as low as 15 nm, using graphene as the immediate source/drain contacts and 10 nm $HfO_2$ as the gate dielectric. As shown in Fig. 3.2, short channel effects were limited in this device, which showed high on/off ratio of $10^6$ and an $SS_{min}$ of 90 mV/dec.

This performance indicated that further scaling to a sub-10-nm channel length might be possible. In a different attempt Yang et al. [10] successfully reduced the $MoS_2$ channel length to 10 nm, in a device with a Ni source/drain contacts and a 7.5-nm $HfO_2$ gate dielectric. The device maintained its low off-current to about 100 pA/μm (Fig. 3.3).

Another approach for aggressive scaling is to extend the channels of transistors in the third dimension, including nanowire gate-all-around (NW GAA) FETs [3, 4], finFETs [16], etc. A surface free of dangling bonds and a low-temperature synthesis method also make $MoS_2$ a promising candidate as the channel material in finFETs.

Chen et al. have demonstrated a CMOS-compatible process for few-layer $MoS_2$/Si hybrid finFETs with improved on-current and good threshold voltage ($V_t$) matching [17]. In subsequent work, the same group improved the process and realized 4-nm-thick ultrathin body $MoS_2$ finFETs in the sub-5 nm technology node with reduced contact resistance and good $V_t$ control with back-gate biasing [18].

In all of the aforementioned devices, standard lithography, including e-beam techniques, was used to define the source/drain electrodes in the $MoS_2$ FETs. Realizing ultra-short channels in $MoS_2$ transistors using lithography can be challenging. Electron beam lithography can potentially provide sub-10-nm patterning resolution; however, it has a low throughput and it is difficult to control at these dimensions. Alternatively, Nourbakhsh et al. [19] used a directed self-assembly (DSA) of block copolymers (BCPs) to push $MoS_2$ channel lengths to the sub-10 nm regime. Unlike conventional lithography methods, DSA-BCP is a bottom-up approach in which smaller building block molecules associate with each other in a coordinated fashion to form more complex supramolecules. Using this fabrication approach to $MoS_2$ FETs, a $MoS_2$ layer was patterned with metallic and semiconducting phases to achieve channel lengths as low as 7.5 nm.

The stable metallic phase of $MoS_2$ can be achieved by chemically treating the semiconducting phase with *n*-butyllithium solution. As shown in Fig. 3.4, the $MoS_2$ channel was first patterned with BCP, then a chemical treatment was used to convert the $MoS_2$ film to a chain of alternating metallic and semiconducting $MoS_2$. The semiconducting regions, 7.5 nm across, acted as the FET channel and the metallic portions acted as the immediate source/drain contacts. This method produced a chain of $MoS_2$ FETs with a record-low channel length of 7.5 nm. This device structure permitted experimental probing of the transport properties of $MoS_2$ in the sub-10-nm channel length regime for the first time. As predicted, $MoS_2$ FETs demonstrated superior subthreshold characteristics with lower off-currents

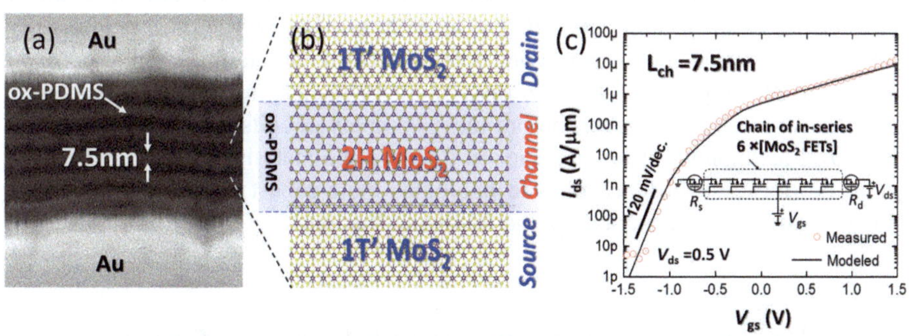


**Fig. 3.4** (**a**) SEM image showing lines of BCP (polystyrene-b-dimethylsiloxane) with a 15 nm pitch formed on a $MoS_2$ film contacted by a pair of Au electrodes. (**b**) Schematic of short channel FET comprising a semiconducting (2H) $MoS_2$ channel contacted to two adjacent metallic (1 T′) $MoS_2$ regions that form internal source/drain contacts. (**c**) $I_d$–$V_g$ of the final $MoS_2$ device (after semiconductor to metallic $MoS_2$ phase transition) the chain transistor was composed of six $MoS_2$ FETs having a channel length of 7.5 nm. Nourbakhsh et al. [19]

than devices based on Si and III-V materials, at the same channel lengths. This $MoS_2$ composite transistor with six FETs in series possessed an off-state current of 100 pA/μm and an $I_{on}/I_{off}$ ratio greater than $10^5$. Modeling of the resulting current-voltage characteristics revealed that the metallic/semiconducting $MoS_2$ junction had a low resistance of 75 Ω μm. These experimental results reveal the remarkable potential of 2D $MoS_2$ for future developments of sub-10 nm technology. Although the structure studied by Nourbaksh et al. was composed of a chain of short channel transistors rather than an individual device, the same short channel effects that occur in single transistors were also active in this series of transistors, because of metallic regions present between any two devices in the chain.

An alternative approach for self-aligned $MoS_2$ transistors was recently demonstrated by English et al. [20] In this work, an $MoS_2$ FET with a self-aligned 10 nm top gate was fabricated by using a self-passivated $Al_2O_3$ layer around an Al gate electrode as the gate dielectric (Fig. 3.5). This allowed for a decrease of the ungated regions to 10 nm.

To reduce the gate length of these devices even further and probe the ultimate limit of scaling, a nanotube-gated $MoS_2$ FET was demonstrated by Desai et al. [21] In their work, a metallic single-wall carbon nanotube (SWCNT) with a diameter of 1 nm was used as the gate electrode enabling a physical gate length down to 1 nm to be achieved (see Fig. 3.6). However, because of the fringing electric field induced by the SWCNTs, the effective channel length in the off-state, calculated by simulation, was 3.9 nm. This ultrashort channel $MoS_2$ FET showed excellent switching characteristics with a subthreshold swing of 65 mV/dec (Fig. 3.6). In this device structure, the SWCNT gate underlapped the source/drain electrodes by some hundreds of nanometers, which caused an extremely large access region. To decrease this resistance, the ungated regions were electrostatically doped by the Si back-gate during electrical measurements.

These initial experimental results show the great promise of $MoS_2$ devices to push Moore's law beyond the scaling limits of silicon. In addition, as all of these devices can be fabricated at low temperature (<400 °C) on top of a fully fabricated Si wafer, they can provide a high performance transistor interposer to be used in 3D chip architectures to drive memory planes, interconnects or power-gate the entire chip. However, before these complex circuits can be fabricated on this novel technology, it is necessary to improve the material, device, and circuit yield. For these novel fabrication technologies more-robust circuit topologies are being investigated as described in the following section.

### *3.2.2 Designing Complex Circuits with Immature Technologies*

Despite its promising material characteristics and initial device performance, $MoS_2$-based circuits to date have been limited to a single or few-transistors [22–25] due to the many challenges associated with the uniformity and yield control in both material growth and device technology. To significantly improve the fabrication

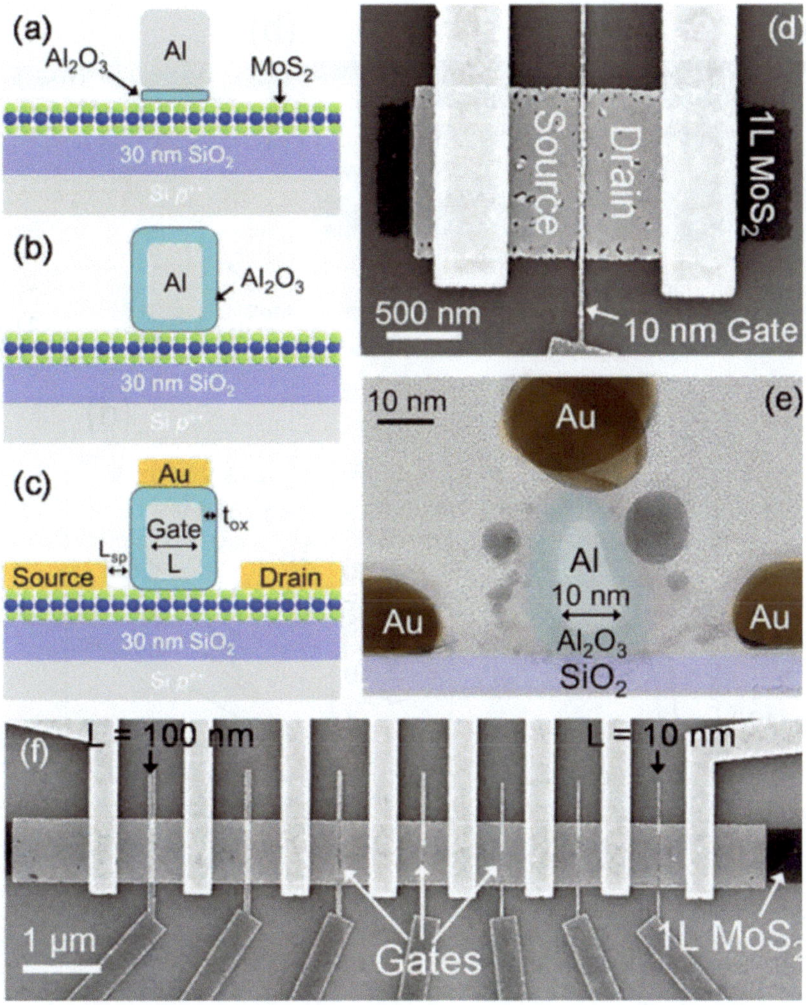


**Fig. 3.5** Fabrication steps of the self-aligned top-gate $MoS_2$ FET with a minimum 10 nm gate length. (**a**) Deposition of Al gate electrode with seed layer, (**b**) formation of self-passivated $Al_2O_3$ gate dielectric, (**c**) self-aligned Au source and drain. (**d**) SEM image (top view) of the device. (**e**) Colorized cross-sectional TEM of the device with the channel length of 10 nm. (**f**) TLM structure used to extract mobility and contact resistance. English et al. [20]

yield and be able to increase the complexity of these circuits, it is necessary to follow an integrated approach where all the main steps to make functional $MoS_2$ circuits, including material synthesis, device technology, compact modeling, circuit design, layout automation, chip fabrication, and circuit measurement are connected (Fig. 3.7) [25–28]. Such an end-to-end design flow allows for the rapid optimization of the material synthesis, device technology, and circuit layout/design to ensure maximum system yield.

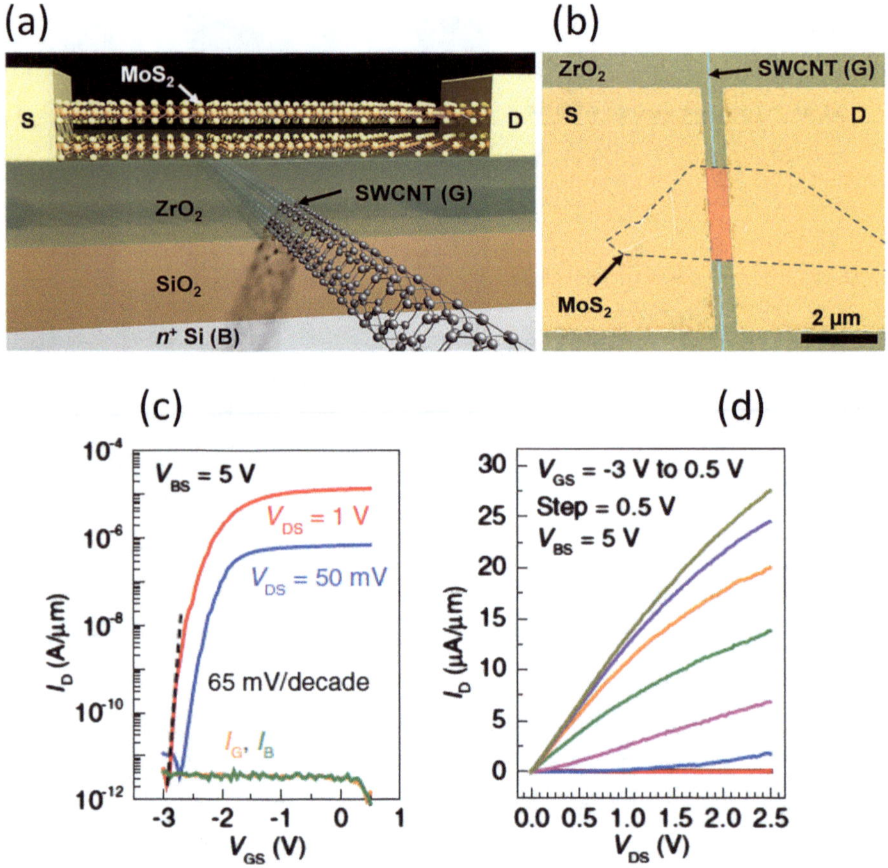


**Fig. 3.6** Schematic of SWCNT-gated $MoS_2$ FETs. (**b**) Optical image showing a device built on a $MoS_2$ flake (indicated by dashed lines) and its SWCNT top gate. (**c**, **d**) $I_D$–$V_{GS}$ and $I_D$–$V_{DS}$ of the SWCNT-gated FET with bilayer $MoS_2$ as a channel. Desai et al. [21]

A main challenge for $MoS_2$ circuit development is that $MoS_2$ is natively n-type, as the material is subject to sulfur vacancies during the CVD growth process resulting in intrinsic electron concentration. Therefore, most $MoS_2$ transistors reported in literature are *n*-MOS with negative threshold voltage [22, 23]. Although tremendous efforts have been made to realize *p*-MOS with $MoS_2$ [29] and new 2D semiconductors have been discovered with improved ambipolar conduction [30, 31], *n*-type $MoS_2$ has so far been the most advanced technology for large-scale integrated circuits. Despite chemical doping of $MoS_2$ in both n type and p type in a controlled manner is key to realize a CMOS technology, it is beyond the scope of this chapter. Nevertheless, enhancement mode FETs with positive threshold voltage are necessary to cascade logic circuits and complete a stand-alone system. Furthermore, FETs made of new materials such as $MoS_2$ often suffer from large process variations and hysteresis, which result in significant performance degradation and poor long-term reliability. This problem should be addressed in both device technology and circuit design.

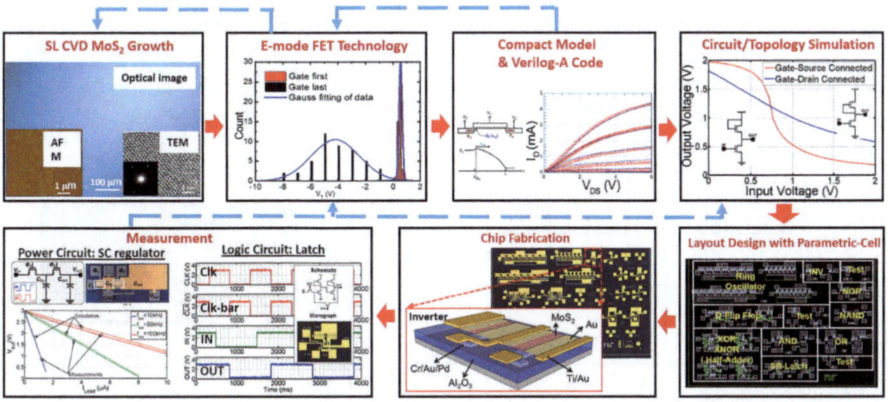


**Fig. 3.7** Design flow for $MoS_2$-based electronics. The flow connects material growth, device technology, compact modeling, circuit design, layout generation, chip fabrication and circuit's measurement. This speeds up the iteration between different development stages of new material technology. The interactions between different developments stages are indicated as dashed arrow [26]

Most 2D transistors today are fabricated with a gate-last process. Dramatic improvement can be achieved by making the gate structure first, and depositing and annealing the gate dielectric before transferring large area, single layer $MoS_2$. This technology, when compared with the conventional gate-last technology, greatly improves the device uniformity. Statistical study of the transistor performance shows that gate-first devices have an average $V_t$ of 2.41 V with a standard deviation of 0.17 V while those from gate-last process have an average $V_t$ of −4.20 V and a much larger standard deviation of 1.75 V [26].

To demonstrate the high uniformity of the new technology, various combinational (NAND, NOR, AND, OR, XOR, XNOR) and sequential (latches, edge-triggered registers) digital circuit as well as switch capacitor voltage regulators based on $MoS_2$ were fabricated on a silicon wafer [25, 26]. Figure 3.8 shows the performance of an edge-triggered register, which is designed using two latches in a master–slave configuration. When the clock is low, the master (connected to the input) is transparent while the slave is in hold mode. When the CLK turns high, the master is in hold mode while the slave latch is transparent. Thus, the output of the edge-triggered register will capture the input value at the positive edge of the CLK. The measurement results of the designed circuits fit well with the simulation, indicating the great promise of our technology and CAD flow for realizing large-scale complex $MoS_2$ systems. The complex four-stage integrated circuits evidence the robustness and scalability of the technology.

The flow as shown in Fig. 3.7 provides an end-to-end guidance to implement functional $MoS_2$ circuits from scratch. However, the availability of more complex circuits is still limited by the poor circuit yield. In the case of $MoS_2$, this yield is limited by variations in the $MoS_2$ material quality, non-uniform doping level across

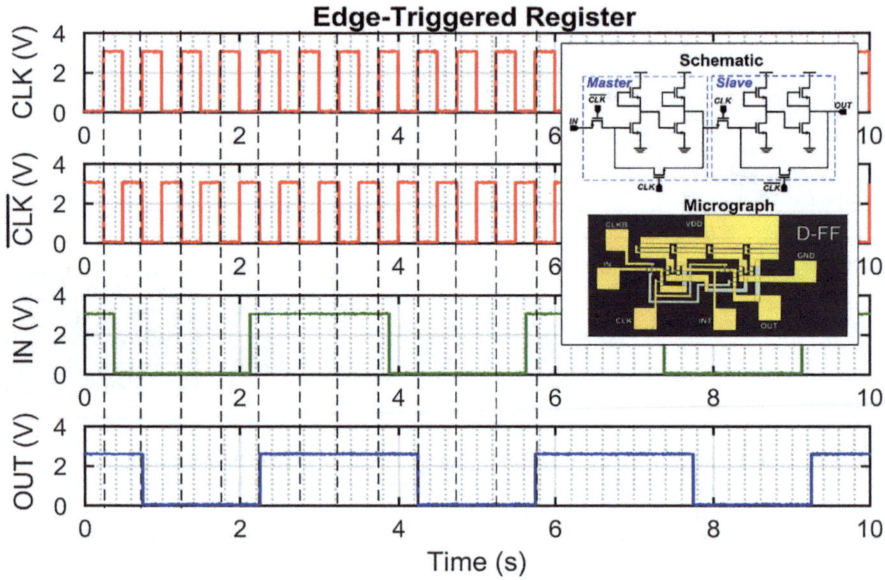


**Fig. 3.8** Schematic, micrograph and measurement results of the fabricated positive edge-triggered register [25, 26]

the sample and scattering from impurities and nucleation centers. Thus, in order to develop a stable technology using $MoS_2$, the sources of variability in the fabrication process need to be understood and captured at the modeling stage and counteracted at the circuit design stage. Moreover, a scalable and flexible yield model, which can quickly evaluate fabrication process and designs as the number of transistors scale, is highly valuable.

Several factors have been identified to be critical to reduce the variation in $MoS_2$ devices. They include: uniformity of material, cleanness of transfer process, robustness of passivation technology and immunity to threshold voltage change in the designed circuits. Figure 3.9 shows that the devices made of high quality CVD $MoS_2$ (Fig. 3.9a, c, e, g) have tighter distribution of current, lower off current, higher on current, compared with low quality CVD $MoS_2$ devices with nucleation particles from growth phase (Fig. 3.9b, d, f, h) [26, 27]. A statistic compact model has been developed to capture the corner performances of the devices for a certain technology as shown in red line in Fig. 3.9c, d, allowing capture variation of devices in circuit design. Very importantly, clean high-quality sample results in very small threshold variation ($\Delta V_{to} = 36$ mV) compared with low quality sample ($\Delta V_{to} = 0.17$ V) [26, 27].

To be able to fabricate large $MoS_2$ circuits on a silicon chip, it is important to develop yield models for those complex logic circuits. This was done by simulating the noise margin (NM) of the $MoS_2$ transistors based on their key parameters [26, 27]. The simulated NM of zero-$V_{GS}$ inverters with different threshold voltages for

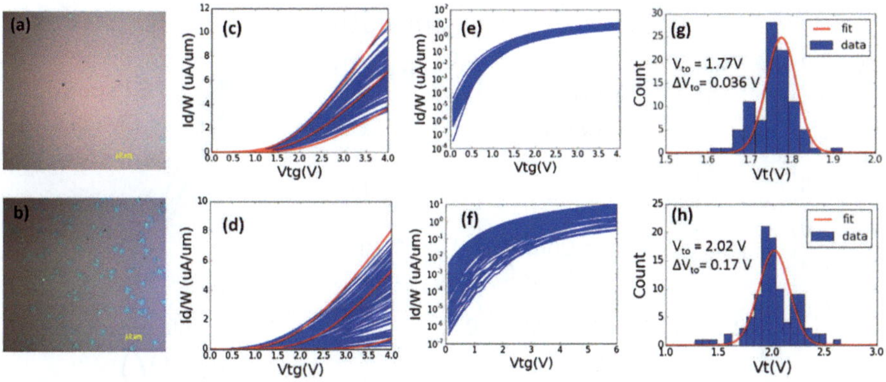


**Fig. 3.9** Performance of the transistor arrays with (**a**) high (sample A) and (**b**) low (sample B) quality $MoS_2$ in linear (**c, d**) and log (**e, f**) scales. The red solid lines are the compact models capturing the slow and fast corners of devices on each sample. Threshold voltage distribution of sample A (**g**) and sample B (**h**). Both threshold voltage distributions are fitted using Gaussian distribution with average value of $V_{\text{to}}$ and standard deviation value of $\Delta V_{t0}$. The fitted parameters for both samples are shown in the images [26, 27]

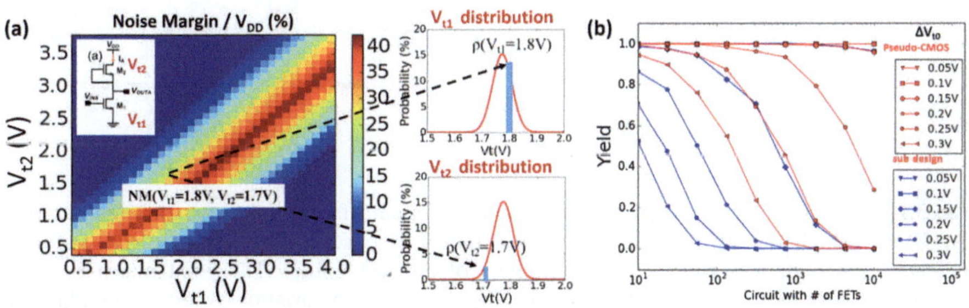


**Fig. 3.10** (**a**) Color map of noise margin/VDD for the zero-$V_{GS}$ inverter as a function of various threshold voltage values for top ($V_{t1}$) and bottom ($V_{t2}$) transistors. The probability of a given noise margin is the joint probability of ($V_{t1}$, $V_{t2}$) pair, e.g., for the probability of the highlighted point, NM ($V_{t1} = 1.7$ V, $V_{t2} = 1.8$ V), $\rho$(NM) $= \rho(V_{t1} = 1.7\text{ V})^* \rho(V_{t2} = 1.8\text{ V})$. (**b**). Yield model for different circuit complexities for both pseudo CMOS and zero-VGS design, with various $V_{t0}$ and constant $V_{t0}$ of 1.8 V. Both topologies show close-to-unit yield for wide range of $V_{t0}$ and $\Delta V_{t0}$ [26, 27]

top and bottom transistors are summarized in the color map shown in Fig. 3.10. The dark blue region indicates NM < 0, that is circuit failure. By combining the NM plot with the statistical distribution of $V_t$'s in the $MoS_2$ technology, it is possible to determine the circuit yield as a function of transistor count ($N$). All transistors in the circuit are assumed to be independent from each other, and their threshold voltages are sampled according to the sample statistical distribution (with measured $V_{t0}$ and $\Delta V_{t0}$). The simulation results for both zero-$V_{GS}$ and pseudo CMOS [32] design are summarized in Fig. 3.10b. As expected, for both topologies, the yield decreases with increasing the circuit complexity ($N$) and global and local variations.

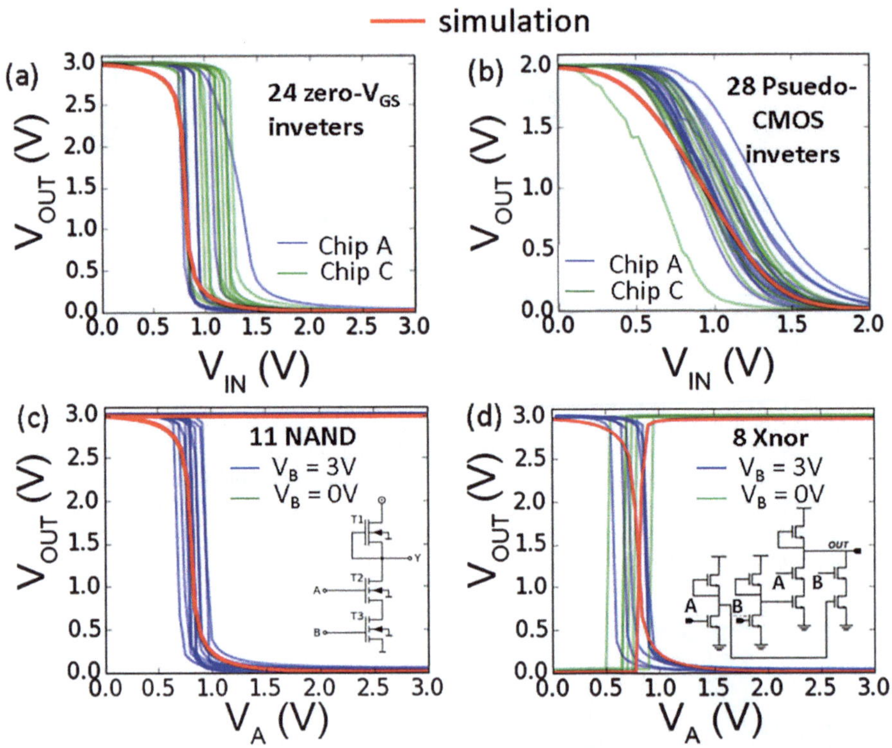


**Fig. 3.11** Experimental and simulated performance of circuits fabricated on two different samples (**A**, **C**) with high quality material. No circuit failure is observed across those two samples, indicating the high performance and robustness of the technology. The good match between the measurement and simulation demonstrates the power of the variation-aware design platform [26, 27]. (**a**) Zero-VGS inverters; (**b**) Psuedo-CMOS inverters; (**c**) NAND gates; (**d**) XNOR gates

Figure 3.10b shows the yield for different standard deviation of the threshold voltage ($\Delta V_{t0}$ from 0.05 V to 0.3 V) for both designs, with $V_{t0}$ of 1.8 V. For $\Delta V_{t0}$ smaller than 0.15 V, both designs show close-to-unity yield and pseudo CMOS design is more robust to larger $\Delta V_{t0}$. Figure 3.11 depicts the simulation and the measurement results of part of the combination circuits and from inverters from two samples (A and C) made with high quality material. No circuit failure is observed across these two samples, indicating the high performance and robustness of the technology. The good match between the measurement and simulation evidences the power of the proposed variation-aware design platform when designing large area complex circuits with new material systems such as $MoS_2$.

## 3.3 2D Materials for IR Detectors on a Si Platform

Infrared (IR) sensing technologies, originally used mainly for night vision, surveillance, and remote controlling in military applications, have gradually shifted to applications for civilian use, including medical, industry, earth resources, and automotive. For example, medical diagnostics can be assisted by IR thermography, in which IR scanning is used to detect spatial temperature abnormality and identify in this way cancers and other trauma [33]. The use of high resolution IR imagers and/or spectrographs on aircrafts and satellites has also allowed the chemical mapping of the surface of earth, with important applications in mineral search [34], as well as geological and environmental survey [35], and inspection of natural hazards and disasters [36]. However, in spite of the large number of applications where IR sensing is key, its use is still limited by the high cost of high performance IR detectors.

The bandgap of 2D materials spans from 0 electron volts (eV) in the case of monolayer graphene, all the way to around 5 eV in the case of hexagonal boron nitride (Fig. 3.12). The energy gaps of graphene and black phosphorus, in particular, lie in the mid- to far-infrared range, and exhibit abundant novel light–matter interaction phenomena, such as thermoelectric effect, photothermoelectric effect, and various other optoelectronic effects related to hot carrier dynamics. The use of these novel phenomena could lead to ultrasensitive and/or fast-response IR detectors with the potential to outperform today's state-of-the-art IR detection technologies.

An IR imaging system is usually made of a large number of detectors, which need to be routed to readout integrated circuits (ROICs) that amplify the signals and implement analog-to-digital conversion. The integration of the detector arrays and the ROICs has traditionally been a major challenge in IR system design, due to material integration and noise issues. As a result, in a conventional IR imaging

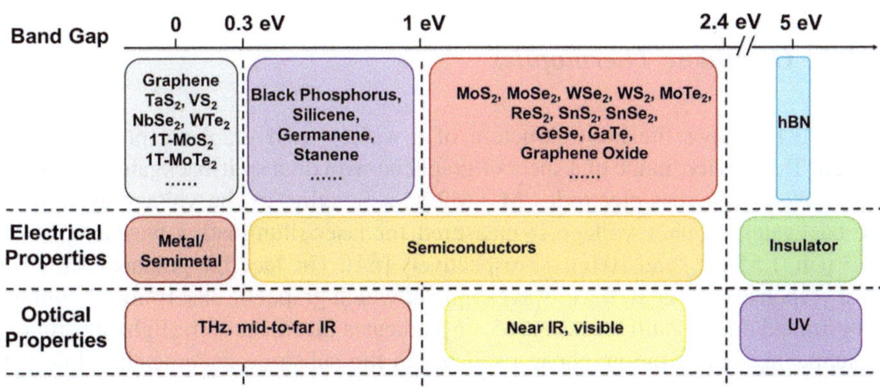


**Fig. 3.12** Summary of band gaps ($E_{bg}$ in eV) of various 2D materials and corresponding wavelength each material is capable of detecting

system, the detector array and the ROIC are usually fabricated on separate wafers and then integrated in a hybrid fashion. However, recent advances of wafer-scale synthesis of graphene and other 2D materials, the easiness of transferring 2D materials onto arbitrary substrates and their low-temperature fabrication technology make it possible to monolithically integrate 2D material-based IR detectors with CMOS integrated circuits. This is expected to reduce the noise and the cost for these systems.

In this section, we will first summarize the current research on 2D material-based IR detectors, and then we will focus on graphene thermopiles as an example to investigate their potential as compared to the state-of-the-art mainstream technologies. Finally, we will discuss the possibility and benefits of monolithically integrating 2D material image arrays and CMOS integrated circuits.

### 3.3.1 2D Material for Infrared Detectors

Due to its linear electronic dispersion relation, monolayer graphene can show ultra-high mobility (up to 200,000 $cm^2/Vs$ [37–39]), which makes graphene an excellent candidate for high-speed electronic or optoelectronic applications. Moreover, the interband optical absorption of suspended, near-intrinsic monolayer graphene is $\pi\alpha \approx 2.3\%$ for a wide range of incident photon energy, determined by the fine structure constant, $\alpha = e^2/\hbar c \approx 1/137$ [40]. As for heavily doped graphene, the interband transition is forbidden by Pauli blocking, whereas the intraband, free-carrier absorption could lead to a resonant absorption, called surface plasmon polariton, in which free electrons and holes in graphene vibrate in-plane with respect to incident light. The resonant absorption can be enhanced efficiently by spatially confining the electromagnetic wave excitation [41–45]. Table 3.1 summarized the key performance metrics of different 2D material infrared detectors.

### 3.3.2 Graphene Thermopiles

Figure 3.13 shows the basic structure of a widely used graphene photodetector design. This device, made of a sheet of graphene with dual split-backgates, develops a photovoltage across electrodes M1–M2 as a function of the voltage applied to the backgates. A photovoltage is measured for laser illumination wavelengths of 0.83 μm, 1.55 μm, and 10.6 μm, respectively [64]. The fact that graphene shows a clear response even at 10.6 μm wavelength at which graphene absorption should be very limited due to Pauli blocking [65, 66] indicates that most of the light absorption is happening not in the graphene itself but in the substrate underneath, while the graphene devices are using the thermoelectric effect to convert the temperature rise in the substrate to a voltage difference [47, 64, 67]. This voltage is described by $V_{TE} = \Delta S \cdot \Delta T$, where $\Delta S$ is the difference of the Seebeck coefficient between the

**Table 3.1** Summary of performances of various 2D material-based infrared detectors

| Device structure | Wavelength | Responsivity | Specific detectivity | Response time | Operating temperature | Ref. |
|---|---|---|---|---|---|---|
| Gr. p–n junct. PTE | Visible, NIR | 10 mA/W | – | 50 fs | Room T | [46–48] |
| Gr.-metal plasmonic | MIR | 0.5 V/W | – | 60 ns | Room T | [49] |
| Gr. thermopile | MIR | 10 V/W | $8 \times 10^8$ cm $Hz^{1/2}$ $W^{-1}$ | 23 ms | Room T | [50] |
| Biased gr. bolometer | Visible, NIR | 0.2 mA/W | – | – | Room T | [51, 52] |
| Dual-gated 2 L gr. bolometer | MIR | $10^5$ V/W | $3 \times 10^{10}$ cm $Hz^{1/2}$ $W^{-1}$ | <1 ns | <10 K | [52] |
| Gr./QDs photo-gating | Visible NIR | $10^8$ A/W | $7 \times 10^{13}$ cm $Hz^{1/2}$ $W^{-1}$ | 10 ms | Room T | [53, 54] |
| Gr./insulator/gr. photo-gating | Visible, NIR, MIR | >1 A/W | $5 \times 10^7$ cm $Hz^{1/2}$ $W^{-1}$ | 1 s | Room T | [55] |
| Gr./insulator/gr. thermionic | NIR | 0.01–0.2 mA/W | – | 10 fs | Room T | [56, 57] |
| Gr./semi. PV | Visible, NIR | 0.7 A/W | $5 \times 10^{13}$ cm $Hz^{1/2}$ $W^{-1}$ | 0.3 ms | Room T | [58, 59] |
| Gr./pyroelectric | MIR | 0.27 mA/W | $6 \times 10^4$ cm $Hz^{1/2}$ $W^{-1}$ | 20 ms | Room T | [60] |
| Gr./NEMS resonator | MIR | – | $1 \times 10^{10}$ cm $Hz^{1/2}$ $W^{-1}$ | 3 ns | Room T | [61] |
| BP photoconductor | Visible, NIR | 5 mA/W | – | 40 μs | Room T | [62, 63] |

*p*- and n-region of graphene, and $\Delta T$ is the temperature difference between the graphene *p–n* junction and the metal contacts. Because the Seebeck coefficient of graphene has an "S" shaped dependence on the Fermi level, the split-gate sweeping maps as shown in Fig. 3.13c, d, e indicate a non-monotonic six-lobe pattern, which is not true if the photovoltaic effect dominated the photocurrent generation.

In a related work, the spectral response of graphene on $SiO_2$ was measured and it was found to match the absorption spectrum of $SiO_2$ very well in the mid-IR. However, the detectivity of these devices was only $10^2$ cm $Hz^{1/2}$ $s^{-1}$. By optimizing the device structure, a follow-up study achieved a detectivity of $10^5$ cm $Hz^{1/2}$ $s^{-1}$ [50]. The observed three orders of magnitude improvement in performance was attributed to two structural changes. First, the absorption in the 8–12 μm spectral range was increased by >40%, as compared to less than 1% absorption in the previous work, thanks to the use of an optimized IR absorption layer made of a $SiO_2/Si_3N_4/SiO_2$ combination instead of a single dielectric layer, which allowed strong resonant absorption at 9 and 11.5 μm, respectively. Secondly, the deposition of the $SiO_2/Si_3N_4/SiO_2$ tri-layer absorber was done with plasma enhanced chemical vapor deposition (PECVD) under an optimized high-frequency

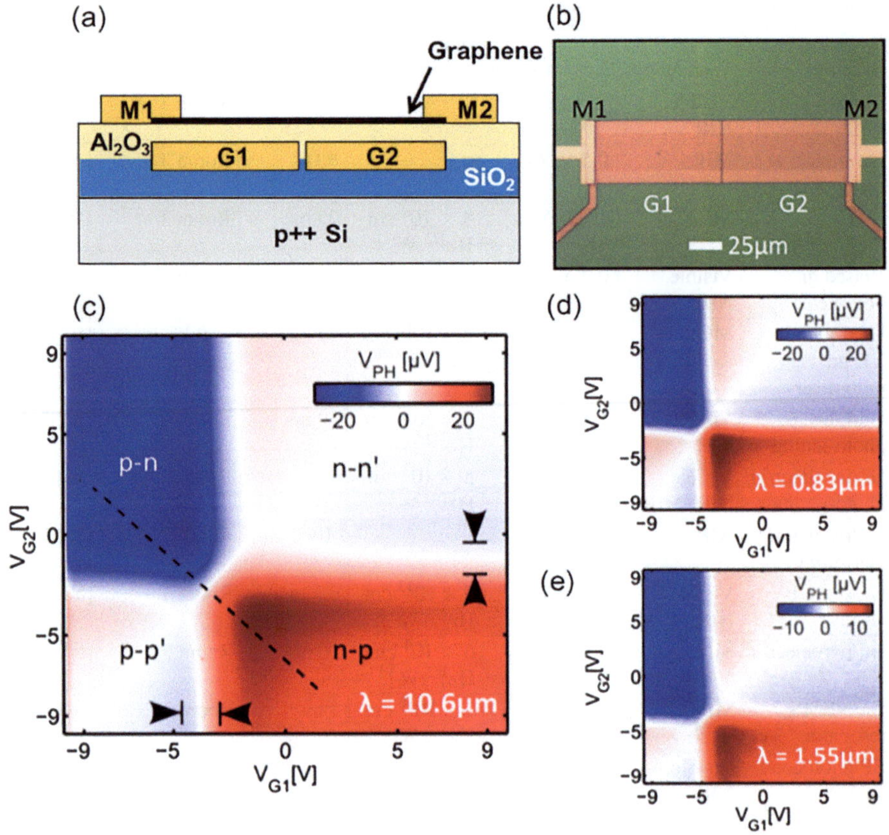


**Fig. 3.13** Graphene split-gate thermopile with supported substrate. (**a**) Schematic and (**b**) microscopic image of the device. M1 and M2 are metal contacts to graphene, and G1 and G2 are split gates that electrostatically dope the graphene channel to form a p-n junction. (**c–e**) Photovoltage ($V_{PH}$) as a function of the split gate voltages $V_{G1}$ and $V_{G2}$, for different wavelength of incident light. (**c**) 10.6 μm, (**d**) 0.83 μm, and (**e**) 1.55 μm. The six-lobe feature indicates that the photoresponse is dominated by thermoelectric effect in graphene [64]

to low-frequency plasma ratio to make the thin film stress-free. This allowed for the fabrication of a free-standing absorber membrane after undercutting the silicon underneath with $XeF_2$ isotropic etching. When measured in vacuum (less than $10^{-2}$ torr), both thermal conduction and convection in the vertical direction were efficiently attenuated. As a result, the temperature generated by the IR irradiance at the absorber is maximized.

Multiple graphene photodetectors can be combined into the thermopile shown in Fig. 3.14a, composed of an infrared absorber that is suspended from the substrate, a series of thermal arms that connect the absorber and the surrounding, with interleaved *p*- and *n*- type graphene channels on top. The schematic of the structure and the key geometrical parameters are shown in Fig. 3.14b. The graphene channels

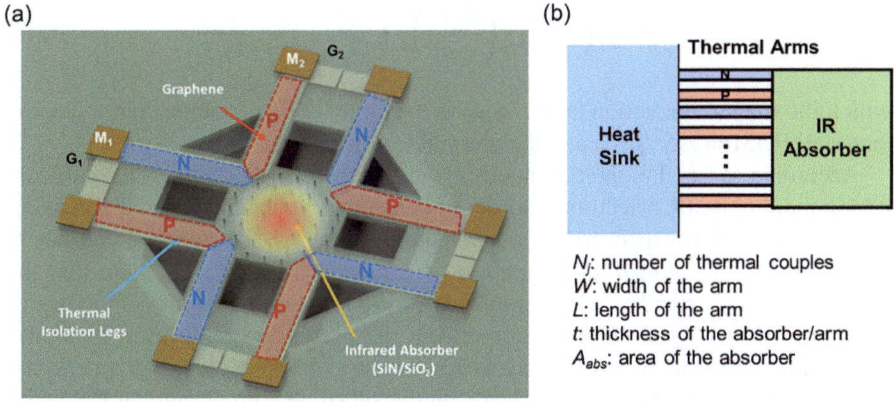


**Fig. 3.14** Graphene thermopile with suspended IR absorber. (**a**) Schematic of a graphene thermopile. The red and blue regions indicate the *p*-type and *n*-type region of graphene, and the square in the center is the dielectric absorber. The whole structure is suspended on the substrate to reduce the thermal conductance in the vertical direction. (**b**) A geometrical abstraction of the graphene thermopile with geometrical parameters listed below

are parallel in terms of temperature gradient, but connected in series electrically. When IR radiation is present, the IR absorber (dielectric multilayer thin film) is heated up, which can then be probed electrically by the graphene p–n junctions due to the thermoelectric effect. The specific detectivity ($D^*$), considering the Johnson–Nyquist noise, can be expressed as

$$D^*=\frac{N_j\Delta S\Delta T}{P_{\text{in}}\sqrt{\overline{v_n^2}}}\sqrt{D_{\text{abs}}^2\Delta f}=\left(\frac{\alpha_{\text{abs}}}{t}\right)\cdot\left(\frac{\Delta S}{\sqrt{\rho_{2D}}}\right)\cdot\left(\frac{1}{k_{\text{th}}}\right)\cdot\left(\frac{L^{1/2}D_{\text{abs}}}{N_j^{1/2}W^{1/2}}\right)\sqrt{\frac{1}{32k_BT}} \tag{3.1}$$

Here the first term ($\alpha_{\text{abs}}/t$) is absorbance per thickness, indicating the capability of IR absorption of the absorber; the second term ($\Delta S/{\rho_{2D}}^{1/2}$), with $\rho_{2D}$ the 2D resistivity of graphene, is determined by the electrical and thermoelectric properties of the sensing material; the third term ($1/\kappa_{\text{th}}$) indicates the quality of thermal isolation, with $\kappa_{\text{th}}$ denoting the thermal conductivity of the absorber; and the fourth term is made of geometrical parameters in the parallel direction, where $N_j$ is the total number of graphene *p–n* junctions at the hot spot. The response time of such a device can be written as

$$\tau = R_{\text{th}}\cdot C_{\text{th}} = \frac{1}{2}\left(\frac{c_V}{k_{\text{th}}}\right)\left(\frac{LD_{\text{abs}}^2}{N_jW}\right) \tag{3.2}$$

with the heat capacitance $C_{\text{th}}$, and the specific heat capacity $c_V$. Notice that ($D^{*2}/\tau$) is independent of the lateral geometries:

$$\frac{(D^*)^2}{\tau} = \left(\frac{\alpha_{\text{abs}}}{t}\right) \cdot \left(\frac{\Delta S^2}{\rho_{2\text{D}}}\right) \cdot \left(\frac{1}{k_{\text{th}} c_V}\right) \cdot \frac{1}{16 k_B T} \tag{3.3}$$

with light–matter interaction factor ($\alpha_{\text{abs}}/t$), the figure of merits of the thermoelectric material (FOM = $\Delta S^2/\rho_{2\text{D}}$) and the thermal transport factor ($1/\kappa_{\text{th}} c_V$).

According to Eq. 3.3, the thermoelectric figure of merit FOM plays a significant role in thermopile IR detectors. In order to benchmark graphene-based thermoelectric detectors with respect to the other material systems, the Seebeck coefficient and the FOM are plotted as a function of resistivity in Fig. 3.15. Note that the FOM for today's standard CVD graphene on $SiO_2$, with the average mobility of 2000 $cm^2\ V^{-1}\ s^{-1}$, can already outperform the performance of any thermopiles made with metals and most of TE materials. The use of higher quality graphene and properly passivating the dangling bonds on the substrate with hexagonal BN could make the FOM two orders of magnitude higher than that of all the other material systems. Also, the FOM of TMDs in 2D form is higher than in their 3D counterparts, which also shows great potential for thermal detection and other thermoelectric applications.

Note from Eq. 3.3 that the optical absorption per thickness ($\alpha_{\text{abs}}/t$) and the thermal transport factor ($1/\kappa_{\text{th}} c_V$) of a graphene thermopile can be improved through properly engineering the IR absorber. Current work used the inherit strong absorption of $SiO_2$/SiN layers [50, 64, 68]. It is also possible to use metal-based metamaterial absorbers in the mid-infrared range [69–71]. The surface plasmon polariton (SPP) in graphene [41, 45, 72, 73] and phonon polariton (PP) in hBN [74, 75] have been studied recently. With appropriate engineering of the light field to enhance the light–matter interaction, it is possible to achieve strong resonant absorption based on these mechanisms with negligible loss.

Figure 3.16 compares the specific detectivity ($D^*$) and the response time ($\tau$) of graphene thermopiles with different types of state-of-the-art thermal detector technologies, including bolometers ($VO_x$, CMOS-MEMS, etc.) [6–8], thermopiles [9, 10] (poly-Si, Al, thermoelectric materials, etc.), and pyroelectric devices (PZT and other piezoelectric materials). Note that photon detectors, such as mercury cadmium telluride (MCT) photoconductors, were not included in this comparison, because they would suffer from high dark current and consequently high noise at room temperature [76]. Here we exploit the dimensionless relation from Eq. 3.2 to represent each technology node of graphene thermopiles. Although current graphene thermopile technology is still not as good as the state-of-the-art thermal detectors, the performance is predicted to be competitive, or even better than today's state-of-the-art technologies. For example, a 100-fold improvement in the FOM = $\Delta S^2/\rho$ could be achieved by encapsulating large area, high-quality CVD graphene with hexagonal boron nitride, which would increase the mobility and Seebeck coefficient to 100,000 $cm^2\ V^{-1}\ s^{-1}$ and 200 $\mu$V/K, respectively. Furthermore, the absorber can be thinned down to 10 nm with good mechanical stability, and 100% perfect absorption can be achieved through nano-photonic structures, which would make graphene thermopiles better than any existing bolometers.

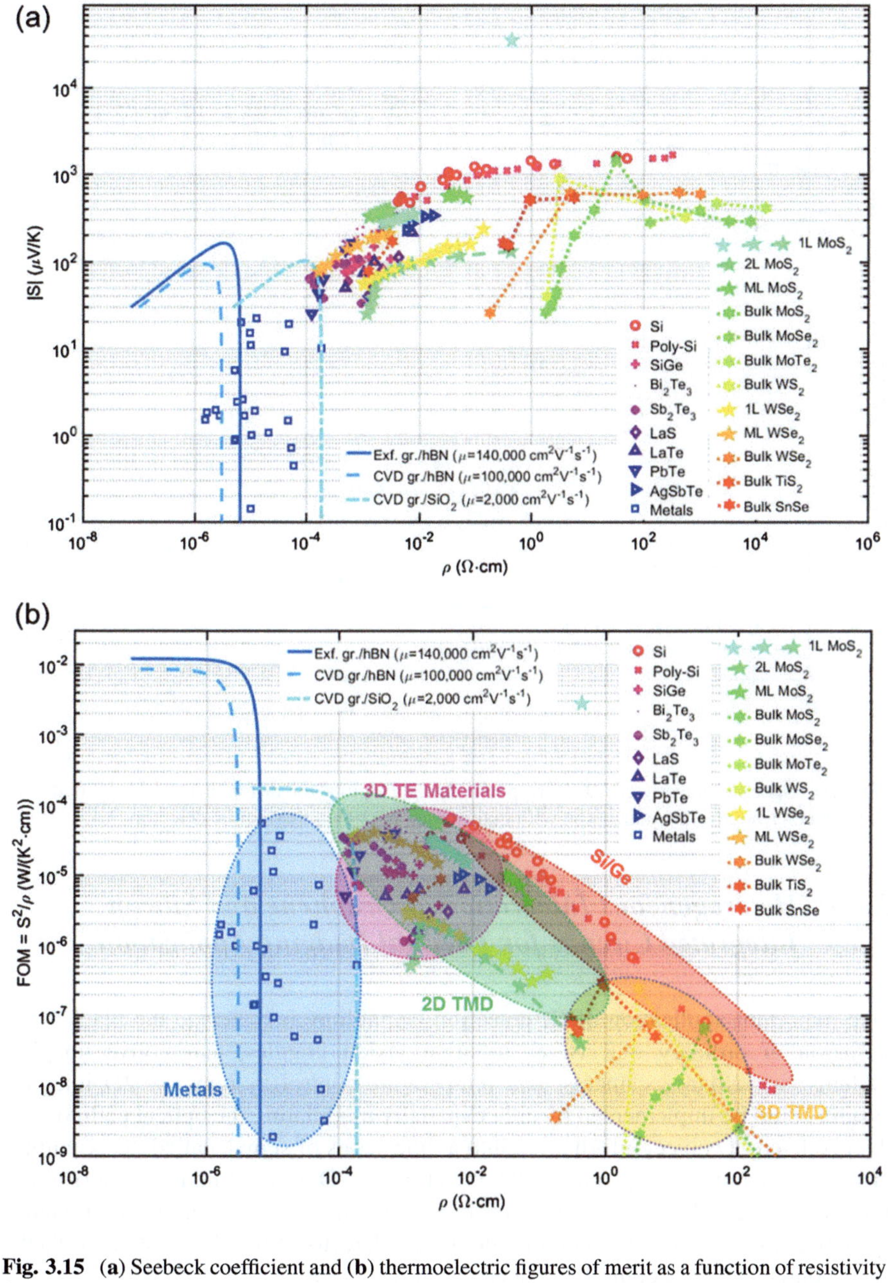


**Fig. 3.15** (**a**) Seebeck coefficient and (**b**) thermoelectric figures of merit as a function of resistivity for various materials

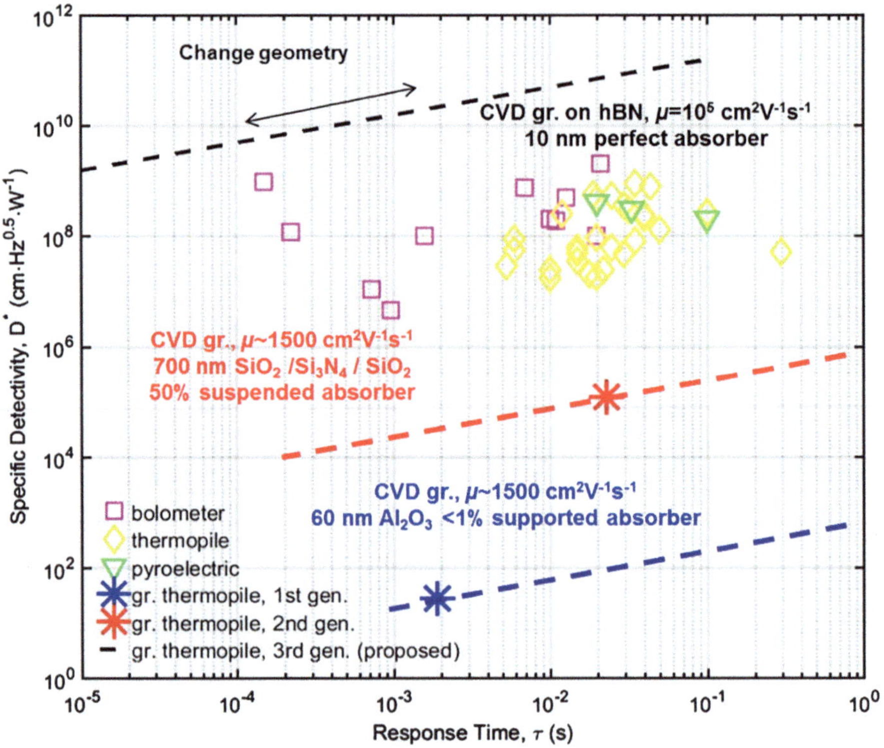


**Fig. 3.16** Specific detectivity ($D^*$)—response time ($\tau$) plots for different technology nodes of graphene thermopiles in comparison with mainstream uncooled thermal IR detectors

### 3.3.3 *Heterogeneous Integration of Graphene and Silicon Integrated Circuits for Thermal Imaging Application*

One important application of IR detection technology is to image the black-body radiation of ordinary objects, as in night vision goggles, automotive imaging systems, surveillance cameras, and temperature-control systems. A 2D imaging system could be implemented by either spatially modulating the optical path to map the pixelated 2D object information into a time series of signal, called 2D scanning, or directly mapping the pixelated 2D object into a 2D array of image sensors at the focal plane, called real-time imaging. Figure 3.17a shows a scanning IR imaging system using a single graphene thermopile IR detector. With this system, a black-body source at 472 K could be imaged, as shown in Fig. 3.17b, c.

A real-time image sensor is made of a focal plane array (FPA), that is a 2D array of IR detectors at the focal plane of the optical system, and a readout integrated

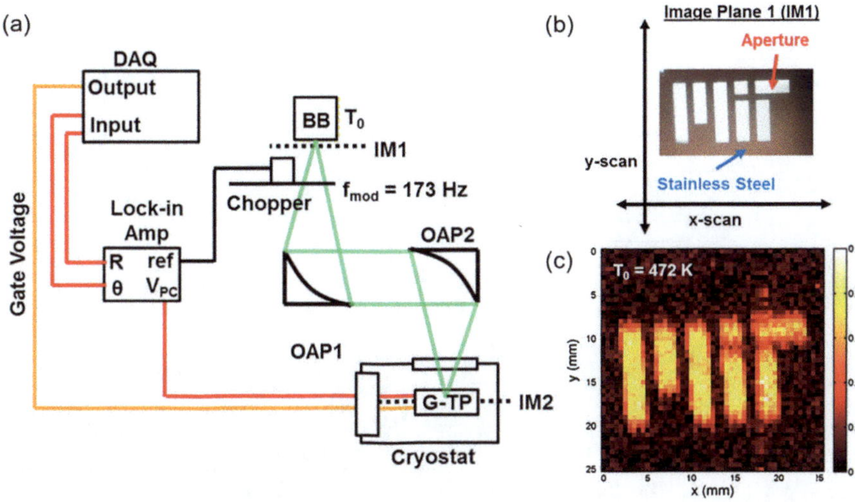


**Fig. 3.17** Scanning thermal imaging system based on a single graphene thermopile. (**a**) Schematics of the scanning imaging setup. (**b**) "MIT" log aperture to be measured. (**c**) Scanning thermal image of the 472 K black-body source passing through the "MIT" log aperture

circuit (ROIC) that controls the operation of the FPAs and converts the analog signals from the FPA into digital signals that can be transmitted to and processed by a microprocessor. The integration between the FPA and the ROIC could be discrete, hybrid, or monolithic. The discrete integration is impractical because the number of connections between the FPA and the ROIC is on the order of thousands or even millions, which could hardly be routed through a PCB board. At present, most of the mainstream IR image sensors have separate FPA and ROIC chips, and integrate them through flip chip bonding. This limits the pixel size, increases the system noise, and reduces the speed of the device. The monolithic integration of the FPA and the ROIC could solve these challenges, however the thermal budget of the ROIC demands that the deposition temperature of the sensing material has to be lower than 400 °C, which degrades the material quality of most 3D semiconductors used in these applications.

Given that 2D materials can be synthesized by chemical vapor deposition on a metal foil, and wet-transferred onto any substrates, it is possible to integrate 2D material-based IR FPAs with CMOS ROICs monolithically without sacrificing the performance of either component. The back-end-of-line process of graphene thermopile FPAs fabricated directly on CMOS ROIC chips is schematically shown in Fig. 3.18a and the optical images of the completed graphene thermopile-CMOS image sensor are shown in Fig. 3.18b, c. This hybrid graphene/Si system enables new opportunities for low cost, high performance IR detection.

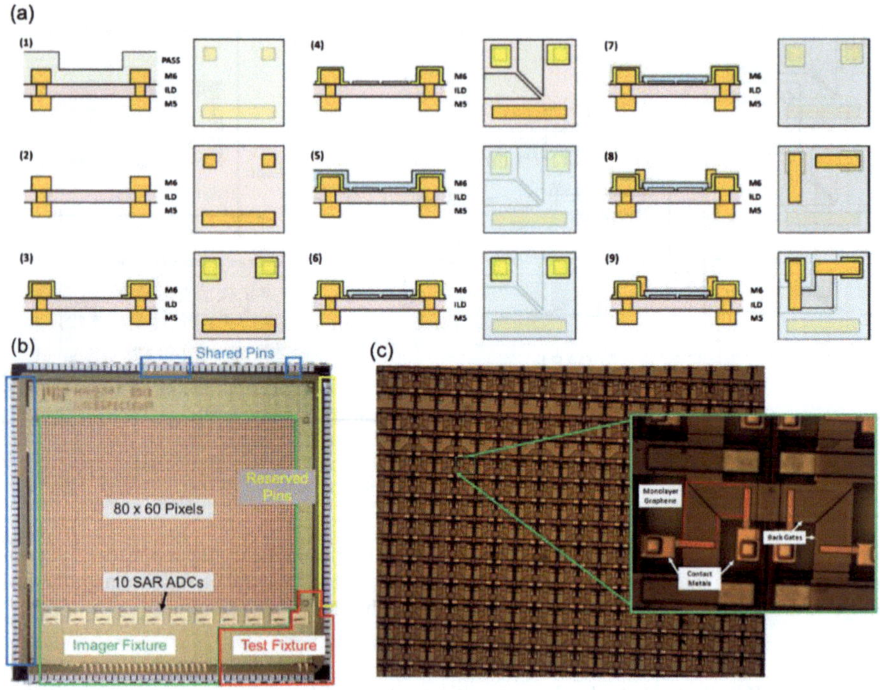


**Fig. 3.18** Graphene thermopile/CMOS monolithic integration. (**a**) Back-end-of-line process of the graphene thermopile FPA fabricated directly onto a CMOS ROIC chip. *PASS* passivation, *ILD* interlayer dielectric, *M6* Metal 6 (Cu), *M5* Metal 5 (Cu), *blue* PECVD SiO2, grey line is graphene. The left and right images are the side and top view of the pixel area, respectively. (**b**) Photograph of the CMOS ROIC chip. (**c**) Microscopic images of the completed Graphene thermopile FPA/CMOS ROIC chip

## 3.4 Heterogeneous Integration of 2D Materials with Si Nanophotonics

Graphene can serve as core materials combined with Si or other conventional semiconductor photonic structures for ultrafast photodetection and optical modulation on a CMOS platform for high speed, low power optical interconnects. Early demonstrated graphene optoelectronics normally rely on a vertical incident configuration to couple light into graphene, therefore resulting in a low absorption coefficient of 2.3% due to the atomic thickness of graphene [77]. Although those devices show promising properties of strong electro-absorptive effect [65, 66] and high-speed photodetection [78, 79], the low total absorption in those devices does not meet the requirement of practical applications. To tackle this issue, nanophotonic devices, including waveguides and cavities, can squeeze optical field in a subwavelength volume, providing a promising platform to enhance light–matter interaction in 2D materials.

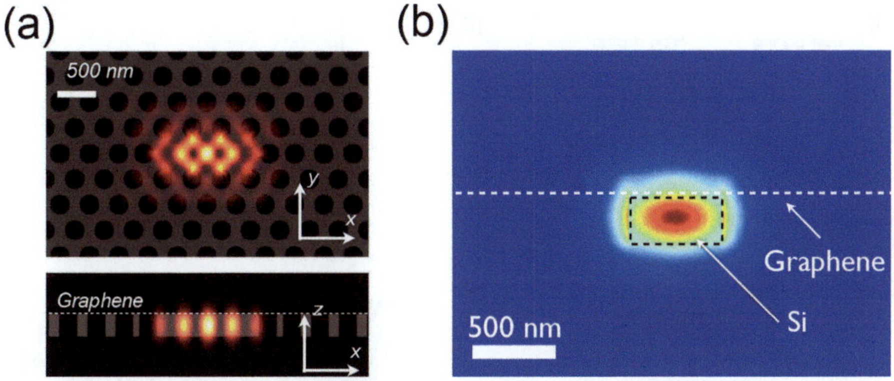


**Fig. 3.19** Finite element simulation (COMSOL) of the waveguide mode indicates that the evanescently coupled bilayer graphene on top of the waveguide introduce a loss of ~0.085 dB/μm. The absorption coefficient suggests that, for example, a 70-μm-long overlapping of graphene with waveguide will absorb ~70% (6 dB) of the incoming light incident from the waveguide, which is consistent with the experimental results and much greater than the 0.1 dB absorption in the normal-incidence configuration

When 2D materials couple to a nanophotonic cavity, enhanced light–matter interaction of graphene in a cavity can enable spectrally selective, order-of-magnitude enhancement in optical absorption and intensity modulation in cavity reflection. Using a temporal coupled mode theory that considers the absorption of graphene coupling to the cavity, Gan et al. showed that the cavity reflection and the absorption strongly depend on the intrinsic cavity loss $\kappa_c$ without 2D materials and the excess loss introduced by absorption of 2D materials $\kappa_{2D}$ in the cavity. The attenuation ratio for the reflectivities with and without 2D materials indicates that the reflectivity of the 2D materials-cavity system can achieve a contrast of 20 dB with a tuning of $\kappa_c/\kappa_{2D}$ in the range of 0.1–1. On the contrary, the on-resonance absorbance of 2D materials has a maximum value of $\eta$, which is the out-coupling efficiency of the cavity field to the environment. The maximum occurs when the two decay rates satisfy $\kappa_{2D} = \kappa_c$, i.e. in the critical coupling condition when excess loss of 2D materials equals to the intrinsic cavity loss.

Figure 3.19a shows an experimental implementation of a coupled graphene-cavity system. The cavity was created by drilling a planar photonic crystal (PPC) air-hole lattice in a suspended silicon membrane. Linear three-missing-hole (L3) defect in the middle of the lattice forms the cavity area. Heterogeneously integrated graphene on top of the membrane can couple to the optical evanescent field and alter the cavity quality factor and out-coupling radiation intensity. Both experimental and calculation results show that such an evanescently coupled graphene-cavity system results in a loss ratio of $\kappa_c/\kappa_{2D} \sim 0.1$, indicating a maximum modulation contrast in reflection can reach more than 20 dB if graphene becomes completely transparent. In this cavity, the out-coupling of the cavity field to the free space radiation mode is symmetric. Therefore, the maximum $\eta$ is 50% if the waveguide-cavity coupling is

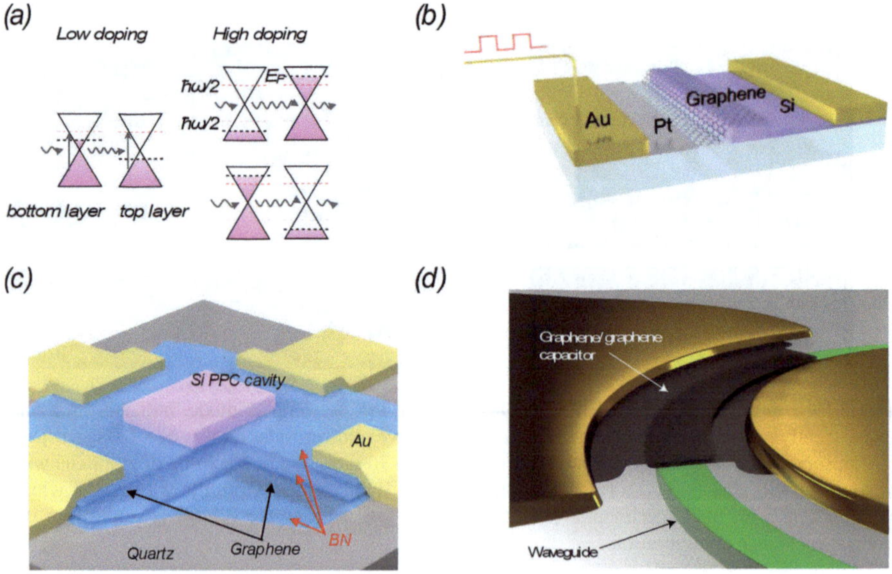


**Fig. 3.20** Schematic of the electro-absorptive effects in graphene. When the Fermi level in graphene raises (descends), the absorption of graphene reduces due to the Pauli-blocking of the interband transition of the electrons in graphene. (**b**) Schematic of a broadband modulator by heterogeneous integration of graphene on top of a silicon waveguide. (**c**) Schematic of the high-speed graphene E–O modulator integrated with a PPC L3 cavity. (**d**) Schematic of a graphene-cladded E–O modulator on top of a silicon nitride ring resonator

responsible for the entire intrinsic loss of the cavity. For cavities with traveling-wave resonant modes, such as ring resonators [80], the $\eta_{\text{maximum}}$ is 100%, enabling 100% absorption in 2D materials.

The enhanced light–matter interaction via optical resonators is inherently narrowband due to the narrow resonant bandwidth of the resonators. On the contrary, coupling 2D materials with a single mode bus waveguide can enhance their interaction with light across a broad spectrum due to the large extension of the interaction length. Figure 3.20b shows a schematic of a graphene layer deposited on top of a silicon waveguide.

### *3.4.1 High-Speed Graphene Electro-Optic Modulators*

Due to the unique linear dispersed band structure [81], graphene exhibits uniform absorption in the spectral ranges from visible to mid-infrared [77]. In addition, the absorption of graphene varies by tuning its Fermi energy ($E_F$) via an electrostatic gate voltage, as schematically shown in Fig. 3.20a. When $E_F$ is tuned away from the Dirac point by more than half of the photon energy $\omega/2$, the interband transitions become forbidden by Pauli blocking, reducing the graphene absorption [65, 66].

Heterogeneously integrated graphene E–O modulators first appear in a structure consisting of a silicon waveguide and ~50-μm-long graphene overlapping [82], as shown in Fig. 3.20b. The enhanced total absorption of the graphene layer on top of the waveguide results in enhanced electro-absorptive effects, enabling broadband intensity modulation for wavelengths from 1.35 to 1.6 μm with 3 dB modulation depths. The 3 dB cut-off speed of this device is about 1 GHz. An improved dual-layer graphene capacitor structure on waveguide shows improved modulation depths of 6 dB [83] and 16 dB [84] with low insertion loss of 3 dB, demonstrating the scalability to acquire higher modulation depth with stacking of multiple single graphene layers.

In these waveguide-based E–O modulators, the footprint of the devices is intrinsically limited by the coupling strength of the waveguide and the graphene layers, which commonly requires a device area of 50 $\mu m^2$. To reduce the footprint of the E–O modulator, therefore reducing the switching energy and increasing the operation speed, a resonator-coupled graphene structure is desired. When graphene strongly absorbs the optical field in a cavity, i.e., $\kappa_c/\kappa_{cg} \ll 1$, the cavity reflection R is strongly attenuated. While the Fermi level in graphene increases (decreases) due to electrostatic gating, the optical absorption in graphene reduces ($\kappa_c/\kappa_{cg}$ increases). Therefore the reflection $R$ of the cavity recovers. This modulation mechanism can in principle provide more than 20 dB modulation depths for a cavity area around 0.5 $\mu m^2$.

We discuss an example of a multilayer graphene/boron nitride heterostructure integrated a PPC L3 cavity. In this architecture, mutually gated dual-layer graphene parallel capacitor could provide high doping strength while operating at high speed due to high carrier mobility of graphene [85]. Figure 3.20c shows the schematic of the high-speed graphene E–O modulators consisting of a dual-layer graphene capacitor and a PPC nanocavity [86]. A BN/Graphene/BN/Graphene/BN five-layer stack was built by the van der Waals (vdW) assembly technique and then transferred onto a quartz substrate [85]. The two graphene sheets were positioned as crossed stripes in order to be contacted individually. The metal contacts to the graphene employ a one-dimensional edge contact technique. In this encapsulated dual-layer graphene structure, each one of the graphene sheets can supply gate voltage to each other.

The device showed in Fig. 3.20c exhibits a maximum modulation depth of 3.2 dB with a high-speed cut-off frequency around 1.2 GHz. The operation frequency response of the device indicates a RC-limited time constant of the dual-layer graphene capacitor, as deduced by the impedance measurement of the device. In this device, the graphene capacitor has an area of ~100 $\mu m^2$ and a capacitance of 320 fF. The switching energy of this device is approximately 1 pJ/bit. For L3 PPC cavities, the overlap between the resonant mode and the graphene capacitor has an area of only ~0.5 $\mu m^2$, corresponding to the three-missing-hole defect region. The graphene capacitor could therefore be reduced in size to match this cavity area to lower the capacitance by approximately 200 times, which should reduce the switching energy to 5 fJ/bit and increase the 3 dB cut-off frequency to 70 GHz

[86]. The cavity bandwidth in this work exceeds 600 GHz for a $Q$ value of 300, i.e., it would be possible to obtain a relatively large modulation contrast without the need for particularly high-$Q$ cavities, as is required in silicon carrier-depletion (injection) modulators. This broader bandwidth would also improve temperature stability, which is a limiting factor in the carrier-modulation of Si modulators [87–89].

The electro-absorptive effect in graphene can not only induce tunable loss of the cavity, but also the cavity coupling efficiency to external optical modes, as the device demonstrated by Phare et al. [90] Fig. 3.20d shows the dual-layer graphene capacitor covering a silicon nitride ring resonator. The ring resonator is intentionally designed to be under-coupled to a bus waveguide when the graphene introduces loss to the cavity. While the absorption in graphene reduces, the resonant field becomes stronger and the coupling of the resonator to the waveguide increases. The tuning of both cavity internal field and out-coupling to the waveguide gives extra tuning of the light transmission, resulting a modulation depth of 15 dB with 10 V voltage swing. The speed of modulator exceeds 30 GHz due to reduced capacitance area of $\sim$ 45 $\mu m^2$. Other resonator-coupled graphene E–O modulators have also shown both high modulation depth and small device footprint, including graphene-silicon ring resonator structures [91, 92], and silicon Mach-Zehnder interferometers-integrated structures [93].

### 3.4.2 *On-Chip Graphene Photodetectors*

On-chip graphene photodetectors could enable on-chip optical interconnects at unprecedented speeds. Although early demonstration of graphene-based photodetector for these applications has shown promising operation speed and broadband response [78, 79], the responsivity of the photodetector is low due to weak absorption in graphene in a vertical incident configuration. To address this, graphene has been integrated with nanocavities [94], microcavities [95], and plasmon resonant structures [49, 72]. In these examples, a range of responsivity of 10–30 mA/W is possible, with a trade-off for limited spectral response due to the narrowband enhancement of the resonant structures. To implemented broadband photodetectors, heterogeneous integrated graphene with waveguides is most promising for broadband spectral response while enhancing the photoresponse of the detector.

Here, we describe a waveguide-coupled graphene photodetector [96, 97], as schematically illustrated in Fig. 3.21a. The silicon waveguides were fabricated in an SOI wafer with a CMOS compatible process. The chip was planarized by backfilling with a thick $SiO_2$ layer, followed by chemical mechanical polishing to reach the top silicon surface. Multilayered hBN/SLG/hBN stack were assembled onto the photonic chip using van der Waals (vdW) assembly [85]. The single layer graphene (SLG) channel spans 40 $\mu$m of the waveguide, inducing $\sim$2.2 dB absorption, consistent with simulation results. One-dimensional edge contacts (Fig. 3.21b) to the encapsulated graphene layer is applied to hBN/SLG/hBN stack and the drain

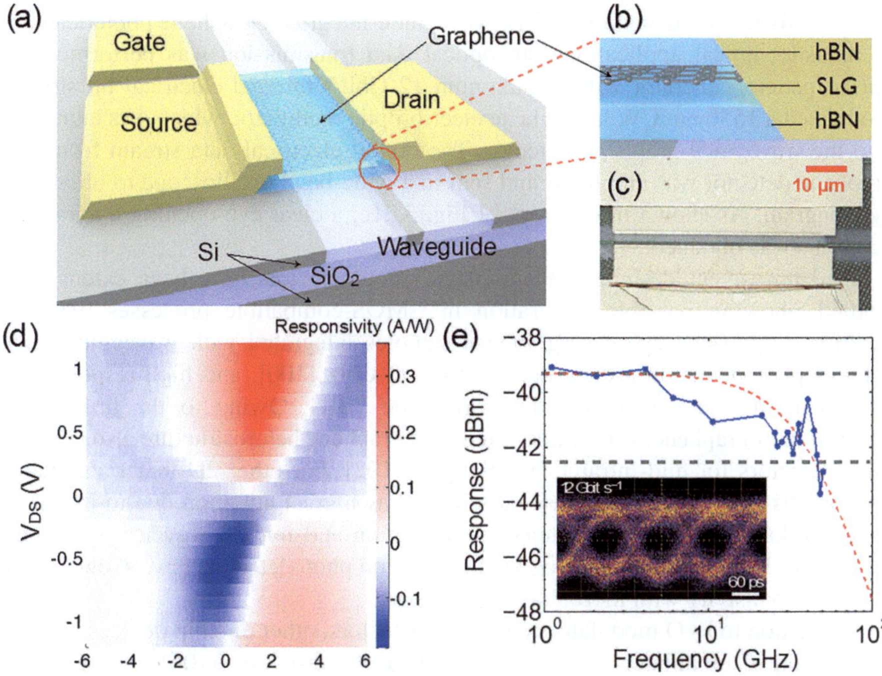


**Fig. 3.21** (**a**) Schematic of the hBN/SLG/hBN photodetector on a buried silicon waveguide. (**b**) Cross section view of the side-contacted hBN/SLG/hBN multilayer stack. (**c**) Optical microscope image of the as-fabricated device. (**d**) Responsivity mapping as a function of gate-source $V_{GS}$ and drain-source $V_{DS}$ voltages. (**e**) High-speed response of the graphene photodetector. The dashed line shows the fitting to the experiment results with a RC low-pass filter model. 12 Gbit $s^{-1}$ optical data link test of the device, showing a clear eye opening

electrode is positioned only 200 nm from the waveguide to induce a pn junction near the optical mode [98]. Figure 3.21c shows the completed structure.

Figure 3.21d shows the photoresponse of the detector via illuminating the waveguide with a continuous-wave (c.w.) laser at 25 μW. The responsivity is defined as the ratio of the short- circuit photocurrent ($I_{ph}$) to the optical power $P_{in}$ in the waveguide, $R = I_{ph}/P_{in}$. Since the metal-induced pn junction near the drain contact serves to separate the photoexcited carrier and generate photocurrent, the doping profile of the pn junction is critical to achieve maximum responsivity of the graphene photodetector. A polymer electrolyte (poly(ethyleneoxide) and $LiClO_4$) layer covering the entire chip serves to independently tune the graphene Fermi level and electric field across the waveguide mode [99]. The responsivity of the detector as a function of $V_{GS}$ and $V_{DS}$ shows a six-fold pattern in the photocurrent, which qualitatively matches the behavior of the photothermoelectric (PTE) effect [47]. The photocurrent reaches a maximum of 0.36 A/W at $V_{GS} = 2$ V and $V_{DS} = 1.2$ V.

Figure 3.21e shows the operation speed of the photodetector measured in the frequency range from 1 to 50 GHz, indicating a 3-dB cut-off frequency at

42 GHz. To gauge the viability of the waveguide-integrated graphene photodetector in realistic optical applications, an optical data transmission was performed. A pulsed pattern generator with a maximum 12 Gbit/s internal electrical bit stream modulated a 1550 nm CW laser via an electro-optic modulator, which was launched into the waveguide-graphene detector. The output electrical data stream from the graphene detector was amplified and sent to a wide-band oscilloscope to obtain an eye diagram. As shown in the inset of Fig. 3.21e, a clear eye-opening diagram at 12 Gbit/s was obtained.

To date, waveguide-integrated graphene photodetectors have been extensively studied, showing possible integration in CMOS-compatible processes [98]. In addition, large-scale epitaxial graphene samples integrated with waveguides can achieve photodetection with a data rate of 50 GHz [100], and high-responsivity by integrating with a silicon slot waveguide [101]. Owing to the broadband absorption of graphene, waveguide-integrated graphene heterostructure also enables photodetectors for mid-infrared wavelengths [102]. For other 2D materials, black phosphorus (BP)-based photodetectors have drawn great attention due to its small bandgap that is promising for telecommunication and mid-IR wavelength ranges [103, 104]. Heterogeneously integrated BP-silicon photodetectors have shown up to 6 A/W responsivity with more than 3 GHz speed.

In addition to E–O modulators and photodetectors, other on-chip devices including all-optical modulators [105], mode-locked ultrafast laser [106], thermo-optic modulators [107], light-emitting diodes [108], and single photon sources [109] could be integrated on a silicon platform in the future, thanks to 2D materials such as transition metal dichalcogenides (TMDs) [110], black phosphorus [104, 111], and superconducting 2D niobium diselenide [112] ($NbSe_2$). The seamless heterogeneous integration enabled by the low processing temperatures of 2D materials will allow an efficient optical interconnect with conventional silicon CMOS technology for inter- and intra-chip communication, back-end deposited silicon photonics [113], mid-infrared photonics [114], and also flexible photonics [115].

## 3.5 2D Material-Based Chemical and Biological Sensors

Chemical and biological sensors are ubiquitous in every aspect of our lives—from health care and environmental monitoring all the way to transportation and industry. Sensors translate a physical, chemical, or biological input signal, for example a gas concentration, into an electrical signal such as a change in voltage or current. In order to benchmark the performance of this transduction process, different metrics exist. One of the most important parameters is the *sensitivity* of a sensor, which is defined as proportionality factor between input signal and electrical output signal. A sensor's *selectivity* is given by its ability to respond to only one input signal and to be insensitive to all other relevant environmental changes. Another important characteristic is the *detection limit,* which marks the lower bound of what can be measured with confidence. Lastly, sensors exhibit a finite *response time* and *recovery time*. They are defined as the delay between a step increase or decrease of the input signal, respectively, and the output reaching a steady state [116, 117].

The integration of chemical and biological sensors with silicon chips could play an important role in the goal of extending the capabilities of the traditional CMOS platform. However this heterogeneous integration has traditionally been difficult due to the instability of $SiO_2$ dielectrics in the presence of many analytes of interest, as well as the large thickness of the interconnect and back-end layers in state-of-the-art silicon chips, which increase the distance between the chip surface and the transistors. 2D materials such as graphene or molybdenum disulfide could change this, and make the seamless integration of chemical and biological sensors with silicon chips a reality. These materials have excellent physical and chemical properties that make them perfectly suited as sensor materials. They are atomically thin, resulting in the highest possible surface to volume ratio and making them extremely sensitive to environmental changes. Furthermore, 2D materials have intrinsically low electronic noise, are biocompatible and mechanically strong and, as mentioned in other sections of these chapter, they can be seamlessly integrated with Silicon microsystems.

This section is organized into two parts. The first half gives an overview of gas and chemical sensors based on 2D materials. The second half focuses on biological sensors made of these materials. The aim of this section is to highlight a few promising sensor concepts rather than to give detailed account of all the extensive research work. In addition, it focuses on sensor solutions that can be eventually integrated with CMOS electronics.

### 3.5.1 Gas and Chemical Sensors

Gas and chemical sensors are used, for example, in our homes to detect carbon monoxide, in security applications to identify explosives or in cars to monitor exhaust and lower hazardous emissions. Sensors can exploit different physical phenomena to detect and quantify a certain gas or chemical, for example changes in the optical, magnetic, or electrical properties of the sensor material. Electrical gas sensors are arguably one of the most common types. They use changes in sensor resistance or capacitance in a variety of materials such as semiconductors, metal thin film, metal oxides, and polymers to facilitate the sensing [118].

Recently, 2D materials have attracted a large amount of interest as gas or chemical sensor due to their high sensitivities and low electronic noise. Both of these properties are rooted in their atomic thickness which yields a very large surface to volume ratio. This makes 2D material very susceptible to the environment with molecules adsorbing on the surface which leads to a charge transfer and dopes the channel with electrons or holes. This increase in carrier concentration strongly alters the channel resistance of the device [119].

In the following paragraphs, a few promising examples of gas sensors using graphene and other 2D materials are given. For a more in-depth discussion, the interested reader is referred to comprehensive literature review papers by Yuan and Shi and Verghese et al. [120, 121] covering graphene-based gas sensors as well as another review paper by Verghese et al. [119] which focuses on gas sensors based on 2D materials (in addition to graphene).

In 2007 Schedin et al. demonstrated the first graphene gas sensor based on a simple Hall bar structure. In this device, gas molecules adsorb on the surface of graphene which dope the material and change its electrical properties. Based on an optimized design and the intrinsically low noise level of graphene, the group was able to detect individual molecule ad- and desorption events which is the ultimate sensitivity limit [122]. More recently, Chen et al. simplified the sensor design to a two-terminal chemically sensitive resistor, or chemiresistor, and added UV-illumination for continuous cleaning of the graphene surface. This way, they were able to demonstrate a detection limit as low as 0.16 parts-per-trillion for nitric oxide gas molecules [123].

Another way to sense gases using graphene is to exploit the influence of specific analytes on the low frequency noise of graphene [124]. Using a back-gated transistor configuration, Rumyantsev et al. found that besides changing the channel resistance, certain gases introduce a Lorentzian noise bulge at a distinctive center frequency (see Fig. 3.22) that differs for each gas and is reproducible across samples. This adds another sensing parameter to chemically sensitive field effect transistors, or chemFETs, which holds the promise of achieving better sensor selectivity through recognizing a distinctive signature of Lorentzian center frequency and change in channel resistance.

Reduced graphene oxide is a graphite-derived material that can be easily solution-processed. Fowler et al. used it to create gas sensors on a chip with an

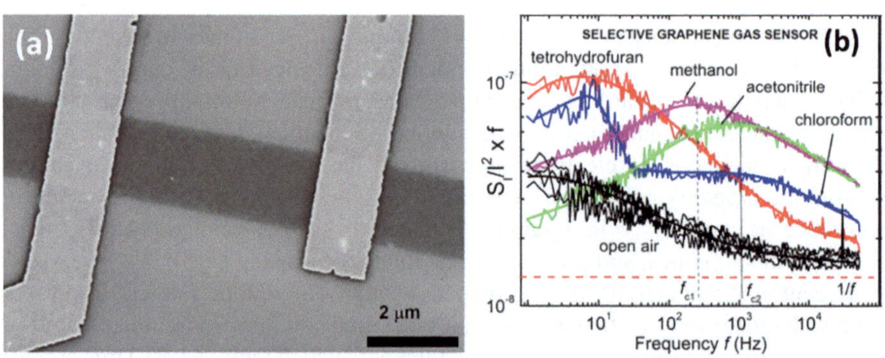


**Fig. 3.22** (**a**) Scanning electron microscope (SEM) image and (**b**) Noise spectral density $S_I/I^2$ multiplied by frequency f versus frequency f for the device in open air and under the influence of various solvent vapors. Different vapors induce noise with different characteristic frequencies fc. The difference in the frequency fc is sufficient for reliable identification of different gases with the same graphene transistor [124]

integrated heater for temperature control by spin coating a graphene oxide solution on the chip and reducing it with anhydrous hydrazine. They found the device to have a detection limit of 52 ppb to 2,4-dintrotoluene (DNT), which is a molecule related to explosive detection. Furthermore, their temperature experiments showed that a substrate temperature of 149 °C helps to decrease the response and recovery time significantly in comparison to room temperature. The benefit however comes as the expense of lowered sensitivity [125]. More recently, reduced graphene oxide sensors were also successfully fabricated by inkjet printing [126, 127].

2D material based gas and chemical sensor are not just limited to graphene. With the exploration of other 2D materials like transition metal dichalcogenides, viable sensors were also shown with molybdenum disulfide ($MoS_2$) [128–131] and phospherene [132] on top of a silicon wafer. Late et al. analyzed the response of $MoS_2$ chemFETs to ammonia and nitrogen dioxide gas while exploring the influence of light illumination, gate bias and number of atomic layers. They found that five-layer $MoS_2$ devices were more sensitive to gas analytes than their bilayer counterparts. Furthermore, applying a gate bias helps to increase the device sensitivity by electrostatically changing the carrier concentration in the channel to an optimal point [129]. In a related study by Perkins et al. the authors demonstrated single-layer $MoS_2$ chemiresistors that are sensitive to different organic compounds such as triethylamine (TEA) or acetone with a detection limit of 10 ppb and 500 ppm, respectively. They also found that the response to these organic vapors is often complementary in polarity to carbon nanotube sensors [130]. Another way to tailor the selectivity of $MoS_2$ to a specific gas is by grafting functional groups onto its surface. Kim et al. demonstrated this approach by attaching a thiolated ligand called mercaptoundecanoic acid (MUA) to chemiresistor devices. The sensors were built with interdigitated electrodes on a layer of vacuum-filtrated $MoS_2$ flakes with or without this functionalization that results in different responses towards volatile organic compounds (see Fig. 3.23a). In particular, the untreated chemiresistors showed an increase in channel resistance being exposed to oxygen-functionalized

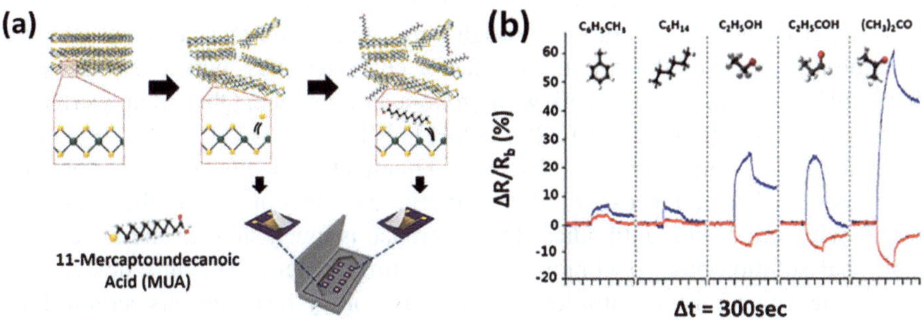


**Fig. 3.23** (**a**) Schematic of $MoS_2$ exfoliation and functionalization with MUA. The chemiresistors were built by vacuum filtration and placed in a test chamber that was exposed to different organic volatile compounds (VOC). (**b**) Relative change in channel resistance of untreated and MUA-treated $MoS_2$ chemiresistors as a result of exposure to different VOCs [131]

VOCs while the MUA treated samples showed a decreasing resistance as shown in Fig. 3.23b. Furthermore, the group showed a detection limit of these sensors below 1 ppm. Considering different available thiolated ligands, this work demonstrates the potential of targeted $MoS_2$ functionalization to create sensor arrays that are sensitive to a variety of gaseous compounds [131].

### 3.5.2 *Biological Sensors*

Biological sensors are another important subgroup of sensors, which is mainly used in medicine to determine blood sugar levels, identify bacteria, cancer cells, and other analytes of interest in the laboratory. There is a large variety of biosensor technologies, although many use optical detection through labeling with fluorescent dyes or electrochemical detection techniques using functionalized electrode surfaces [133, 134].

Recently, 2D materials such as graphene have been extensively explored as biosensors. Besides their low intrinsic noise and large surface to volume ratio, 2D materials are especially compelling to use as biosensors because of their chemical inertness and easier handling compared to other new materials such as carbon nanotubes. The paragraphs below summarize some of the approaches to use graphene, $MoS_2$, and other 2D materials in monitoring cells or pH levels and detect DNA, biomolecules and other proteins. Most of these devices were fabricated on top of silicon wafers and could easily be integrated with silicon electronic chips. This could eventually enable small, inexpensive diagnostic tests for laboratory and personal use. For a more detailed literature review, the interested reader is referred to more comprehensive review papers by Liu et al. and Moldvan et al. [135].

Being able to sense pH levels is important in biology and medicine, for example for cell monitoring or diagnosing disease in humans [133]. Ang et al. demonstrated that graphene electrolyte gated FETs (EGFETs) can be used to measure pH levels. In particular, they found that the hydroxyl ($OH^-$) and hydronium ($H_3O^+$) change the channel carrier concentration through capacitive coupling and hence change the neutrality point of the chemFET transfer characteristics with a sensitivity of 99 mV/pH [136]. In a similar work with monolayer graphene, Ohno et al. were able to improve the sensitivity down to 25 mV/pH [137].

In the realm of biological cell detection, Huang et al. demonstrated a graphene chemiresistor that can sense *E. coli* bacteria by functionalizing the graphene surface with anti-*E. coli* antibodies. This approach is much simpler and faster than traditional solution-based techniques for detecting bacteria. Huang's device was able to detect bacteria concentrations as low as 10 colony-forming cells per milliliter (cfu/mL) and also showed high selectivity against other bacteria [138]. Another interesting graphene biosensor application was highlighted by Ang et al. who fabricated a microfluidic channel equipped with protein-functionalized graphene chemFETs that could detect malaria infected red blood cells. The group showed the selective capture of infected blood cells to a transistor channel, with a much simpler setup compared to traditional optical detection methods. This capture event changes the channel conductance when in contact with a cell due to capacitive charging. By analyzing the dwell time and magnitude of the conductivity change, they could also distinguish between two different stages of the infected blood cells [139]. Graphene transistors can also be used to monitor the action potential of cells for neural prosthesis application [140]. Hess et al., for example, demonstrated graphene EGFET arrays that are very biocompatible and are able to monitor the action potential of cells grown on top of the channel [140]. In their experiments, the graphene arrays proved to be more chemical resistant and exhibit far higher gate sensitivities than traditional silicon FET arrays. Lastly, Jiang et al. fabricated hemin-functionalized graphene chemFETs with integrated microfluidic delivery system to sense nitric oxide (NO), an important by-product of many cell reactions. With their devices, they demonstrated nanomolar NO sensitivity with sensor areas of 0.25–1 $\mu m^2$, which is similarly sensitive but much smaller than state-of-the-art electrochemical sensors. Hence, this approach paves the way to a specially resolved cell monitoring [141].

DNA sensors are becoming a vital tool for genetic screening and pathogen detection that help diagnose diseases. However, the current technology for DNA sensing relies on marking it with fluorescent dyes and optically detecting it. Recently, 2D materials have been intensively studied for this application due to their easy functionalization and electronic integration, which could lead to compact biochips that can detect DNA in real time. In 2008, Mohanty and Berry used reduced graphene oxide (rGO) functionalized with single-stranded DNA to build highly sensitive and selective DNA chem-FETs that changed between 60–200% in channel conductance after DNA hybridization [142]. A slightly different design by Stine et al. improved the selectivity of this sensor concept by introducing a second reference transistor without functionalization that could compensate for sensor drift and non-specific binding [143]. In an effort to make graphene DNA sensors more scalable, Ping et al. recently reported on graphene transistor arrays with 52 devices functionalized with single-stranded DNA as depicted in Fig. 3.24a. The sensors measure DNA concentration by a positive shift Dirac voltage, see Fig. 3.24b, which is ascribed to an increase in positive carrier concentration induced by the negatively charged phosphate groups of the target DNA molecules. In their experiments, the transistor arrays had a device yield of over 90%, good reproducibility and could achieve a detection limit of down to 1 fM for 60mer DNA strands. Furthermore, the sensor selectivity was tested by analyzing the response to DNA strands that had one or two mismatching base pairs either at the center or the end of the stand. The group found that a mismatch at the end or the center of a sequence resulted in a 20% reduction or 90% reduction in Dirac voltage shift, respectively, which highlights the good selectivity of the sensor to even small mismatches [144].

Detecting proteins and biomolecule such as glucose or immunoglobulin has also been demonstrated with 2D materials by using different kinds of functionalization [145–149]. Hunag et al., for example, used graphene chemFETs decorated with glucose oxidase to sense glucose in solution down to a level of 100 $\mu M$ which is comparable to common electrochemical sensors. Using glutamic dehydrogenase as

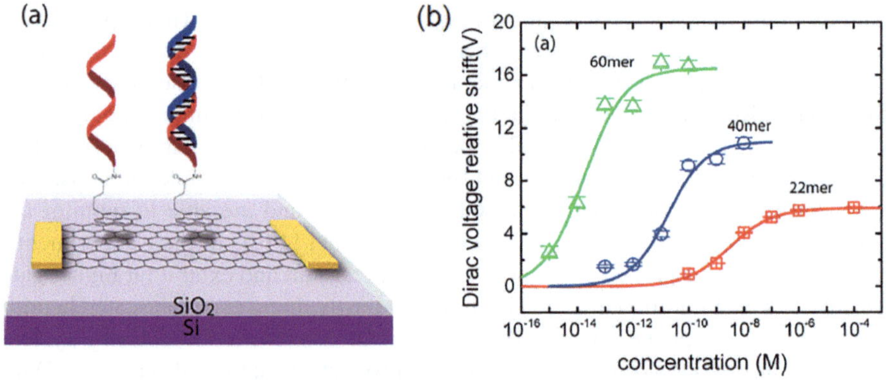


**Fig. 3.24** (**a**) Schematic of back-gated graphene transistor with single-strand DNA functionalization and hybridized DNA strand. (**b**) Relative Dirac voltage shift of transistor as a function of DNA concentration and DNA strand length [144]

immobilized enzyme, they could furthermore measure glutamate with a detection limit of 5 μM [145, 147]. A group around Mao et al. chose a different approach to detect immunoglobulin G (IgG). They build reduced graphene oxide chemiresistors and deposited gold nanoparticles by a combination of electrospraying and electrostatic assembly that were decorated with matching antibodies.

After exposing the devices to an IgG solution for 1 h, washing and drying the chemiresistors, the group was able to show statistically significant changes in channel resistance at IgG concentrations as low as 2 ng/mL while maintaining a good selectivity against other types of immunoglobulin [147]. Using exfoliated $MoS_2$, Lee et al. recently demonstrated a prostate specific antigen biosensor which is an important tool to detect prostate cancer. The group functionalized the $MoS_2$ transistors by using PSA antibodies that nonspecifically physisorb on the sensor surface. Compared to the pristine devices this leads to an increased baseline off-current of the transistor due to doping of the positively charged antibodies. Once the biosensors are exposed to a PSA solution, the negatively charged PSA antigens selectively bind to their respective antibodies, which decreases the transistor off-current again and is illustrated in Fig. 3.25. With this approach, the researchers were able to demonstrate a detection limit of 1 pg/mL which is three orders of magnitude below the clinical cut-off level [149].

Many of the aforementioned biosensors are based on electrolyte gated field effect transistors (EGFETs). To better understand the variations that are associated with building these arrays, large-scale sensor systems based on graphene EGFETs have been fabricated [150]. These systems contain up to 256 individual devices that can be analyzed by DC-measurements within minutes, which enables meaningful statistical analysis. Given the low thermal budget involved in the fabrication of these sensor arrays, their integration with fully fabricated silicon chips is expected to be imminent, and very similar to what has already happened with infrared detectors.

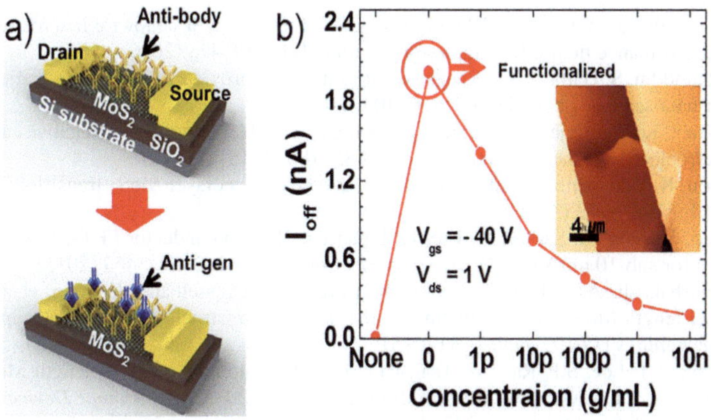


**Fig. 3.25** (**a**) Schematic of a $MoS_2$ biosensor functionalized with prostate specific antigen (PSA) antibody on $MoS_2$ surface (top) and subsequent binding of PSA antigen with antibody receptors (bottom). (**b**) Transistor current in off-state ($V_{gs}$ − 40 V and $V_{ds}$ + 1 V) as a function of PSA concentration. The off-current increases with the functionalization and then decreases as antigens connect to the immobilized antibodies [149]

## 3.6 Conclusion

2D materials, thanks to their atomically thin nature and diverse electronic properties, are promising candidates for a variety of applications. Semiconducting $MoS_2$ is, for example, an ideal channel material to both provide power gating in future Si chips and replace Si in sub-10 nm technology. At the same time, graphene, a gapless semiconductor, can serve as a core material combined with a Si photonic structure for fast photodetection and optical modulation on a CMOS platform. The transferable nature of atomically thin 2D materials and the low thermal budget of their processing technology allows seamless integration of these and other 2D devices, circuits and systems on a Si platform to develop complex 3D systems. Such 3D integration will significantly increase the performance of future Si microsystems, enabling exciting new opportunities for hybrid systems.

## References

1. J.A. del Alamo, Nanometre-scale electronics with III-V compound semiconductors. Nature **479**, 317–323 (2011)
2. J.A.d. Alamo, D. Antoniadis, A. Guo, D.H. Kim, T.W. Kim, J. Lin et al., InGaAs MOSFETs for CMOS: Recent advances in process technology, in *Electron Devices Meeting (IEDM), 2013 IEEE International* (2013), pp. 2.1.1–2.1.4
3. J.P. Colinge, C.W. Lee, A. Afzalian, N.D. Akhavan, R. Yan, I. Ferain, et al., Nanowire transistors without junctions. Nat. Nanotechnol. **5**, 225–229 (2010)

4. J. Xiang, W. Lu, Y.J. Hu, Y. Wu, H. Yan, C.M. Lieber, Ge/Si nanowire heterostructures as high-performance field-effect transistors. Nature **441**, 489–493 (2006)
5. S. Salahuddin, S. Dattat, Use of negative capacitance to provide voltage amplification for low power nanoscale devices. Nano Lett. **8**, 405–410 (2008)
6. A.I. Khan, K. Chatterjee, B. Wang, S. Drapcho, L. You, C. Serrao, et al., Negative capacitance in a ferroelectric capacitor. Nat. Mater. **14**, 182–186 (2015)
7. Y. Yoon, K. Ganapathi, S. Salahuddin, How good can monolayer $MoS_2$ transistors be? Nano Lett. **11**, 3768–3773 (2011)
8. W. Cao, J. Kang, D. Sarkar, W. Liu, K. Banerjee, 2D semiconductor FETs: Projections and Design for sub-10 nm VLSI. IEEE Trans Electron Devic **62**, 3459–3469 (2015)
9. A. Nourbakhsh, A. Zubair, S. Huang, X. Ling, M.S. Dresselhaus, J. Kong, et al., 15-nm channel length $MoS_2$ FETs with single- and double-gate structures, in *2015 symposium on Vlsi Technology (Vlsi Technology)* (2015)
10. L. Yang, R.T. Lee, S.P. Rao, W. Tsai, P.D. Ye, 10 nm nominal channel length $MoS_2$ FETs with EOT 2.5 nm and 0.52 mA/μm drain current, in *2015 73rd Annual Device Research Conference (DRC)* (2015), pp. 237–238
11. G. Fiori, F. Bonaccorso, G. Iannaccone, T. Palacios, D. Neumaier, A. Seabaugh, et al., Electronics based on two-dimensional materials. Nat. Nanotechnol. **9**, 768 (2014)
12. X.L. Chen, Z.F. Wu, S.G. Xu, L. Wang, R. Huang, Y. Han, et al., Probing the electron states and metal-insulator transition mechanisms in molybdenum disulphide vertical heterostructures. Nat. Commun. **6**, 6088 (2015)
13. E.J.G. Santos, E. Kaxiras, Electrically driven tuning of the dielectric constant in $MoS_2$ layers. ACS Nano **7**, 10741–10746 (2013)
14. Z. Ni, M. Ye, J. Ma, Y. Wang, R. Quhe, J. Zheng, L. Dai, J.S. Dapeng Yu, J. Yang, S. Watanabe, J. Lu, Performance upper limit of sub-10 nm monolayer $MoS_2$ transistors. Adv Electron Mater **2**, 1600191 (2016)
15. H. Liu, A.T. Neal, P.D.D. Ye, Channel length scaling of $MoS_2$ MOSFETs. ACS Nano **6**, 8563–8569 (2012)
16. D. Hisamoto, L. Wen-Chin, J. Kedzierski, H. Takeuchi, K. Asano, C. Kuo, et al., FinFET-a self-aligned double-gate MOSFET scalable to 20 nm. IEEE Trans Electron Devic **47**, 2320–2325 (2000)
17. M.C. Chen, C.Y. Lin, L. Kai-Hsin, L.J. Li, C.H. Chen, C. Cheng-Hao, et al., Hybrid Si/TMD 2D electronic double channels fabricated using solid CVD few-layer-$MoS_2$ stacking for Vth matching and CMOS-compatible 3DFETs, in *2014 IEEE International Electron Devices Meeting* (2014), pp. 33.5.1–33.5.4
18. M.C. Chen, K.S. Li, L.J. Li, A.Y. Lu, M.Y. Li, Y.H. Chang, et al., TMD FinFET with 4 nm thin body and back gate control for future low power technology, in *2015 IEEE International Electron Devices Meeting (IEDM)* (2015), pp. 32.2.1–32.2.4
19. A.Z.A. Nourbakhsh, A. Tavakkoli, R. Sajjad, X. Ling, M. Dresselhaus, J. Kong, K.K. Berggren, D. Antoniadis, T. Palacios, Serially connected monolayer $MoS_2$ FETs with channel patterned by a 7.5 nm resolution directed self-assembly lithography, in *2015 Symposium on Vlsi Technology (VLSI Technology)* (2016)
20. C.D. English, K.K. Smithe, R.L. Xu, E. Pop, Approaching ballistic transport in monolayer $MoS_2$ transistors with self-aligned 10 nm top gates, in *2016 IEEE International Electron Devices Meeting (IEDM)* (2016), p. 131
21. S.R.M. Sujay, B. Desai, A.B. Sachid, J.P. Llinas, Q. Wang, G.H. Ahn, G. Pitner, M.J. Kim, J. Bokor, H. Chenming, H.-S. Philip Wong, A. Javey, $MoS_2$ transistors with 1-nanometer gate lengths. Science **354**, 99–102 (2016)
22. L. Yu, Y.-H. Lee, X. Ling, E.J.G. Santos, Y.C. Shin, Y. Lin, et al., Graphene/$MoS_2$ hybrid technology for large-scale two-dimensional electronics. Nano Lett. **14**, 3055–3063 (2014)
23. H. Wang, L. Yu, Y.-H. Lee, Y. Shi, A. Hsu, M.L. Chin, et al., Integrated circuits based on bilayer $MoS_2$ transistors. Nano Lett. **12**, 4674–4680 (2012)
24. R. Kappera, D. Voiry, S.E. Yalcin, B. Branch, G. Gupta, A.D. Mohite, et al., Phase-engineered low-resistance contacts for ultrathin $MoS_2$ transistors. Nat. Mater. **13**, 1128–1134 (2014)

25. L. Yu, D. El-Damak, S. Ha, X. Ling, Y. Lin, A. Zubair, et al., Enhancement-mode single-layer CVD $MoS_2$ FET technology for digital electronics, in *2015 IEEE International Electron Devices Meeting (IEDM)* (2015)
26. L. Yu, D. El-Damak, U. Radhakrishna, X. Ling, A. Zubair, Y. Lin, et al., Design, modeling, and fabrication of chemical vapor deposition grown $MoS_2$ circuits with E-mode FETs for large-area electronics. Nano Lett. **16**, 6349–6356 (2016)
27. L. Yu, D. El-Damak, U. Radhakrishna, A. Zubair, D. Piedra, X. Ling, et al., High-yield large area $MoS_2$ technology: Material, device and circuits co-optimization, in *2016 IEEE International Electron Devices Meeting (IEDM)* (2016), pp. 5.7.1–5.7.4
28. L. Yu, D. El-Damak, U. Radhakrishna, X. Ling, A. Zubair, Y. Lin, et al., Design, modeling and fabrication of CVD grown $MoS_2$ circuits with E-mode FETs for large-area electronics. Nano Lett. **16**, 6349–6356 (2016)
29. S. Chuang, C. Battaglia, A. Azcatl, S. McDonnell, J.S. Kang, X. Yin, et al., $MoS_2$ P-type transistors and diodes enabled by high work function MoOx contacts. Nano Lett. **14**, 1337–1342 (2014)
30. L. Yu, A. Zubair, E.J.G. Santos, X. Zhang, Y. Lin, Y. Zhang, et al., High-performance WSe2 complementary metal oxide semiconductor technology and integrated circuits. Nano Lett. **15**, 4928–4934 (2015)
31. G. Long, D. Maryenko, S. Pezzini, S. Xu, Z. Wu, T. Han, et al., Ambipolar quantum transport in few-layer black phosphorus. Phys. Rev. B **96**, 155448 (2017)
32. T.C. Huang, K. Fukuda, C.M. Lo, Y.H. Yeh, T. Sekitani, T. Someya, et al., Pseudo-CMOS: A design style for low-cost and robust flexible electronics. IEEE Trans Electron Devic **58**, 141–150 (2011)
33. B.B. Lahiri, S. Bagavathiappan, T. Jayakumar, J. Philip, Medical applications of infrared thermography: a review. Infrared Phys. Technol. **55**, 221–235 (2012)
34. J.L. Thomson, J.W. Salisbury, The mid-infrared reflectance of mineral mixtures (7–14 μm). Remote Sens. Environ. **45**, 1–13 (1993)
35. Q. Weng, Thermal infrared remote sensing for urban climate and environmental studies: Methods, applications, and trends. ISPRS J. Photogramm. Remote Sens. **64**, 335–344 (2009)
36. K.E. Joyce, S.E. Belliss, S.V. Samsonov, S.J. McNeill, P.J. Glassey, A review of the status of satellite remote sensing and image processing techniques for mapping natural hazards and disasters. Prog. Phys. Geogr. **33**, 183–207 (2009)
37. K.I. Bolotin, K.J. Sikes, Z. Jiang, M. Klima, G. Fudenberg, J. Hone, et al., Ultrahigh electron mobility in suspended graphene. Solid State Commun. **146**, 351–355 (2008)
38. C.R. Dean, A.F. Young, L.C. MericI, L. Wang, S. Sorgenfrei, et al., Boron nitride substrates for high-quality graphene electronics. Nat Nanotechnol **5**, 722–726 (2010)
39. L. Banszerus, M. Schmitz, S. Engels, J. Dauber, M. Oellers, F. Haupt, et al., Ultrahigh-mobility graphene devices from chemical vapor deposition on reusable copper. Sci. Adv. **1**, e1500222 (2015)
40. R.R. Nair, P. Blake, A.N. Grigorenko, K.S. Novoselov, T.J. Booth, T. Stauber, et al., Fine structure constant defines visual transparency of graphene. Science **320**, 1308–1308 (2008)
41. F.H.L. Koppens, D.E. Chang, F.J. García de Abajo, Graphene plasmonics: A platform for strong light–matter interactions. Nano Lett. **11**, 3370–3377 (2011)
42. L. Ju, B. Geng, J. Horng, C. Girit, M. Martin, Z. Hao, et al., Graphene plasmonics for tunable terahertz metamaterials. Nat Nanotechnol **6**, 630–634 (2011)
43. Z. Fei, A.S. Rodin, G.O. Andreev, W. Bao, A.S. McLeod, M. Wagner, et al., Gate-tuning of graphene plasmons revealed by infrared nano-imaging. Nature **487**, 82–85 (2012)
44. J. Chen, M. Badioli, P. Alonso-Gonzalez, S. Thongrattanasiri, F. Huth, J. Osmond, et al., Optical nano-imaging of gate-tunable graphene plasmons. Nature **487**, 77–81 (2012)
45. A.N. Grigorenko, M. Polini, K.S. Novoselov, Graphene plasmonics. Nat Photon **6**, 749–758 (2012)
46. M.C. Lemme, F.H.L. Koppens, A.L. Falk, M.S. Rudner, H. Park, L.S. Levitov, et al., Gate-activated photoresponse in a graphene p–n junction. Nano Lett. **11**, 4134–4137 (2011)

47. N.M. Gabor, J.C.W. Song, Q. Ma, N.L. Nair, T. Taychatanapat, K. Watanabe, et al., Hot carrier–assisted intrinsic photoresponse in graphene. Science **334**, 648–652 (2011)
48. M. Freitag, T. Low, P. Avouris, Increased responsivity of suspended graphene photodetectors. Nano Lett. **13**, 1644–1648 (2013)
49. Y. Yao, R. Shankar, P. Rauter, Y. Song, J. Kong, M. Loncar, et al., High-responsivity mid-infrared graphene detectors with antenna-enhanced photocarrier generation and collection. Nano Lett. **14**, 3749–3754 (2014)
50. A.L. Hsu, P.K. Herring, N.M. Gabor, S. Ha, Y.C. Shin, Y. Song, et al., Graphene-based thermopile for thermal imaging applications. Nano Lett. **15**, 7211–7216 (2015)
51. M. Freitag, T. Low, F. Xia, P. Avouris, Photoconductivity of biased graphene. Nat Photon **7**, 53–59 (2013)
52. J. Yan, M.H. Kim, J.A. Elle, A.B. Sushkov, G.S. Jenkins, H.M. Milchberg, et al., Dual-gated bilayer graphene hot-electron bolometer. Nat Nanotechnol **7**, 472–478 (2012)
53. G. Konstantatos, M. Badioli, L. Gaudreau, J. Osmond, M. Bernechea, F.P.G. de Arquer, et al., Hybrid graphene-quantum dot phototransistors with ultrahigh gain. Nat Nanotechnol **7**, 363–368 (2012)
54. Z. Sun, Z. Liu, J. Li, G.-A. Tai, S.-P. Lau, F. Yan, Infrared photodetectors based on CVD-grown graphene and PbS quantum dots with ultrahigh responsivity. Adv. Mater. **24**, 5878–5883 (2012)
55. C.-H. Liu, Y.-C. Chang, T.B. Norris, Z. Zhong, Graphene photodetectors with ultra-broadband and high responsivity at room temperature. Nat Nanotechnol **9**, 273–278 (2014)
56. Q. Ma, T.I. Andersen, N.L. Nair, N.M. Gabor, M. Massicotte, C.H. Lui, et al., Tuning ultrafast electron thermalization pathways in a van der Waals heterostructure. Nat. Phys. **12**, 455–459 (2016)
57. M. Massicotte, P. Schmidt, F. Vialla, K. Watanabe, T. Taniguchi, K.J. Tielrooij, et al., Photo-thermionic effect in vertical graphene heterostructures. Nat. Commun. **7**, 12174 (2016)
58. X. An, F. Liu, Y.J. Jung, S. Kar, Tunable graphene–silicon heterojunctions for ultrasensitive photodetection. Nano Lett. **13**, 909–916 (2013)
59. X. Li, M. Zhu, M. Du, Z. Lv, L. Zhang, Y. Li, et al., High detectivity graphene-silicon heterojunction photodetector. Small **12**, 595–601 (2016)
60. U. Sassi, R. Parret, S. Nanot, M. Bruna, S. Borini, D. De Fazio, et al., Graphene-based mid-infrared room-temperature pyroelectric bolometers with ultrahigh temperature coefficient of resistance. Nat. Commun. **8**, 14311 (2017)
61. Z. Qian, Y. Hui, F. Liu, S. Kang, S. Kar, M. Rinaldi, Graphene–aluminum nitride NEMS resonant infrared detector. Microsyst Nanoeng **2**, 16026 (2016)
62. H. Yuan, X. Liu, F. Afshinmanesh, W. Li, G. Xu, J. Sun, et al., Polarization-sensitive broadband photodetector using a black phosphorus vertical p–n junction. Nat Nanotechnol **10**, 707–713 (2015)
63. T. Low, M. Engel, M. Steiner, P. Avouris, Origin of photoresponse in black phosphorus phototransistors. Phys. Rev. B **90**, 081408 (2014)
64. P.K. Herring, A.L. Hsu, N.M. Gabor, Y.C. Shin, J. Kong, T. Palacios, et al., Photoresponse of an electrically tunable ambipolar graphene infrared thermocouple. Nano Lett. **14**, 901–907 (2014)
65. Z.Q. Li, E.A. Henriksen, Z. Jiang, Z. Hao, M.C. Martin, P. Kim, et al., Dirac charge dynamics in graphene by infrared spectroscopy. Nat. Phys. **4**, 532–535 (2008)
66. F. Wang, Y.B. Zhang, C.S. Tian, C. Girit, A. Zettl, M. Crommie, et al., Gate-variable optical transitions in graphene. Science **320**, 206–209 (2008)
67. J.C.W. Song, M.S. Rudner, C.M. Marcus, L.S. Levitov, Hot carrier transport and photocurrent response in graphene. Nano Lett. **11**, 4688–4692 (2011)
68. M. Badioli, A. Woessner, K.J. Tielrooij, S. Nanot, G. Navickaite, T. Stauber, et al., Phonon-mediated mid-infrared photoresponse of graphene. Nano Lett. **14**, 6374–6381 (2014)
69. Y. Yao, R. Shankar, M.A. Kats, Y. Song, J. Kong, M. Loncar, et al., Electrically tunable metasurface perfect absorbers for ultrathin mid-infrared optical modulators. Nano Lett. **14**, 6526–6532 (2014)

70. I.J. Luxmoore, P.Q. Liu, P. Li, J. Faist, G.R. Nash, Graphene-metamaterial photodetectors for integrated infrared sensing. ACS Photonics **3**(6), 936–941 (2016)
71. S. Song, Q. Chen, L. Jin, F. Sun, Great light absorption enhancement in a graphene photodetector integrated with a metamaterial perfect absorber. Nanoscale **5**, 9615–9619 (2013)
72. T.J. Echtermeyer, L. Britnell, P.K. Jasnos, A. Lombardo, R.V. Gorbachev, A.N. Grigorenko, et al., Strong plasmonic enhancement of photovoltage in graphene. Nat. Commun. **2**, 458 (2011)
73. T. Low, P. Avouris, Graphene plasmonics for terahertz to mid-infrared applications. ACS Nano **8**, 1086–1101 (2014)
74. P. Li, M. Lewin, A.V. Kretinin, J.D. Caldwell, K.S. Novoselov, T. Taniguchi, et al., Hyperbolic phonon-polaritons in boron nitride for near-field optical imaging and focusing. Nat. Commun. **6**, 7507 (2015)
75. S. Dai, Z. Fei, Q. Ma, A.S. Rodin, M. Wagner, A.S. McLeod, et al., Tunable phonon polaritons in atomically thin van der Waals crystals of boron nitride. Science **343**, 1125–1129 (2014)
76. A. Rogalski, HgCdTe infrared detector material: History, status and outlook. Rep. Prog. Phys. **68**, 2267 (2005)
77. K.F. Mak, M.Y. Sfeir, Y. Wu, C.H. Lui, J.A. Misewich, T.F. Heinz, Measurement of the optical conductivity of graphene. Phys. Rev. Lett. **101**, 196405 (2008)
78. T. Mueller, F.N.A. Xia, P. Avouris, Graphene photodetectors for high-speed optical communications. Nat. Photonics **4**, 297–301 (2010)
79. F. Xia, T. Mueller, Y.M. Lin, A. Valdes-Garcia, P. Avouris, Ultrafast graphene photodetector. Nat. Nanotechnol. **4**, 839–843 (2009)
80. C. Manolatou, M.J. Khan, S.H. Fan, P.R. Villeneuve, H.A. Haus, J.D. Joannopoulos, Coupling of modes analysis of resonant channel add-drop filters. IEEE J. Quantum Electron. **35**, 1322–1331 (1999)
81. A.H. Castro Neto, F. Guinea, N.M.R. Peres, K.S. Novoselov, A.K. Geim, The electronic properties of graphene. Rev. Mod. Phys. **81**, 109–162 (2009)
82. M. Liu, X. Yin, E. Ulin-Avila, B. Geng, T. Zentgraf, L. Ju, et al., A graphene-based broadband optical modulator. Nature **474**, 64–67 (2011)
83. M. Liu, X.B. Yin, X. Zhang, Double-layer graphene optical modulator. Nano Lett. **12**, 1482–1485 (2012)
84. M. Mohsin, D. Schall, M. Otto, A. Noculak, D. Neumaier, H. Kurz, Graphene based low insertion loss electro-absorption modulator on SOI waveguide. Opt. Express **22**, 15292–15297 (2014)
85. L. Wang, I. Meric, P.Y. Huang, Q. Gao, Y. Gao, H. Tran, et al., One-dimensional electrical contact to a two-dimensional material. Science **342**, 614–617 (2013)
86. Y. Gao, R.J. Shiue, X. Gan, L. Li, C. Peng, I. Meric, et al., High-speed electro-optic modulator integrated with graphene-boron nitride heterostructure and photonic crystal nanocavity. Nano Lett. **15**, 2001–2005 (2015)
87. J. Teng, P. Dumon, W. Bogaerts, H.B. Zhang, X.G. Jian, X.Y. Han, et al., Athermal silicon-on-insulator ring resonators by overlaying a polymer cladding on narrowed waveguides. Opt. Express **17**, 14627–14633 (2009)
88. G.T. Reed, G. Mashanovich, F.Y. Gardes, D.J. Thomson, Silicon optical modulators. Nat. Photonics **4**, 518–526 (2010)
89. S. Manipatruni, R.K. Dokania, B. Schmidt, N. Sherwood-Droz, C.B. Poitras, A.B. Apsel, et al., Wide temperature range operation of micrometer-scale silicon electro-optic modulators. Opt. Lett. **33**, 2185–2187 (2008)
90. C.T. Phare, Y.H.D. Lee, J. Cardenas, M. Lipson, Graphene electro-optic modulator with 30 GHz bandwidth. Nat. Photonics **9**, 511 (2015)
91. Y.H. Ding, X.L. Zhu, S.S. Xiao, H. Hu, L.H. Frandsen, N.A. Mortensen, et al., Effective electro-optical modulation with high extinction ratio by a Graphene-Silicon Microring Resonator. Nano Lett. **15**, 4393–4400 (2015)

92. C.Y. Qiu, W.L. Gao, R. Vajtai, P.M. Ajayan, J. Kono, Q.F. Xu, Efficient modulation of 1.55 mu m radiation with gated graphene on a silicon microring resonator. Nano Lett. **14**, 6811–6815 (2014)
93. N. Youngblood, Y. Anugrah, R. Ma, S.J. Koester, M. Li, Multifunctional graphene optical modulator and photodetector integrated on silicon waveguides. Nano Lett. **14**, 2741–2746 (2014)
94. R.J. Shiue, X.T. Gan, Y.D. Gao, L.Z. Li, X.W. Yao, A. Szep, et al., Enhanced photodetection in graphene-integrated photonic crystal cavity. Appl. Phys. Lett. **103**, 241109 (2013)
95. M. Furchi, A. Urich, A. Pospischil, G. Lilley, K. Unterrainer, H. Detz, et al., Microcavity-integrated graphene photodetector. Nano Lett. **12**, 2773–2777 (2012)
96. X.T. Gan, R.J. Shiue, Y.D. Gao, I. Meric, T.F. Heinz, K. Shepard, et al., Chip-integrated ultrafast graphene photodetector with high responsivity. Nat. Photonics **7**, 883–887 (2013)
97. R.J. Shiue, Y.D. Gao, Y.F. Wang, C. Peng, A.D. Robertson, D.K. Efetov, et al., High-responsivity graphene-boron nitride photodetector and autocorrelator in a silicon photonic integrated circuit. Nano Lett. **15**, 7288–7293 (2015)
98. A. Pospischil, M. Humer, M.M. Furchi, D. Bachmann, R. Guider, T. Fromherz, et al., CMOS-compatible graphene photodetector covering all optical communication bands. Nat. Photonics **7**, 892–896 (2013)
99. C.G. Lu, Q. Fu, S.M. Huang, J. Liu, Polymer electrolyte-gated carbon nanotube field-effect transistor. Nano Lett. **4**, 623–627 (2004)
100. D. Schall, D. Neumaier, M. Mohsin, B. Chmielak, J. Bolten, C. Porschatis, et al., 50 GBit/s photodetectors based on wafer-scale graphene for integrated silicon photonic communication systems. ACS Photonics **1**, 781–784 (2014)
101. J.Q. Wang, Z.Z. Cheng, Z.F. Chen, X. Wan, B.Q. Zhu, H.K. Tsang, et al., High-responsivity graphene-on-silicon slot waveguide photodetectors. Nanoscale **8**, 13206–13211 (2016)
102. X.M. Wang, Z.Z. Cheng, K. Xu, H.K. Tsang, J.B. Xu, High-responsivity graphene/silicon-heterostructure waveguide photodetectors. Nat. Photonics **7**, 888–891 (2013)
103. Q.S. Guo, A. Pospischil, M. Bhuiyan, H. Jiang, H. Tian, D. Farmer, et al., Black phosphorus mid-infrared photodetectors with high gain. Nano Lett. **16**, 4648–4655 (2016)
104. N. Youngblood, C. Chen, S.J. Koester, M. Li, Waveguide-integrated black phosphorus photodetector with high responsivity and low dark current. Nat. Photonics **9**, 247–252 (2015)
105. W. Li, B.G. Chen, C. Meng, W. Fang, Y. Xiao, X.Y. Li, et al., Ultrafast all-optical graphene modulator. Nano Lett. **14**, 955–959 (2014)
106. Z.P. Sun, T. Hasan, F. Torrisi, D. Popa, G. Privitera, F.Q. Wang, et al., Graphene mode-locked ultrafast laser. ACS Nano **4**, 803–810 (2010)
107. S. Gan, C.T. Cheng, Y.H. Zhan, B.J. Huang, X.T. Gan, S.J. Li, et al., A highly efficient thermo-optic microring modulator assisted by graphene. Nanoscale **7**, 20249–20255 (2015)
108. F. Withers, O. Del Pozo-Zamudio, A. Mishchenko, A.P. Rooney, A. Gholinia, K. Watanabe, et al., Light-emitting diodes by band-structure engineering in van der Waals heterostructures. Nat. Mater. **14**, 301–306 (2015)
109. T.T. Tran, C. Elbadawi, D. Totonjian, C.J. Lobo, G. Grosso, H. Moon, et al., Robust multicolor single photon emission from point defects in hexagonal boron nitride. ACS Nano **10**, 7331–7338 (2016)
110. K.F. Mak, J. Shan, Photonics and optoelectronics of 2D semiconductor transition metal dichalcogenides. Nat. Photonics **10**, 216–226 (2016)
111. L.K. Li, Y.J. Yu, G.J. Ye, Q.Q. Ge, X.D. Ou, H. Wu, et al., Black phosphorus field-effect transistors. Nat. Nanotechnol. **9**, 372–377 (2014)
112. M.M. Ugeda, A.J. Bradley, Y. Zhang, S. Onishi, Y. Chen, W. Ruan, et al., Characterization of collective ground states in single-layer NbSe2. Nat. Phys. **12**, 92–U126 (2016)
113. Y.H.D. Lee, M. Lipson, Back-end deposited silicon photonics for monolithic integration on CMOS. IEEE J Sel Top Quantum Electron **19**, 8200207 (2013)
114. R. Shankar, R. Leijssen, I. Bulu, M. Loncar, Mid-infrared photonic crystal cavities in silicon. Opt. Express **19**, 5579–5586 (2011)

115. L. Li, H.T. Lin, S.T. Qiao, Y. Zou, S. Danto, K. Richardson, et al., Integrated flexible chalcogenide glass photonic devices. Nat. Photonics **8**, 643–649 (2014)
116. J.S. Wilson, *Sensor technology handbook* (Newnes, New South Wales, 2004)
117. J. Fraden, *Handbook of modern sensors*, 4th edn. (Springer, Berlin, 2010)
118. G. Korotcenkov, *Handbook of gas sensor materials* (Springer, Berlin, 2013)
119. S.S. Varghese, S.H. Varghese, S. Swaminathan, K.K. Singh, V. Mittal, Two-dimensional materials for sensing: graphene and beyond. Electronics **4**, 651–687 (2015)
120. W. Yuan, G. Shi, Graphene-based gas sensors. J. Mater. Chem. A **1**, 10078–10091 (2013)
121. S.S. Varghese, S. Lonkar, K.K. Singh, S. Swaminathan, A. Abdala, Recent advances in graphene based gas sensors. Sensors Actuators B Chem. **218**, 160–183 (2015)
122. F. Schedin, A.K. Geim, S.V. Morozov, E.W. Hill, P. Blake, M.I. Katsnelson, et al., Detection of individual gas molecules adsorbed on graphene. Nat. Mater. **6**, 652–655 (2007)
123. G. Chen, T.M. Paronyan, A.R. Harutyunyan, Sub-ppt gas detection with pristine graphene. Appl. Phys. Lett. **101**, 053119 (2012)
124. S. Rumyantsev, G. Liu, M.S. Shur, R.A. Potyrailo, A.A. Balandin, Selective gas sensing with a single pristine graphene transistor. Nano Lett. **12**, 2294–2298 (2012)
125. J.D. Fowler, M.J. Allen, V.C. Tung, Y. Yang, R.B. Kaner, B.H. Weiller, Practical chemical sensors from chemically derived graphene. ACS Nano **3**, 301–306 (2009)
126. V. Dua, S.P. Surwade, S. Ammu, S.R. Agnihotra, S. Jain, K.E. Roberts, et al., All-organic vapor sensor using inkjet-printed reduced graphene oxide. Angew. Chem. Int. Ed. **49**, 2154–2157 (2010)
127. F. Ricciardella, B. Alfano, F. Loffredo, F. Villani, T. Polichetti, M.L. Miglietta, et al., Inkjet printed graphene-based chemi-resistors for gas detection in environmental conditions, in *The AISEM Annual Conference* (2015), p. XVIII
128. H. Li, Z. Yin, Q. He, H. Li, X. Huang, G. Lu, et al., Fabrication of single- and multilayer $MoS_2$ film-based field-effect transistors for sensing NO at room temperature. Small **8**, 63–67 (2012)
129. D.J. Late, Y.-K. Huang, B. Liu, J. Acharya, S.N. Shirodkar, J. Luo, et al., Sensing behavior of atomically thin-layered $MoS_2$ transistors. ACS Nano **7**, 4879–4891 (2013)
130. F.K. Perkins, A.L. Friedman, E. Cobas, P.M. Campbell, G.G. Jernigan, B.T. Jonker, Chemical vapor sensing with monolayer $MoS_2$. Nano Lett. **13**, 668–673 (2013)
131. J.-S. Kim, H.-W. Yoo, H.O. Choi, H.-T. Jung, Tunable volatile organic compounds sensor by using thiolated ligand conjugation on $MoS_2$. Nano Lett. **14**, 5941–5947 (2014)
132. S. Cui, H. Pu, S.A. Wells, Z. Wen, S. Mao, J. Chang, et al., Ultrahigh sensitivity and layer-dependent sensing performance of phosphorene-based gas sensors. Nat. Commun. **6**, 8632 (2015)
133. X. Zhang, H. Ju, J. Wang, *Electrochemical sensors, biosensors and their biomedical applications* (Elsevier, New York, 2008)
134. D. Grieshaber, R. MacKenzie, J. Vörös, E. Reimhult, Electrochemical biosensors – sensor principles and architectures. Sensors (Basel) **8**, 1400–1458 (2008)
135. O. Moldovan, B. Iniguez, M.J. Deen, L.F. Marsal, Graphene electronic sensors – review of recent developments and future challenges. IET Circuits Devices Syst **9**, 446–453 (2015)
136. P.K. Ang, W. Chen, A.T.S. Wee, K.P. Loh, Solution-gated epitaxial graphene as pH sensor. J. Am. Chem. Soc. **130**, 14392–14393 (2008)
137. Y. Ohno, K. Maehashi, K. Matsumoto, Chemical and biological sensing applications based on graphene field-effect transistors. Biosens. Bioelectron. **26**, 1727–1730 (2010)
138. Y. Huang, X. Dong, Y. Liu, L.-J. Li, P. Chen, Graphene-based biosensors for detection of bacteria and their metabolic activities. J. Mater. Chem. **21**, 12358–12362 (2011)
139. P.K. Ang, A. Li, M. Jaiswal, Y. Wang, H.W. Hou, J.T.L. Thong, et al., Flow sensing of single cell by graphene transistor in a microfluidic channel. Nano Lett. **11**, 5240–5246 (2011)
140. L.H. Hess, M. Seifert, J.A. Garrido, Graphene transistors for bioelectronics. Proc. IEEE **101**, 1780–1792 (2013)
141. S. Jiang, R. Cheng, X. Wang, T. Xue, Y. Liu, A. Nel, et al., Real-time electrical detection of nitric oxide in biological systems with sub-nanomolar sensitivity. Nat. Commun. **4**, 2225 (2013)

142. N. Mohanty, V. Berry, Graphene-based single-bacterium resolution biodevice and DNA transistor: interfacing graphene derivatives with nanoscale and microscale biocomponents. Nano Lett. **8**, 4469–4476 (2008)
143. R. Stine, Real-time DNA detection using reduced graphene oxide field effect transistors. Adv. Mater. **22**, 5297–5300 (2010)
144. J. Ping, Scalable production of high sensitivity, label-free DNA biosensors based on back-gated graphene field effect transistors. ACS Nano **10**, 8700–8704 (2016)
145. Y. Huang, X. Dong, Y. Shi, C.M. Li, L.-J. Li, P. Chen, Nanoelectronic biosensors based on CVD grown graphene. Nanoscale **2**, 1485–1488 (2010)
146. S. Mao, G. Lu, K. Yu, Z. Bo, J. Chen, Specific protein detection using thermally reduced graphene oxide sheet decorated with gold nanoparticle-antibody conjugates. Adv. Mater. **22**, 3521–3526 (2010)
147. S. Mao, G. Lu, K. Yu, Z. Bo, J. Chen, Specific protein detection using thermally reduced graphene oxide sheet decorated with gold nanoparticle-antibody conjugates. Adv. Mater. **22**, 3521–3526 (2010)
148. Y. Ohno, K. Maehashi, K. Inoue, K. Matsumoto, Label-free aptamer-based immunoglobulin sensors using graphene field-effect transistors. Jpn. J. Appl. Phys. **50**, 070120 (2011)
149. J. Lee, P. Dak, Y. Lee, H. Park, W. Choi, M.A. Alam, et al., Two-dimensional layered $MoS_2$ biosensors enable highly sensitive detection of biomolecules. Sci. Rep. **4**, 7352 (2014)
150. C. Mackin, T. Palacios, Large-scale sensor systems based on graphene electrolyte-gated field-effect transistors. Analyst **141**, 2704–2711 (2016)
151. X. Gan, K. F. Mak, Y. Gao, Y. You, F. Hatami, J. Hone, et al., Strong enhancement of light-matter interaction in graphene coupled to a photonic crystal nanocavity. Nano Lett. 12, 5626–5631 (2012).

# Chapter 4
# Emerging NVM Circuit Techniques and Implementations for Energy-Efficient Systems

**Win-San Khwa, Darsen Lu, Chun-Meng Dou, and Meng-Fan Chang**

## 4.1 Introduction

With ever-increasing amount of data being transferred and stored in today's data centers, it is imminent to develop low-cost, low-power, and high-speed storage hardware. Due to the rapidly diminishing cost per bit, NAND flash memory is beginning to replace hard disk drive as large-capacity nonvolatile storage. The programming speed of NAND flash is still limited due to the fundamental write mechanism based on quantum mechanical tunneling [1]. Emerging nonvolatile memory (NVM) devices such as phase change memory (PCM), spin-torque transfer memory (STT-MRAM), and resistive memory (ReRAM) are being intensively studied to overcome the challenges faced by current mainstream memories. The figure-of-merit of various emerging NVMs, NAND flash, and DRAM are summarized in Table 4.1. Generally, the emerging NVMs exhibit superior write performance with fast speed and low power compared to conventional ones. Among them, while STT-MRAM shows particularly fast write operation and good endurance, PCM and ReRAM show advantages on the storage density. Based on their different characteristics, emerging NVMs have a broad range of applications and further enrich the memory hierarchy, as shown in Fig. 4.1. In particular, emerging NVMs are expected to serve as embedded memory for low-power SoC (system-on-chip) designs due to their fast speed and high-energy efficiency. Besides, the recent progress of emerging NVMs toward high density and low cost has further enabled their applications on the storage class memory and high-density storage. Furthermore, these new storage

W.-S. Khwa · C.-M. Dou · M.-F. Chang (✉)
National Tsing-Hua University, Hsinchu, Taiwan
e-mail: mfchang@ee.nthu.edu.tw

D. Lu
National Cheng-Kung University, Tainan, Taiwan



R. O. Topaloglu, H.-S. P. Wong (eds.), *Beyond-CMOS Technologies for Next Generation Computer Design*, https://doi.org/10.1007/978-3-319-90385-9_4

**Table 4.1** Device metric comparison for emerging memory (PCM, STT-MRAM, ReRAM), NAND flash, and DRAM

| Device metric | | PCM | STT-MRAM | ReRAM | NAND | DRAM |
|---|---|---|---|---|---|---|
| Power | Write energy/bit | 18 pJ [2][a] | 1.0 pJ [3] | 0.1 pJ [3] | 100 pJ [4] | <1.0 pJ [5] |
| Performance | Write current | 100 μA [2] | 50 μA [6] | 1.0 μA [7] | | |
| | Write latency | 150 ns [2] | 5 ns [8] | 50 ns [3] | >100 μs | 5 ns [9] |
| Reliability | Read latency | 50 ns [10] | 10 ns [11] | <10 ns [12] | 15–50 μs [13] | 20–80 ns [13] |
| | Program window | 3 bit/cell [14] | Good [15] | Variable [16] | 4 bit/cell [17] | Good |
| Density | Endurance | $10^8$–$10^9$ [5, 18] | Unlimited [19] | $10^5$–$10^{10}$ [19] | $10^5$–$10^6$ [20] | Unlimited |
| | Retention | R-drift [14] | Good [8] | RTN [21] | Good | 64 ms |
| | Cell size | 4 $F^2$ [2] | 12 $F^2$ [6] | 4–6 $F^2$ [22] | <4 $F^2$ [2] | 7 $F^2$ [2] |

[a]Write energy estimated using $I_{RESET} \times V_{DD} \times t_{RESET}$

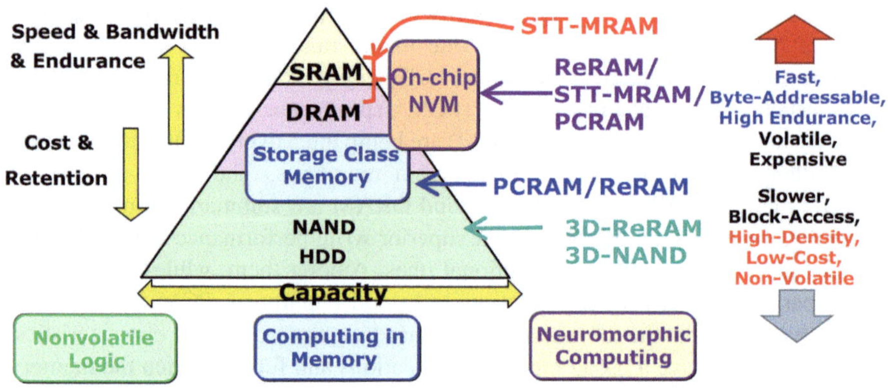


**Fig. 4.1** Applications of emerging nonvolatile memories

devices can be basis of novel applications such as nonvolatile logic, computing-in-memory, neuromorphic circuits, or hardware security.

Several bit cell configurations of emerging NVMs have been developed for different applications. Typical one-transistor-one-resistor (1T1R) for PCRAM, STT-MRAM, and RRAM is shown in Fig. 4.2, in which the memory cells are integrated in the drain sides of transistors though the back-end-of-line (BEOL) process. Here, the cell access is controlled by word lines (WLs), and the write/read operations can be done by controlling bit lines (BLs) and source lines (SLs). The 1T1R configuration is commonly adopted for embedded memory systems because of its (1) high compatibility with CMOS process, (2) good immunity of write/read disturb,

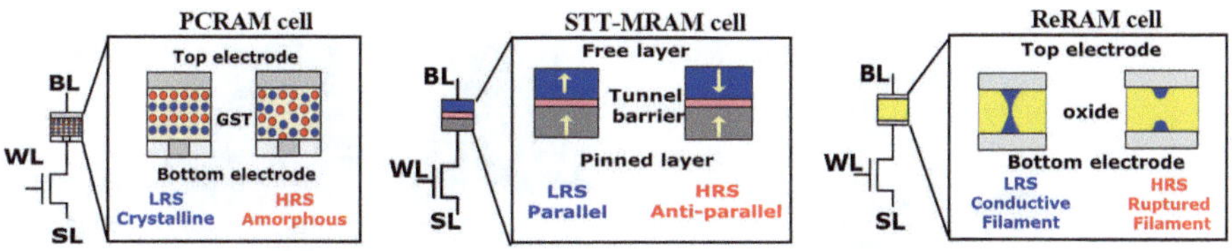


**Fig. 4.2** 1T + 1R bit cell configurations of emerging NVMs

and (3) effective suppression of leakage currents. Other configurations are also under developments. For example, one-diode-one-resistor (1S1R) and one-selector-one-resistor (1S1R) [9] can benefit from high-density storage, though specific diodes/selectors are required. In this chapter, we mainly focus on the 1T1R array for energy-efficient systems, but it is noteworthy that the circuit techniques introduced here can be extended to other configurations.

In the rest of this section, we begin with discussing the physical principles, figure of merits, and key technological challenges for PCM, STT-MRAM, and ReRAM devices. Subsequently, in the next section, we will analyze these emerging memory technologies from a circuit perspective for storage as well as various other innovative applications.

### *4.1.1 Phase Change Memory Device*

$Ge_2Sb_2Te_5$ (germanium-antimony-telluride), or GST, is a type of phase change material often used in re-writable DVDs. By applying a short laser pulse to melt the material locally and have it quickly cooled, it enters the amorphous phase (darker in color). On the other hand, when heated above its crystallization temperature ($T_{cryst}$) but below the melting point ($T_{melt}$) for a longer duration, the material enters the crystalline phase (lighter in color). Information is stored in the form of a sequence of GST material phase, or dark/light spots on the DVD disk. Since the change in phase of GST material can be repeated many times, these DVD disks are "re-writable."

The operation of PCM is based on similar principles. As shown in Fig. 4.3a, the PCM device consists of two conductors (electrodes) that are connected to the top and bottom parts of the GST material, respectively. The $SiO_2$ serves the purpose of reducing bottom contact dimensions for reasons that will be discussed later in this subsection. GST material has higher electrical resistivity in its amorphous phase as compared to the crystalline phases. Therefore, with the dome-shaped amorphous region shown in Fig. 4.3a, the PCM device is in its high-resistance state. Figure 4.3b illustrates real PCM devices in the "SET" and "RESET" states, respectively [23]. GST material in amorphous and crystalline phases can be clearly distinguished by optical contrast. Figure 4.3c, d shows temperature as a function of time for "SET," "RESET," and read operations for PCM devices. When a large current pulse is applied between the top and bottom electrodes, the device is heated above its

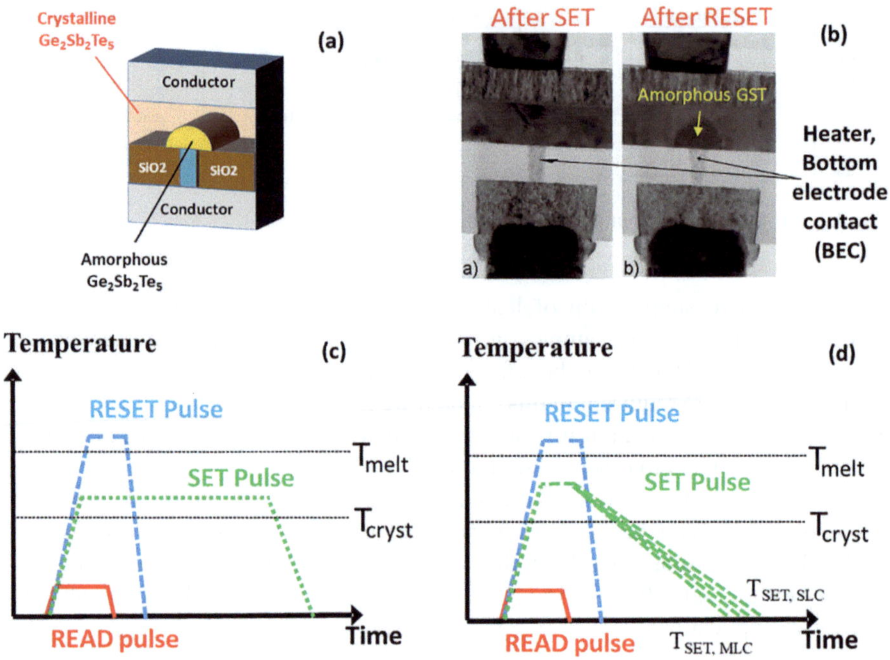


**Fig. 4.3** PCM device (**a**) Schematic illustration of device structure (**b**) Cross-sectional micrograph of a real PCM cell in the "SET" (crystalline) or "RESET" (amorphous) states, after [23], and temperature cycle applied to program (set/reset) and read the PCM with (**c**) rectangular set pulse and (**d**) slow-quench pulse

melting temperature for a short duration, and the amorphous region is formed. This brings the device to its high-resistance "RESET" state. Subsequently, a medium current pulse is applied to heat the device only slightly above its crystallization temperature. The amorphous region crystallizes and the device is back to its low-resistance "SET" state. Note that by using slow-quench set pulses, multilevel cell (MLC) operation can be achieved by writing the cell to different resistance levels. For read operation, an even lower current is adopted to sense the cell resistances.

One of the key challenges for PCM for memory applications is that very large current is required to melt the GST material to "RESET" it to amorphous state. This current is often referred to as RESET current ($I_{RESET}$). At 40 nm dimensions, $I_{RESET}$ is about 100 μA [2]. In a memory array, one select device is required for each memory cell to reduce leakage current through unselected cells. For best isolation, we typically use the MOS transistor as the select device. However, for PCM, it is an immense challenge to design a MOS transistor that delivers 100 μA at 40 nm. An alternative is to use a diode selector device [2]. However, for density's sake, nonvolatile memory typically requires the cells to be programmed at four or more levels. This is known as MLC. In such cases, leakage requirement is even more stringent, and diode may not provide sufficient isolation. Fortunately, $I_{RESET}$ is

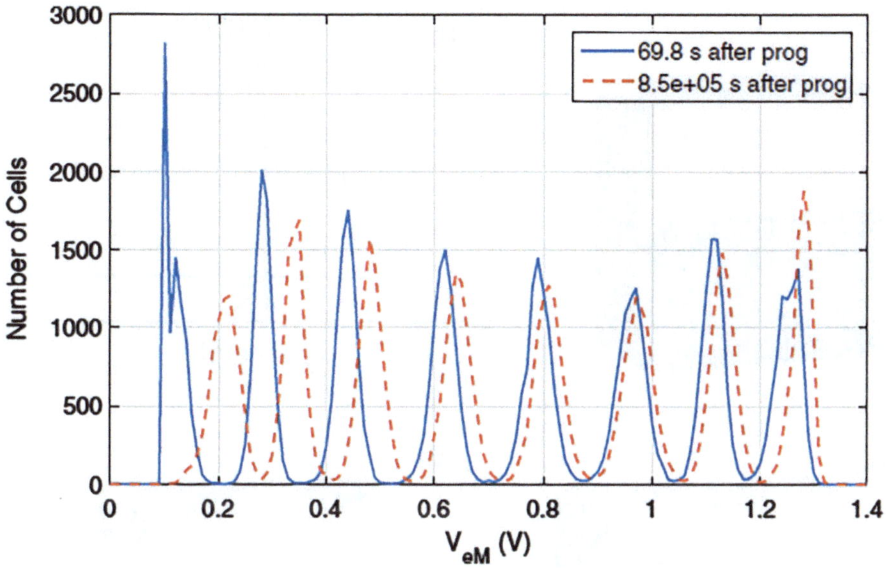


**Fig. 4.4** eM-metric distribution after a PCM device is programmed at eight different levels for 3-bits-per-cell storage [14]. Even though resistance drift causes the eight levels to shift from its initial value 10 days ($5 \times 10^5$ s) after programming, the spacing is sufficient to distinguish one level from another

reduced as we scale the contact dimension, making use of MOS selector device possible, provided contact area can be scaled [24]. It has been demonstrated that with a 3 nm contact $I_{RESET}$ can be reduced to 5 μA [25].

Another issue of PCM is retention problem, known as resistance drift. As shown in Fig. 4.4, the programmed voltage distribution is shifted from its original one about 10 days after programming. Physically, crystalline state is the lower energy configuration; therefore, PCM material gradually shifts toward that state. To overcome resistance drift, a novel programming scheme known as eM-metric [14] allows PCM to be programmed and read properly in spite of resistance drift.

PCM is superior compared to NAND or NOR flash in terms of endurance. In general, PCM can be written around $10^8$–$10^9$ cycles. There are even reports of PCM cells that can be programmed $10^{10}$–$10^{11}$ times [26, 27]. Lower $I_{RESET}$ is one of the key contributing factors to further improved endurance. Hence, PCM can be expected to be used as storage class memory or on-chip embedded NVM.

### *4.1.2 Spin-Torque Transfer Memory Device*

Ferromagnetic materials have been widely used as information storage medium since the invention of magnetic tapes, magnetic core memories, and hard disk drives. The operation principle of the magnetic tunneling junction (MTJ) is similar to hard

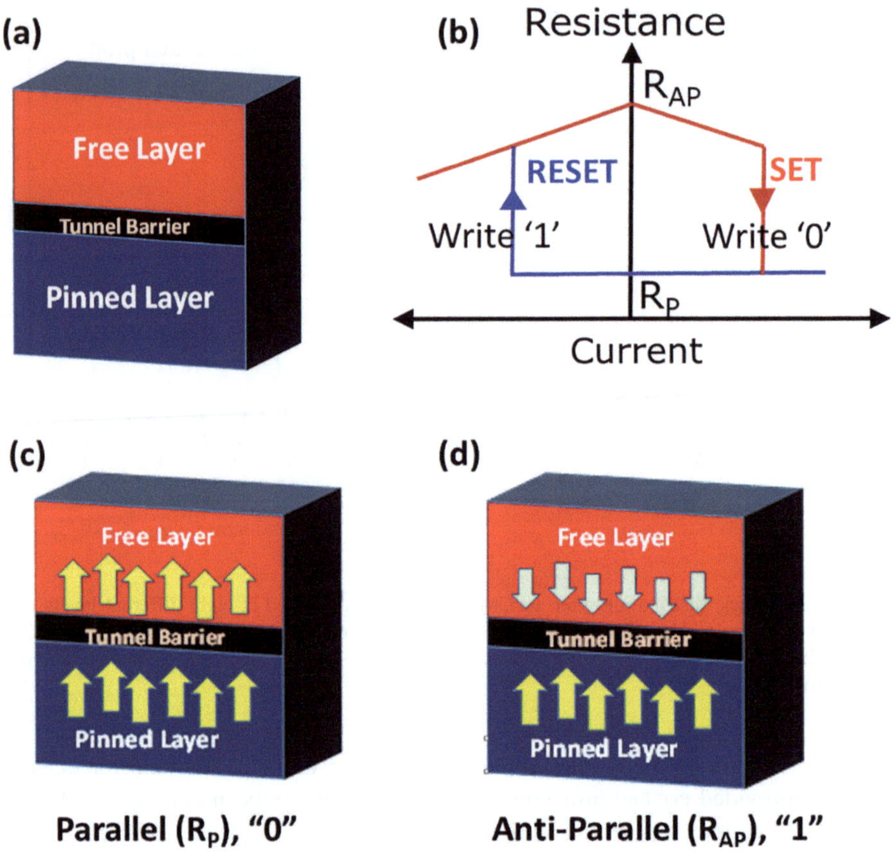


**Fig. 4.5** (**a**) STT-MRAM cell structure consists of a tunnel barrier (insulator) and "free" and "pinned" ferromagnetic layers, (**b**) MTJ resistance versus injected current, (**c**) the low resistive state with parallel polarization, and (**d**) the high resistive state with antiparallel polarization

disk drives. It consists of a layer of tunnel barrier (insulator) sandwiched between two ferromagnetic layers. The insulator is thin enough to allow quantum mechanical tunneling of electrons. Tunneling current, or effectively resistance across the MTJ depends on the relative spin polarization of the top and bottom ferromagnetic layer. If the two layers have the same spin polarization, resistance is small. On the other hand, if the two layers are polarized oppositely, resistance is larger. The relative resistance change is defined as tunneling magnetoresistance (TMR).

STT-MRAM cell consists of an MJT junction with free and pinned (reference) ferromagnetic layers, as shown in Fig. 4.5a. The free layer's magnetization can be changed, whereas the pinned layer's magnetization is fixed. Most of successful demonstrations of STT-MRAM use CoFeB as free/pin layer material and MgO as the tunnel barrier. Besides, additional layers are usually required to fix the polarization of the pin layer.

Writing of information is done by simply passing current across the MTJ with different directions, as illustrated in Fig. 4.5b. Figure 4.5c, d shows the low (logic "0") and high (logic "1") resistive states with parallel and antiparallel magnetic polarizations. In order to write "0," the electrons are injected from the side of the pinned layer. The spin-polarized electrons that aligned with the pin layer are then generated, which will force the magnetization of the free layer to be parallel with the pin layer. On the other hand, the current is passed from the free layer to the pinned layer for writing "1." While the electrons whose spin is aligned with the pinned layer's magnetization direction pass through, the oppositely-polarized electrons are accumulated in the free layer. These oppositely-polarized electrons flip the magnetization of the free layer, and the magnetic polarizations of the two layers become opposite.

One of the key advantages of STT-MRAM is its endurance—STT-MRAM can be written $10^{15}$ times or more. For this reason, STT-MRAM may be suitable for DRAM applications, embedded DRAM in SoC (system-on-chip) applications, or even SRAM due to the comparable access time [28] and smaller cell size. Non-volatility is an added benefit. Commercial DRAM products based on STT-MRAM are available [29].

In general, the size of MRAM cell is larger than that of any other nonvolatile memory or DRAM. Non-STT MRAM, where write operation is conducted by passing current through a wire to induce magnetic field, requires large cell-to-cell spacing to avoid interference and prevent electromigration [30]. The aforementioned spin-torque transfer writing mechanism had already reduced MRAM cell size significantly. However, STT-MRAM is still larger than NAND or DRAM [6] (Table 4.1). Most successful demonstrations show cell size larger than 12 $F^2$ [6, 8, 11], where F (feature size) is the minimum feature size of the process technology. Besides, STT-MRAM usually has a relatively small current difference between high and low resistance values compared to other emerging NVMs (Table 4.1). Moreover, the cell characteristics are sensitive to variations due to process and temperature, which further degrades the sensing margin (Fig. 4.6). Sensing circuits with small offset are necessary to tackle this issue.

Another drawback of STT-MRAM is process complexity. Typically, multiple layers of magnetic materials are required to construct the pinned layer. In addition, magnetically hard material is required on top of the pinned layer to limit external interference and magnetic field leakage. Such process complexity may limit the scaling of STT-MRAM into three dimensions.

### *4.1.3 Resistive Memory Device*

ReRAM is a two-terminal device with programmable resistance. Bipolar resistive switching has been widely observed in transition metal oxide [9]. Basic physical mechanism for bipolar-switching ReRAM is illustrated in Fig. 4.7a. By applying a large voltage across the device, conductive filament (CF) forms, which links the top

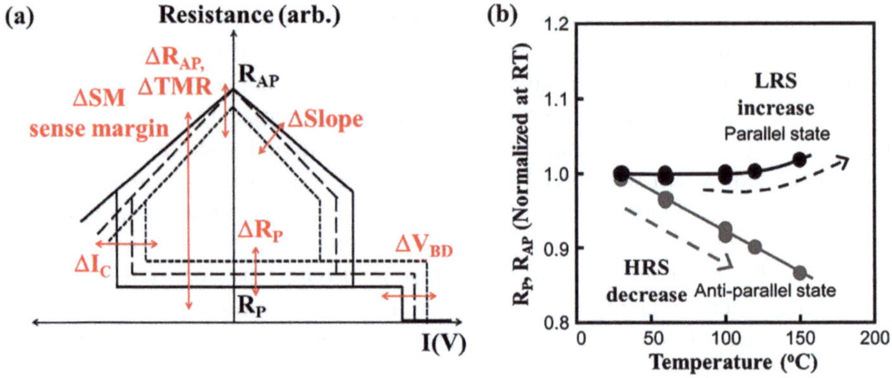


**Fig. 4.6** (**a**) Resistance variations of the MJT junction [31] and (**b**) its temperature dependence [32]

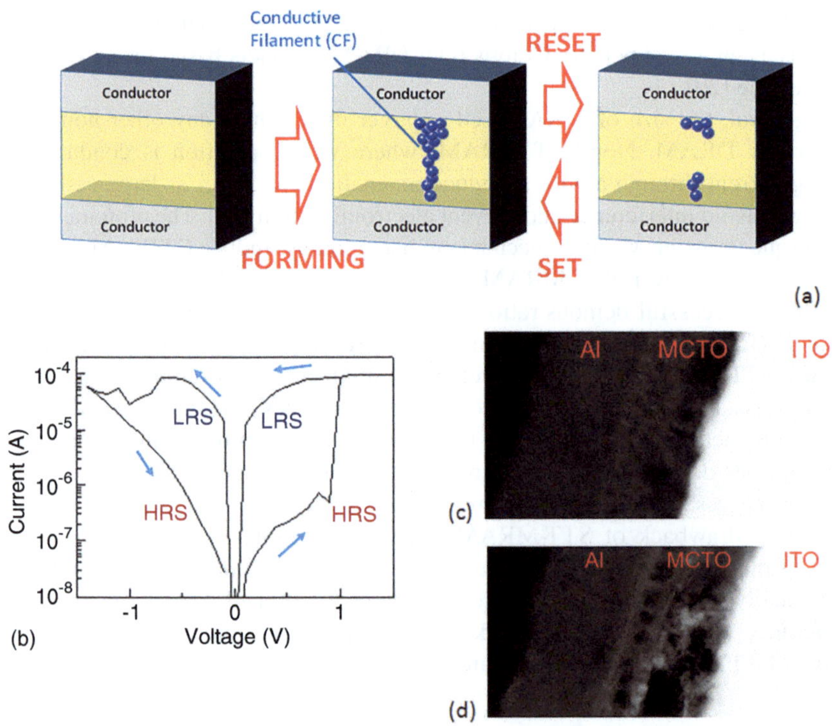


**Fig. 4.7** Illustration of ReRAM (**a**) programming mechanism (**b**) I–V curve for an $HfO_X$-based bipolar-switching ReRAM device [12] (**c**) high-resistance state and (**d**) low-resistance state of an MCTO-based ReRAM in STEM-HAADF images [33]

and bottom electrodes, dramatically reducing device resistance. This step is known as "Forming." Subsequently, if voltage with opposite polarity is applied, CF will be broken and resistance will become large again ("RESET operation). If we switch voltage polarity again and apply intermediate voltage (lower than the "Forming" voltage), CF will re-form, bringing ReRAM back to the low-resistance state ("SET" operation).

Figure 4.7b shows the typical I–V characteristics of a bipolar-switching ReRAM device, with arrow indicating sweep direction. Starting from 0 V, by increasing the voltage past 1.0 V, ReRAM switches from the high-resistance state (HRS) to the low-resistance state (LRS). Afterward the voltage is decreased from 1.0 to −1.0 V, and the I–V follows the LRS trajectory. At −1.0 V, the ReRAM switches back to HRS. Subsequent forward sweep now follows the HRS trajectory. Fig. 4.7c, d illustrates the observation of CF formation in an MCTO-based ReRAM device [33]. Crystalline $In_2O_3$ CF in the low-resistance state is clearly visible from the STEM (scanning transmission electron microscopy) image.

ReRAM requires the lowest write energy among all memory technologies (Table 4.1). Unlike PCM where a significant portion of the device material must switch states, ReRAM relies on the formation of conductive filament with dimension as low as 2–3 nm [34]. As a result, write current for ReRAM can be 1 μA or less with write energy less than 0.1 pJ [3, 7]. For NAND, assuming Fowler–Nordheim tunneling current of 1.0 A/cm$^2$ [35] is required to program a 20 nm by 20 nm floating gate transistor, the total current is only 4 pA. However, with large programming voltage and relatively long programming duration, the average energy to write a bit is on the order of 100 pJ including external overhead, which is much larger than ReRAM.

The low switching energy comes at an expense of retention problems and large device variability. As shown in [16], cell-to-cell variation makes it difficult to distinguish LRS from HRS unless we use a large current to program the ReRAM. In addition, even for a single cell resistance distribution varies from one programming cycle to another. Iterative sense-and-write mechanism may not work if random telegraphic noise (RTN) causes single-event driven current fluctuation [21]. In fact, the same programmed resistance value may change overtime due to RTN. This is no surprise, as the conductive path in the LRS is typically a very narrow path. The use of conductive-bridge RAM (CBRAM), a special type of ReRAM, design helps overcome the variability problem by enlarging the window between LRS and HRS [36]. A 4 Mb single-level ReRAM array has been demonstrated successfully [12]. Multiple-bits-per-cell is still challenging due to the variability problem.

In summary, unlike PCM or STT-MRAM, ReRAM is a technology for which material exploration is still underway. Given the wide variety of materials that can be used as ReRAM [9], significant research opportunity exists for both academia and industry.

## 4.2 Read Circuits for Emerging Resistive NVM

Voltage-mode sense amplifier (VSA) and current-mode sense amplifier (CSA) are two frequently used configurations in NVM. Here we provide an introduction on the circuit structure and the operation waveform of both SA types. Afterward, we

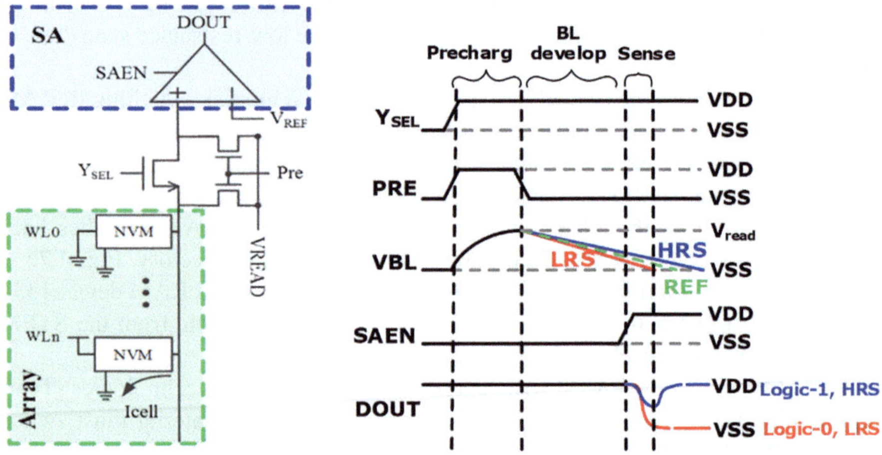


**Fig. 4.8** Circuit schematic and conceptual waveform of a typical voltage-mode sensing scheme

discuss some of the challenges associated with each sensing scheme, and review examples of advanced circuit techniques.

## *4.2.1 Voltage-Mode and Current-Mode Sensing Schemes*

Figure 4.8 illustrates the circuit schematic and the conceptual waveform of a typical VSA. The read operation could be separated into three phases: (1) precharge, (2) voltage development, and (3) sensing. First, precharge transistors are switched on to charge the bit-line voltage ($V_{BL}$) from 0 V to a read voltage ($V_{READ}$). In the second phase, the word-line (WL) is turned on, and the $V_{BL}$ starts to discharge through the ReRAM device. If the ReRAM cell is in the logic-1 (HRS) state, the cell read current ($I_{READ}$) is small and the $V_{BL}$ is maintained near $V_{PRE}$. On the contrary, the $V_{BL}$ will decrease at a more rapid pace if the ReRAM cell is in the logic-0 (LRS) state, in which the $I_{READ}$ is larger. When a considerable voltage swing is developed for the LRS cell, we entered the sensing phase. The VSA is used to compare $V_{BL}$ with a predefined reference voltage ($V_{REF}$), the digital comparison result is generated at data-out (DOUT).

Figure 4.9 illustrates the circuit schematic and conceptual waveform of a typical current-mode sensing scheme. Unlike the VSA scheme, in which the sensing result depends on the discharging of $V_{BL}$, the CSA employs a fixed $V_{BL}$ at the clamping voltage ($V_{CLP}$) to induce an $I_{READ}$ through the ReRAM device for reading. When the WL is turned on, the ReRAM device will induce an $I_{READ}$ according to its resistance state. The CSA result is obtained by comparing the $I_{READ}$ with a predefined reference current ($I_{REF}$). Clamping the $V_{BL}$ allows the $I_{READ}$ to be transferred to the CSA without discharging the large BL capacitance. As a result, for longer BL length, the current-mode sensing scheme usually has smaller sensing delay and power.

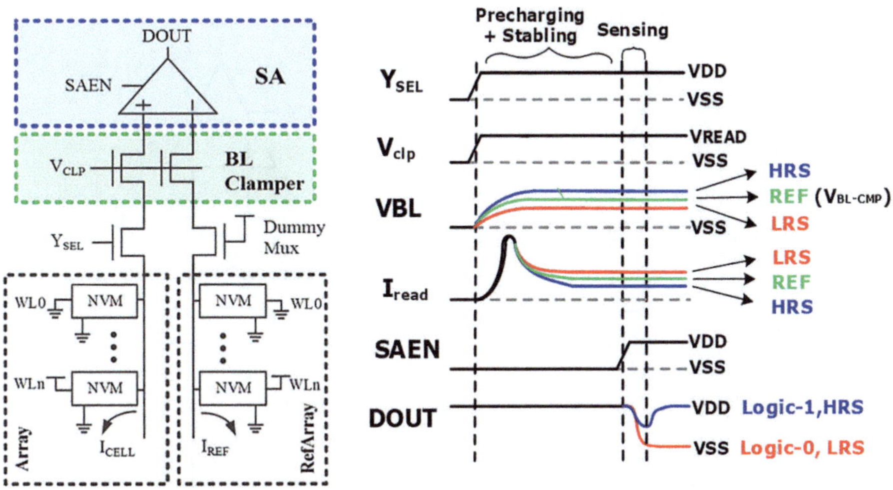


**Fig. 4.9** Circuit schematic and conceptual waveform of a typical current-mode sensing scheme

Two commonly used bit-line voltage clamping approaches are static clamping and dynamic clamping. Figure 4.10 illustrates their circuit structures. In static clamping, the clamping transistor ($N_{CLP}$) is controlled by supplying a constant voltage ($V_{CLP}$) at the gate terminal. In dynamic bias clamping, the $V_{BL}$ fluctuation is suppressed by the negative feedback to the gate of $N_{CLP}$. The latter approach is capable of achieving a larger $N_{CLP}$ gate swing and a more rapid BL bias speed, at the cost of area and power overhead.

Choosing the appropriate sensing scheme is a design decision that depends on both the ReRAM device resistance and the cell array structure. The rule of thumb is to choose VSA over CSA for shorter BL, because smaller BL loading favors the $V_{BL}$ discharge phase used in VSA. Figure 4.11 compares the ReRAM read access time of VSA and CSA schemes with different BL lengths. When reading a 0.18 μm ReRAM LRS cell ($I_{READ} = 20$ μA) on a BL with over 128 cells, the CSA provides a faster read speed than the VSA.

### 4.2.2 Challenges for Voltage-Mode Sensing Scheme

The VSA scheme faces design challenges associated with sensing margin and low $V_{DD}$ read latency. In conventional memory technologies, such as SRAM and logic-ROM, the bit-line voltage ($V_{BL}$) swing for the read-0 operation is significantly larger than that for the read-1 operation. This advantage is attributed to their high on–off current ratio ($I_{RATIO} = I_{CELL\text{-}0}/I_{CELL\text{-}1}$). This characteristics allows SRAM and logic-ROM to employ traditional voltage-mode sense amplifiers (VSA) for differential sensing or single-ended sensing using a reference voltage ($V_{REF}$). In

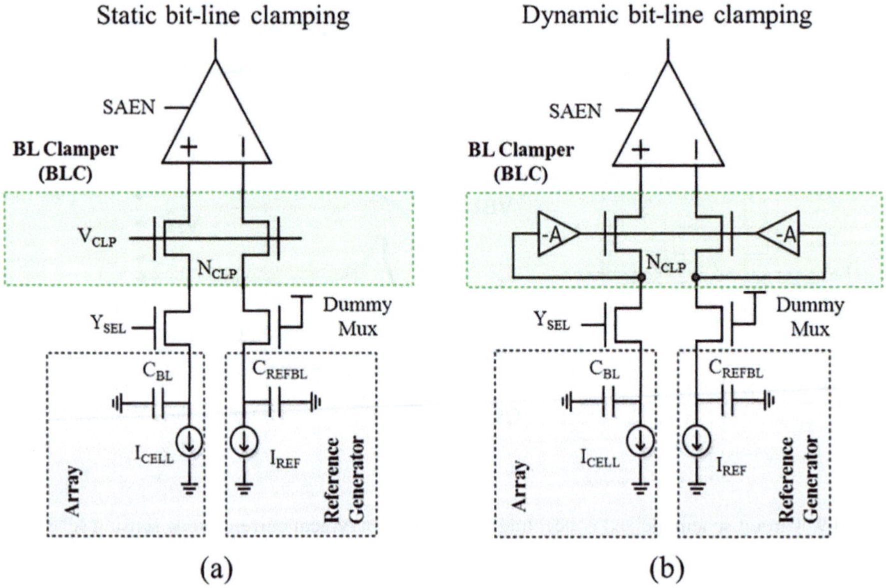


**Fig. 4.10** Circuit structure of (**a**) static and (**b**) dynamic BL voltage clamping schemes

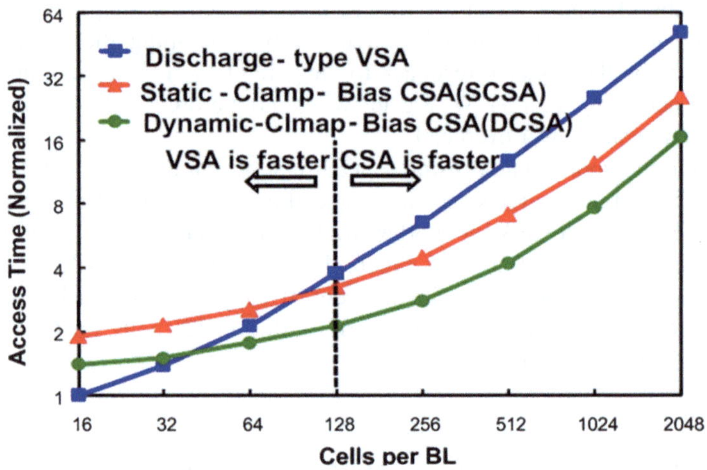


**Fig. 4.11** Read access time comparison among various sensing schemes with different BL lengths

VSA, the $V_{REF}$ is generally chosen to allocate equal sensing margin ($V_{SM}$) for read-0 and read-1 operations. Therefore, $V_{REF}$ is often chosen to be the midpoint between the $V_{BL}$ when reading logic-0 and logic-1 tail cells. Consequently, this provides a $V_{SM}$ that only equals (or less than) half of the tail $V_{BL}$ swing.

The situation is more challenging for emerging memories where process or temperature variation further reduces sensing margin. As shown in Fig. 4.12a,

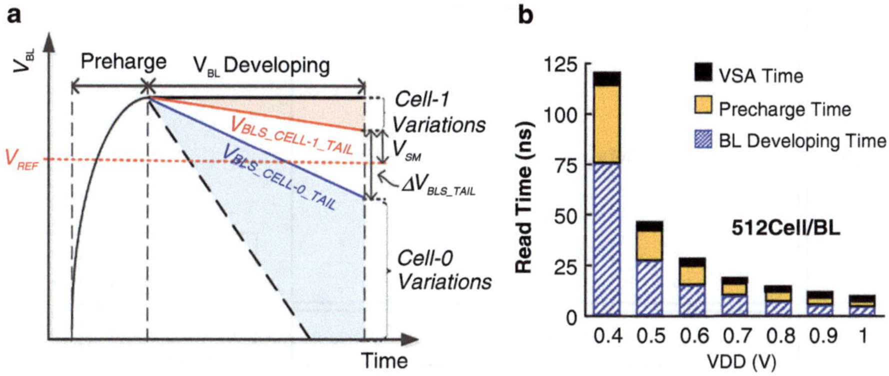


**Fig. 4.12** (**a**) Variations in BL voltage swing during read operation due to tailing cells. (**b**) Read time breakdown of a typical VSA under various $V_{DD}$

emerging memories suffer from smaller $V_{BL}$ difference between tail cells when reading logic-0 and logic-1 cells ($\Delta V_{BLS\_TAIL}$). In particular, both read-1 and read-0 operations experience dynamic $V_{BL}$ swings during the $V_{BL}$ development phase. This prevents emerging memories from employing traditional fixed-value $V_{REF}$ sensing scheme [37–42] or other novel VSAs that requires the read-1 $V_{BL}$ to be maintained near $V_{PRE}$ or $V_{DD}$ [43–46].

Second, lowering the $V_{DD}$ has a profound impact on read time. The dynamic $V_{BL}$ swings for both read-1 and read-0 require a longer $V_{BL}$ development time to ensure sufficient $\Delta V_{BLS\_TAIL}$ for accurate sensing. The decrease in $V_{DD}$ reduces the transistor driving strength and $V_{SM}$, which further increases the macro-level read time. Figure 4.12b breaks down the read time of a conventional voltage-mode sensing scheme using a 65 nm emerging memory macro with a bit-line length of 512 cells. Under low $V_{DD}$, the precharge time and the BL developing time dominate the read time, while the VSA response time remains relatively unchanged. This trend becomes more prominent in macros with longer BL lengths. Therefore, speed and yield improvements of low $V_{DD}$ voltage-mode sensing should emphasize on bit-line development time reduction, while maintaining a $V_{SM}$ sufficient to tolerate the input offset voltage of the VSA.

### *4.2.3 Challenges for Current-Mode Sensing Scheme*

The current-mode sensing scheme has three design challenges associated with (1) BL clamping voltage, (2) reference current, and (3) voltage head room. Conventional current-mode sensing requires the BL voltage clamping. To ensure robust and high-speed read operations, it is desirable to have $V_{BL}$ clamped as high as possible to compensate for the bit-line unselected cells leakage current, and the CSA input

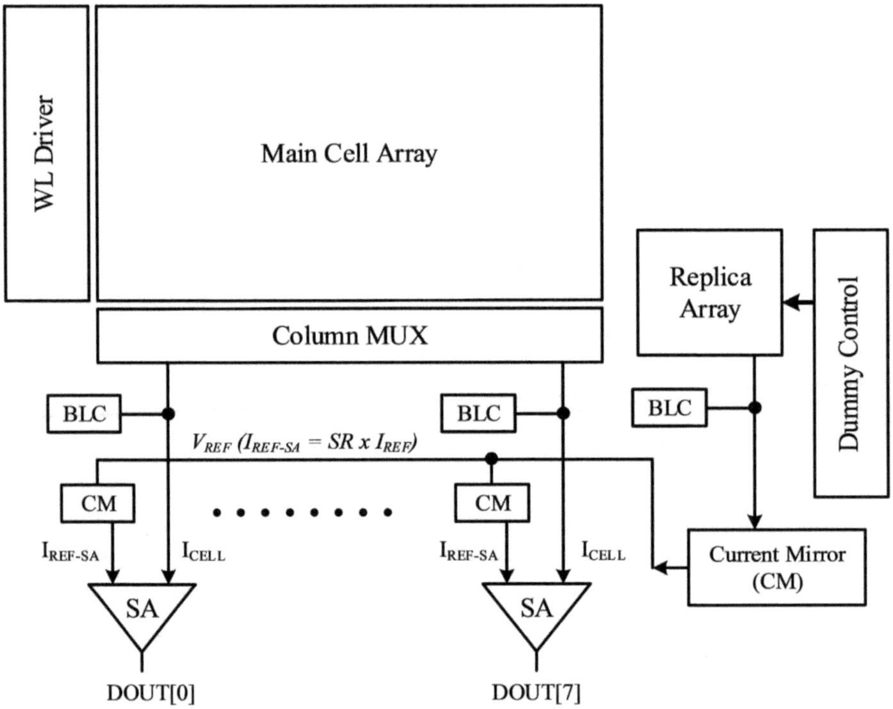


**Fig. 4.13** An example of the replica cell current sensing circuit for emerging NVM cell array

offset. However, the $V_{BL}$ must be kept considerably smaller than the typical program voltage to prevent read disturbance, which might modify the existing cell content. This imposes a trade-off between read speed and read disturbance.

As previously mentioned, current-mode sensing usually involves comparing the cell current ($I_{CELL}$) with a predefined reference current ($I_{REF}$). However, using a fixed $I_{REF}$ cannot accommodate for the $I_{CELL}$ fluctuations across various process, voltage, and temperature (PVT) variations. Several publications address this issue using the replica cell schemes [47–53]. In the replica cell schemes, because both the memory cells and the replica cells experience similar conditions, allowing the generated $I_{REF}$ to fluctuate along with $I_{CELL}$ across different PVT variations. Figure 4.13 presents an example of the replica cell current sensing circuit for emerging NVM cell array. The replica array includes both active and inactive replica cells. The active replica cell generates replica current ($I_{CELL\_REPLICA}$) matches that from regular main cell array. The purpose of inactive replica cells is to minimize mismatch between $I_{CELL}$ and $I_{CELL\_REPLICA}$ by providing similar neighboring geometric patterns and parasitic loads to what a regular memory cell experiences in the main cell array.

If a hard defect or a significant process variation occurs in the replica array, the $I_{REF}$ could be inaccurate and susceptible to high die-to-die variation. This could lead to read failure or increase in read access time and power consumption. Therefore,

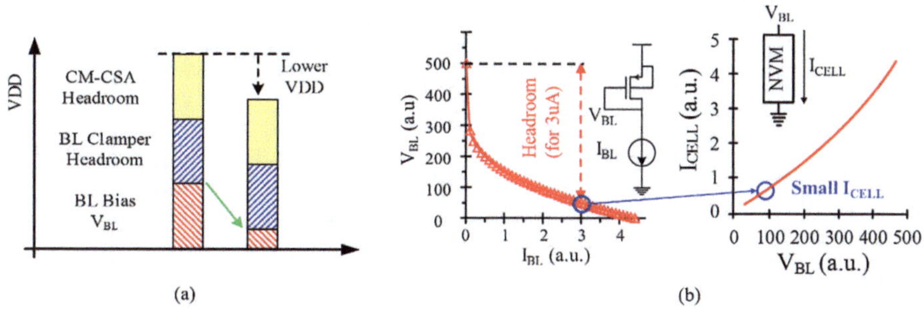


**Fig. 4.14** (**a**) Concept of voltage budget and (**b**) bit-line voltage vs. bit-line current in a typical 65 nm current-mode sensing scheme

read reference current generation is a crucial challenge in the design of emerging NVM.

Figure 4.14a outlines the concept of voltage budge in a current-mode sensing scheme, which consists of current-mirror (CM) headroom, BL clamper headroom, and BL bias $V_{BL}$. The amount of supply voltage ($V_{DD}$) reduction is limited by the voltage headroom of the diode-connected CM and the BL voltage clamping circuitry. As illustrated in Fig. 4.14b, a 3 μA bit-line current would require a voltage headroom of 400 mV for a typical 65 nm CM circuit [54]. Lowering $V_{DD}$ suppresses the effective $V_{BL}$ as well as the voltage across the emerging memory device. This leads to smaller cell current during read operation, which further reduces the sensing margin and read speed.

## 4.2.4 Advanced Circuit Design Techniques for VSA and CSA

### 4.2.4.1 Low-$V_{DD}$ Swing-Sample-and-Couple VSA (SSC-VSA)

Figure 4.15 illustrates the concept of a low-$V_{DD}$ swing-sample-and-couple (SSC) VSA [55]. While conventional VSA utilizes only half the bit-line voltage swing of the tailing cells (½ $\Delta V_{BLS\_TAIL}$), this SSC-VSA is capable of exploiting the full $\Delta V_{BLS\_TAIL}$ for sensing margin ($V_{SM}$). The SSC circuit comprises of one PMOS transistor (T1), two switches (SW1 and SW2), and one capacitor (C1). Unlike midpoint $V_{REF}$ in conventional VSA, SSC-VSA places the $V_{REF}$ below the midpoint. In phase-1 ($V_{BLS}$ sampling), the BL develops a voltage swing $V_{BLS}$ (=$V_{DD} - V_{BL}$) during a time ($T_{BL}$), and SW1 is turned on to transfer $V_{BL}$ to node IN1 of VSA and node A of C1, while node B of C1 is maintained at $V_{REF}$.

In phase-2 ($V_{BLS}$ coupling), SW1 is turned off for isolation between node A and BL. Then T1 is turned on to pull node A up from $V_{BL}$ to $V_{DD}$. Node A's voltage shift boosts the node IN2 voltage from $V_{REF}$ to ($V_{REF}$ + $V_{BLS}$). In phase-3 (comparison), the SAEN signal enables the

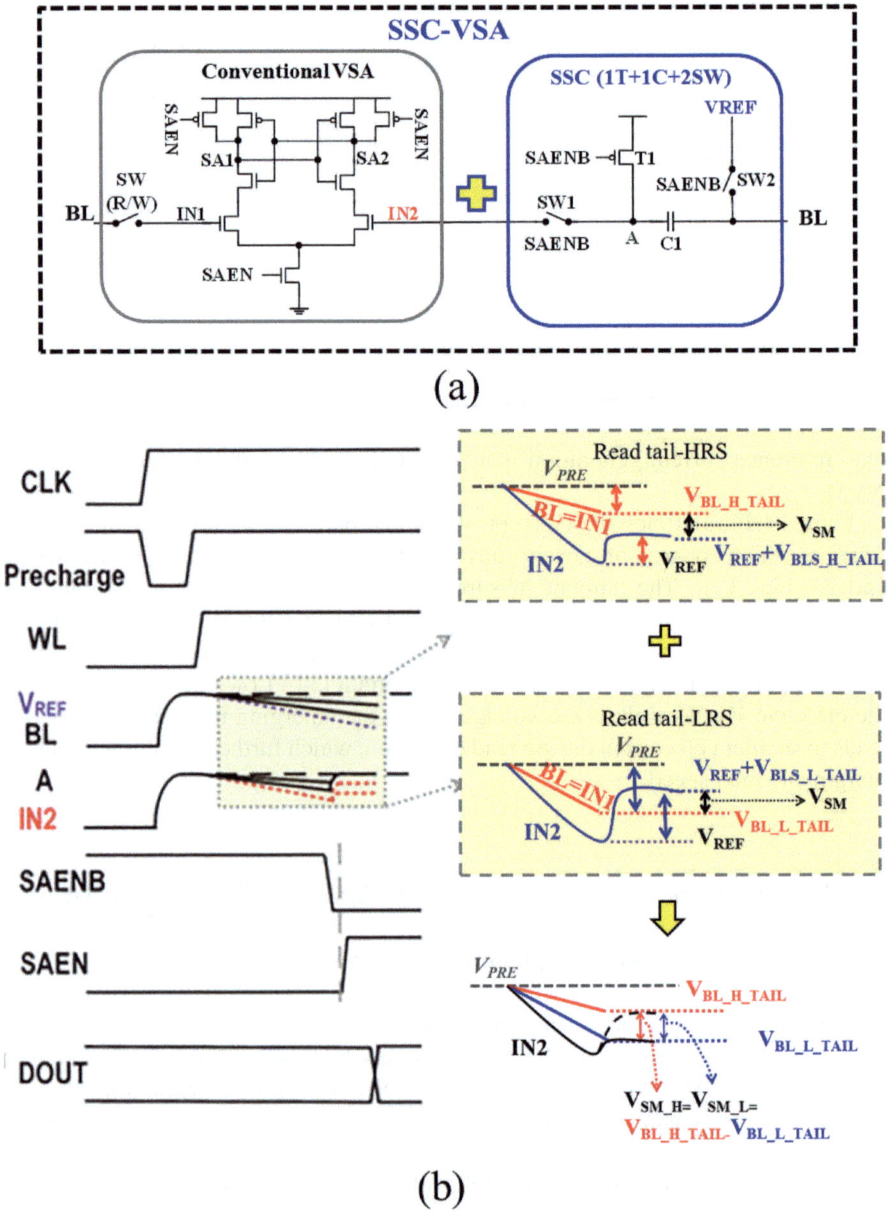


**Fig. 4.15** (**a**) Circuits and (**b**) waveforms of a low-$V_{DD}$ swing-sample-and-couple (SSC) VSA

VSA to detect the differential voltage between its two inputs, which is $\Delta V_{SA} = (V_{REF} + V_{BLS}) - (V_{DD} - V_{BLS}) = 2V_{BLS} - (V_{BLS\text{-}H\_TAIL} + V_{BLS\text{-}L\_TAIL})$. The increase in $V_{SM}$ (nearly double) lowers $V_{DD}$ minimal requirement and achieves read speed exceeding those of conventional VSA at low $V_{DD}$.

#### 4.2.4.2 Reference Current Generation

Most floating gate (FG)-based replica cells connect selected floating gates to their control gate at the edge of replica array. This avoids the need for additional erase cycle and allows the replica cell FG voltage potential to be maintained throughout multiple access cycles. However, the sharp corners and the special shapes of the FG layer in split-gate flash technology make it difficult to maintain uniformity and lower resistances in metal-to-FG contact and FG poly. Therefore, the resistance ($R_{DWL\_FG}$) between control-gate (dummy word-line) and FG of the accessed replica cells often exhibits significant variation across rows/dies. The $R_{DWL\_FG}$ variation affects the selected replica cell turn-on time and produces $I_{REF}$ settling time fluctuation. This could potentially lead to many issues, including data-out (DOUT) ringing, long-address access times, and increased power consumption. To overcome these issues, FG-always-high and prestable current sensing (PSCS) schemes have been proposed.

The FG-always-high scheme is a straightforward concept such that the FG replica cells are kept high at all times [56]. This makes it a non-switching approach and eliminates the $I_{REF}$ settling time completely. However, it exposes the replica cell FG oxide to longer voltage-stress time than the regular flash cells. Consequently, the replica cells in FG-always-high scheme are more susceptible to aging degradation. Moreover, the $I_{REF}$ generation in FG-always-high scheme is more power consuming, because one of the replica cells is always on, even during non-read operations (erase and program).

Figure 4.16 illustrates the conceptual waveform of the PSCS scheme [53]. The $I_{REF}$ generation and the replica cell switching occur in the final phase of an erase/program cycle or during the device power-up, instead of the beginning of every read operation. The time required to complete the erase/program operation is usually much longer than the worst-case $I_{REF}$ settling time in eFlash macros provided by foundries [57, 58]. Therefore, it is feasible to hide the replica cell FG switching time and the $I_{REF}$ settling time near the end of an erase/program operation prior to subsequent read operations. Because the replica cells are not always turned on, they experience less voltage-stress and consume less power than those in the FG-always-high scheme. Moreover, narrow $I_{REF}$ distribution could be achieved with multiple replica cells turning on simultaneously.

Several $I_{REF}$ generation schemes based on replica cells have been proposed for emerging resistive NVM. Figure 4.17 illustrates three $I_{REF}$ generation schemes: $k$-parallel-cell, high-low-average, and parallel-series reference cell (PSRC). The $k$-parallel-cell scheme provides the average $I_{CELL}$ of k replica LRS cells. The high-low-average schemes provide the average $I_{CELL}$ of LRS and HRS cells, as long as their R-ratio is small. In the presence of large-cell resistance variation, both $k$-parallel-cell and high-low-average schemes suffer from wide $I_{REF}$ distribution, because $I_{REF}$ will be dominated by the tailing LRS replica cell. The PSRC scheme eliminates this concern by having two parallel LRS sets connected in series. This enables the PSRC scheme to generate a narrower $I_{REF}$ distribution even with HRS and LRS cell variations.

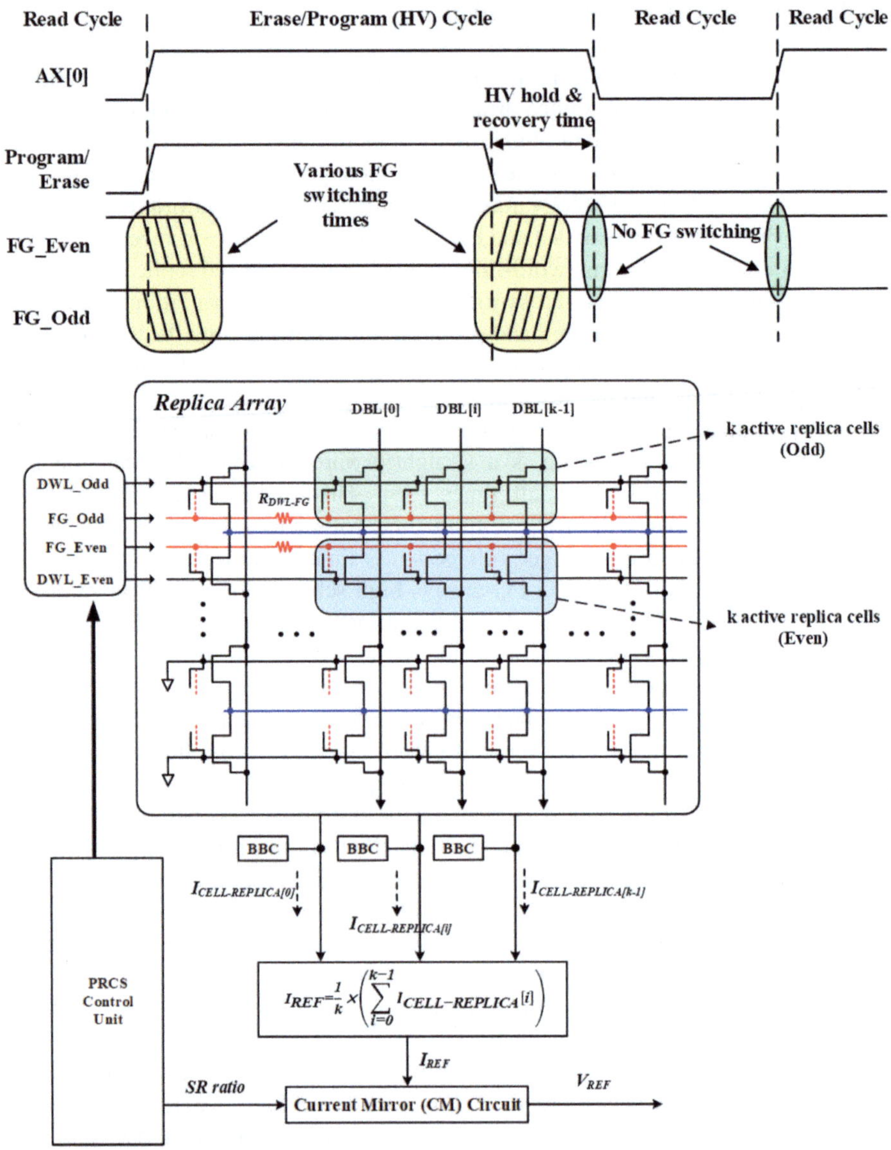


**Fig. 4.16** Conceptual illustration of the prestable current sensing (PSCS) scheme

#### 4.2.4.3 Current-Mode Sensing with Small Input-Offsets

Several CSA schemes with small input-offset were reported for high-speed or small cell-current emerging resistive NVM sensing. Figure 4.18 presents a small-offset CSA sampling scheme using the threshold voltage ($V_{TH}$) of M1 and M2 [59]. The MOS latch (M1/M2) $V_{TH}$ is stored in capacitors C1/C2 to provide offset

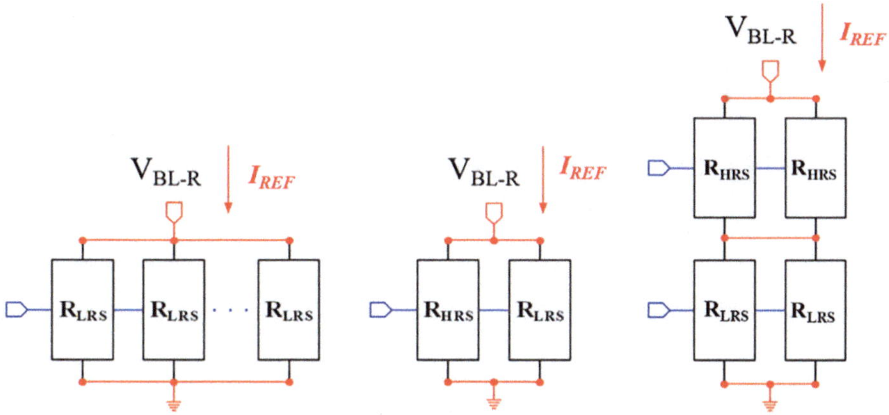


**Fig. 4.17** $I_{REF}$ generation scheme based on replica cells for resistive emerging NVM: (**a**) *k*-parallel-cell, (**b**) high-low-average, and (**c**) parallel-series reference cell (PSRC) schemes

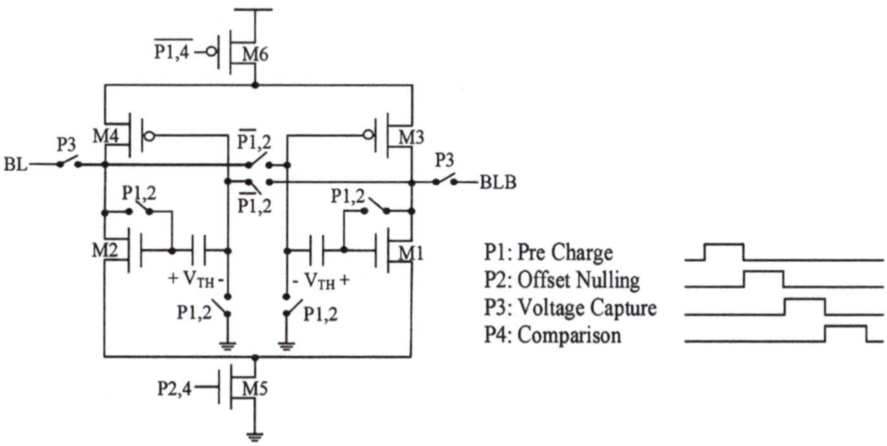


**Fig. 4.18** Small-offset CSA using threshold-voltage ($V_{TH}$) sampling scheme

compensation. This scheme achieves small input offset but requires considerable area overhead due to its use of numerous switches.

Figure 4.19 presents two CSA schemes based on current-sampling (IS-CSA). As shown in Fig. 4.19a, IS-CSA [60] employs two capacitors (C1/C2) to store the gate-source voltage ($V_{GS}$) of critical transistors (M1/M2) to generate $I_{CELL}$ and $I_{REF}$. IS-CSA then uses the sampled $I_{CELL}$ and $I_{REF}$ for second-stage amplification operations, which are insensitive to the $V_{TH}$ variations of M1/M2. Figure 4.19b shows the time-differential IS-CSA (TD-IS-CSA) [61] uses a single capacitor for the sampling of $I_{REF}$. In the subsequent timing phase, TD-IS-CSA compares the sampled $I_{REF}$ with $I_{CELL}$ in order to achieve robust sensing with small input offset.

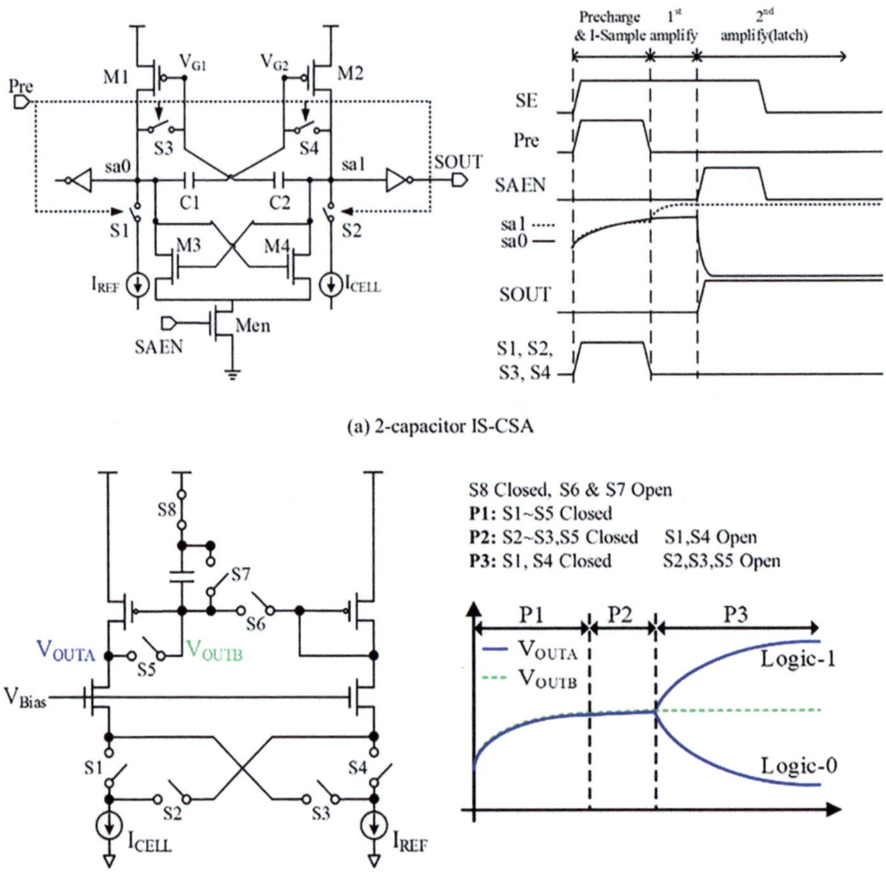


(a) 2-capacitor IS-CSA

(b) single-capacitor time-differential IS-CSA

**Fig. 4.19** Two current-sampling CSA: (**a**) 2-capacitor IS-CSA and (**b**) single-capacitor time-differential IS-CSA

Figure 4.20 shows two digital-calibration-based CSAs (DC-CSA) designed for high-speed sensing without the requirement of run-time offset-cancellation operation. A digital offset cancellation CSA (DOC-CSA) [62] implements additional offset-cancel current ($I_{OC}$) on the differential input (LBLL or LBLR) of the comparator. The amount of $I_{OC}$ is digitally calibrated with respect to the comparator mismatch in the wafer testing stage. In the calibration-based asymmetric-voltage-biased CSA (AVB-CSA) [63], the differential nodes (CP and RP) of a latch-type comparator are biased at different precharge voltages ($V_{AP\_CP}$ and $V_{AP\_RP}$) prior to the sensing operation. $V_{AP\_CP}$ and $V_{AP\_RP}$ are calibrated during wafer testing in accordance with the offset voltage of the read-path detected at nodes CP and RP. As a result, AVB-CSA suppresses the offset caused by the comparator and the entire read-path ($I_{OS\_SUM}$).

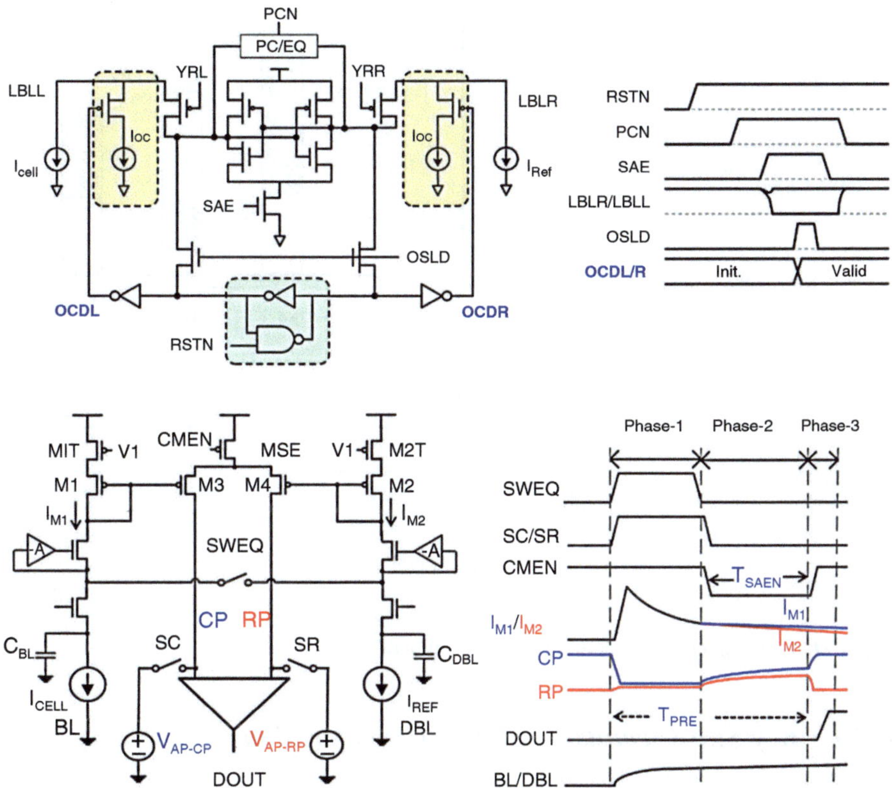


**Fig. 4.20** Two digital-calibration-based CSAs (DC-CSA): (**a**) digital offset cancellation CSA (DOC-CSA), and (**b**) asymmetric-voltage-biased CSA (AVB-CSA)

#### 4.2.4.4 Low-Voltage Current-Mode Sensing Schemes

Figure 4.21 illustrates a low-voltage, body-drain-driven (BDD) CSA [54]. The BDD-based current-mirror (CM) circuit needs a smaller voltage headroom than typical diode-connected current mirrors. As a result, the BDD-CSA is capable of allocating more voltage budget to the voltage across NVM devices. This generates sufficient $I_{CELL}$ that enables sensing even at low $V_{DD}$.

## 4.3 Write Circuits for Emerging Resistive NVM

In this section, we introduce the basic operations of key sub-circuits used in the write operations of emerging resistive NVM. Afterward, we describe the various circuit design challenges and review some of the advanced circuit techniques.

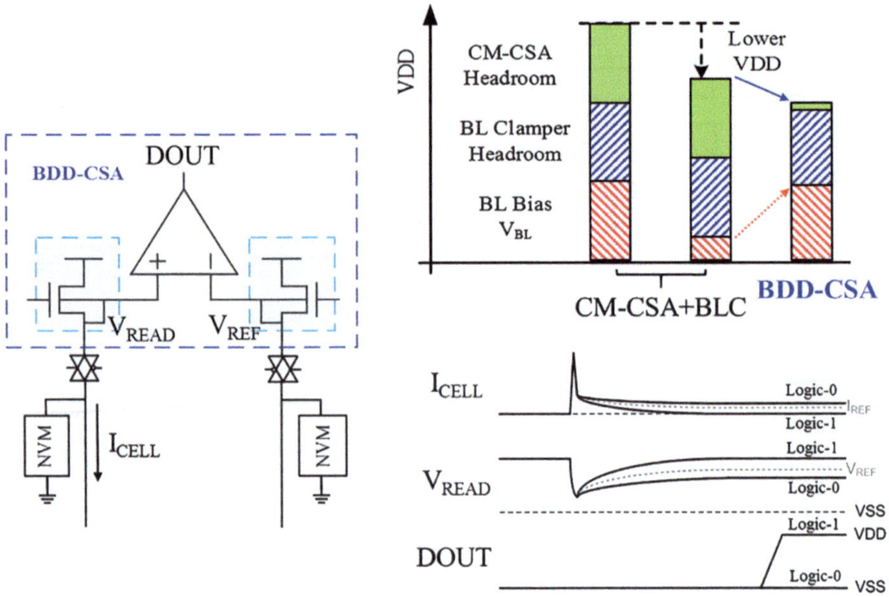


**Fig. 4.21** Low-voltage, body-drain-driven CSA (BDD-CSA)

### *4.3.1 Key Sub-Circuits Used in Write Operations*

Figure 4.22 illustrates functional blocks along the write-path in a typical emerging resistive NVM macro. The key sub-circuits used for write operation include level shifters, write drivers, and charge-pump circuits. Depending on the memory device, some might require voltages higher than the supply voltage ($V_{DD}$). In such cases, a charge-pump (CP) circuit is implemented for voltage conversion. Write drivers provide the required write voltage to the cell array. For read and write operations, write drivers supply different output voltage levels to word-lines (WL) or source-lines (SL). As a result, level shifters are commonly employed along with WL/SL drivers.

Because the generation of high voltage and high current is generally inefficient, the high-voltage path current load should be minimized to reduce power consumption. As a result, low-power emerging resistive NVM generally requires advanced circuit techniques to reduce the DC/leakage current of level shifters and write drivers. In addition, designing an energy-efficient CP with small peak current is also important when these low-power emerging NVM macros are embedded on a chip.

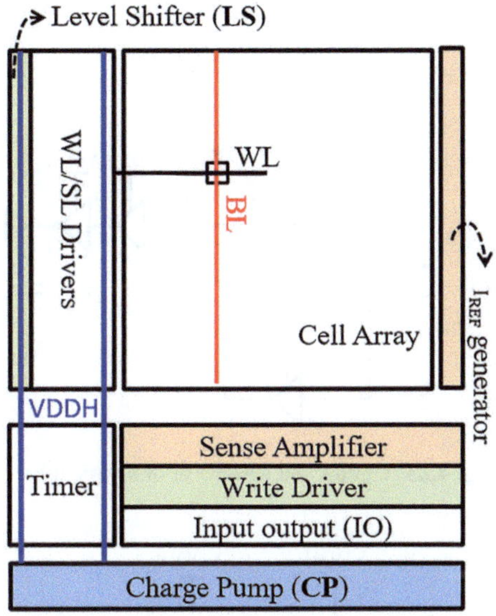


**Fig. 4.22** Illustration of functional blocks along write-path in a typical emerging resistive NVM macro

#### 4.3.1.1 Challenges and Advanced Circuits in Level-Shifters

Level shifters are implemented for converting the row decoder logic signals (0 V or $V_{DD}$) to voltage higher than $V_{DD}$ ($V_{DDH}$). $V_{DDH}$ could be the write voltage required for NVM devices. Generally, level shifters (LS) have two challenges: (1) generating a high $V_{DDH}$ for write operation from a low input voltage ($V_{DDL}$), and (2) consuming a large DC current due to the $V_{DDH}$ path charge loss for conventional low-$V_{DDL}$ level shifters in unselected rows. These challenges potentially lead to an increase of current load and power consumption, particularly in NVM macros with a large number of rows. In the following texts, we will examine a number of existing level shifters.

Figure 4.23 illustrates the circuit and waveform of a common half-latch level shifter (HL-LS) [64, 65]. The HL-LS comprises of two pull-down NMOS transistors ($M_{NM}$, $M_{NR}$), a PMOS cross-coupled transistor pair ($M_{PL}$, $M_{PR}$), an input inverter (INV1), and an output buffer (BUF). When SEL = 1, $M_{NL}$ is turned on to pull the node OB voltage ($V_{OB}$) down to ground. This turns on $M_{PR}$ and raises the voltage at node OT ($V_{OT}$) to $V_{DDH}$. Thereafter, the cross-coupled feedback ensures the $M_{PR}$ remains turned off and the output ($V_O$) of HL-LS is $V_{DDH}$. When SELb = 1, $M_{NR}$ turns on and creates a discharge current ($I_D$) to lower $V_{OT}$ to below $V_{DDH}$. If $I_D$ is large, the current contention between $M_{PR}$ and $M_{NR}$ can cause $M_{NR}$ to pull $V_{OT}$ down to 0 V. When $V_{DD}$ is low (i.e., below or near the threshold voltage of $M_{NL}$/$M_{NR}$), $I_D$ is too small to win the current contention over $M_{PL}$/$M_{PR}$. This challenge prevents the HL-LS from being implemented in applications requiring low $V_{DDL}$. In addition, HL-LS also requires the ultra-large sizing for $M_{NL}$ and $M_{NR}$ transistors.

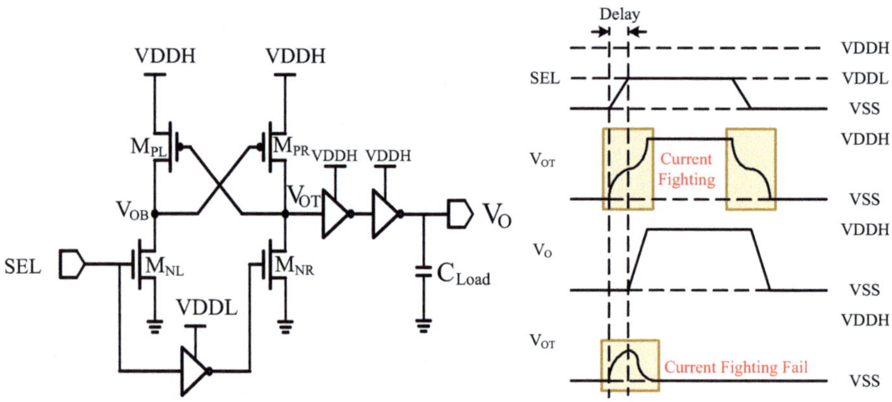


**Fig. 4.23** Circuit and waveform of a typical half-latch level shifter (HL-LS)

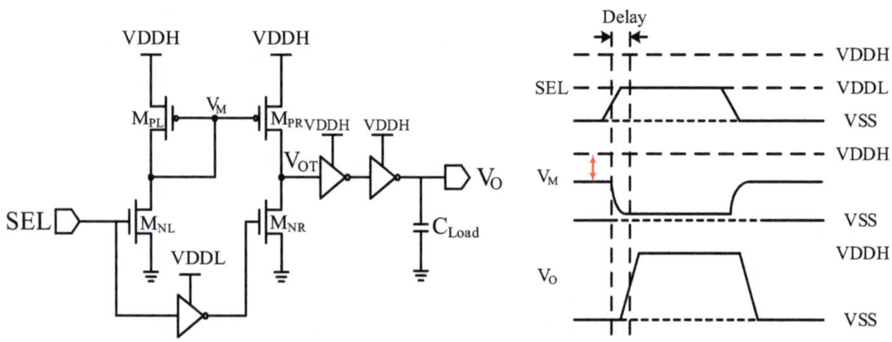


**Fig. 4.24** Circuit and waveform of a typical current-mirror-based level shifter (CM-LS)

Figure 4.24 shows typical current-mirror-based level shifters (CM-LS) were developed to reduce $V_{DDL}$ for low-$V_{DD}$ applications [64]. CM-LS comprises two pull-down NMOS transistors ($M_{NL}$ and $M_{NR}$), a pair of PMOS current-mirror transistors ($M_{PL}$ and $M_{PR}$), an inverter, and an output buffer (BUF). When SEL = 1, $M_{NL}$ is enabled to turn on $M_{PR}$ by lowering the voltage of node M ($V_M$). The voltage of node OT ($V_{OT}$) is then raised to $V_{DDH}$. When SEL = 0, $M_{NL}$ is off and $M_{PL}$ slowly pulls down $V_M$ to $V_{DDH}$ minus the threshold voltage of $M_{PL}$ ($V_{DDH}-V_{TH\_MPL}$). This reduces the current through $M_{PR}$ significantly and enables $M_{NR}$ to pull down $V_{OT}$ beyond the trip-point of the output buffer without incurring series current contention. As a result, CM-LS is capable of achieving a low $V_{DDL}$ using smaller $M_{NL}/M_{NR}$ compared to HL-LS design.

However, there is still DC current in selected and unselected CM-LSs. When SEL = 1 (selected row), we have DC current flowing through $M_{PL}$ and $M_{NL}$ ($I_{DC\_PLNL}$). When SEL = 0, the weakly turned-off $M_{PR}$ ($I_{LEAK\_PR}$) conducts a DC leakage current, because its gate is biased at $V_{DDH}-V_{TH\_MPL}$. In the case of an emerging NVM macro with k rows, the DC leakage current on the $V_{DDH}$ path of the CM-LSs is $I_{DC\_PLNL} + (k - 1)\, I_{LEAK\_PR}$. For large k, the high current load

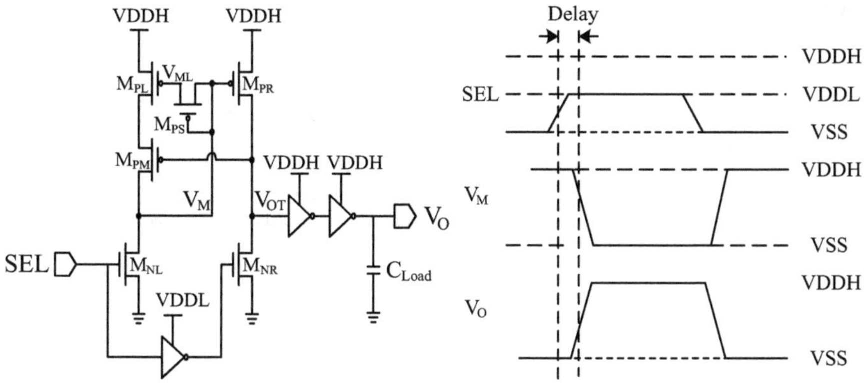


**Fig. 4.25** Circuit and waveform of a pseudo-diode-mirrored level shifter (PDM-LS)

on $V_{DDH}$ path weakens the charge pump and prevents it from providing the current and voltage required in write operations. A Wilson CM-LS [66] has been proposed to suppress the DC current when SEL = 1, but the DC current leakage issue when SEL = 0 remains unsolved.

Figure 4.25 shows the pseudo-diode-mirrored level shifter (PDM-LS) proposed to accommodate lower $V_{DDL}$, while suppressing DC current [67]. The PDM-LS structure comprises of two pull-down NMOS transistors ($M_{NL}$ and $M_{NR}$), a pseudo-diode PMOS transistor ($M_{PS}$), a current-cutoff PMOS transistor ($M_{PM}$), and pseudo PMOS current-mirror transistors ($M_{PL}$ and $M_{PR}$), an inverter, and a buffer. While one source/drain terminal of $M_{PS}$ is connected to the gate of $M_{PL}$, the gate and the other source/drain terminal of $M_{PS}$ are connected to the gate of $M_{PR}$ and the drain of $M_{NL}$, respectively. $M_{PM}$ is controlled by $V_{OT}$ and placed between $M_{PL}$ and $M_{NL}$.

For SEL = 1, $M_{NL}$ is turned on to pull $V_M$ down to 0 V. This consequently turns on $M_{PR}$ and charges the voltage of node OT ($V_{OT}$) to $V_{DDH}$. When $V_M = 0$ V, $M_{PS}$ forms a diode-connected structure, causing $V_{ML}$ to be equal to the threshold voltage of $M_{PS}$ ($V_{TH\_MPS}$). Low $V_M$ also turns on $M_{PR}$ pulling $V_{OT}$ to $V_{DDH}$, which turns $M_{PM}$ off. For SEL = 0, $M_{NR}$ is turned on to pull $V_{OT}$ down, this slightly turns on $M_{PM}$ and charges $V_M$. Consequently, this causes $M_{PS}$ to enter cutoff mode. $V_M$ then approaches $V_{DDH}$, which is higher than the level of CM-LS and suppresses the DC leakage current flowing through $M_{PR}$. By cutting of the DC current at SEL = 1 and suppressing DC leakage current at SEL = 0, the PDM-LS is able to operate under lower minimum $V_{DDL}$ and consume less DC current than the conventional CM-LS.

#### 4.3.1.2 Challenges and Advanced Circuits in Write Drivers

In a typical emerging NVM macro, multiple cells on the same row are selected for write operation. Consequently, these cells experienced the same voltage bias conditions on word-line (WL) and source-line (SL). Figure 4.26a, b illustrate

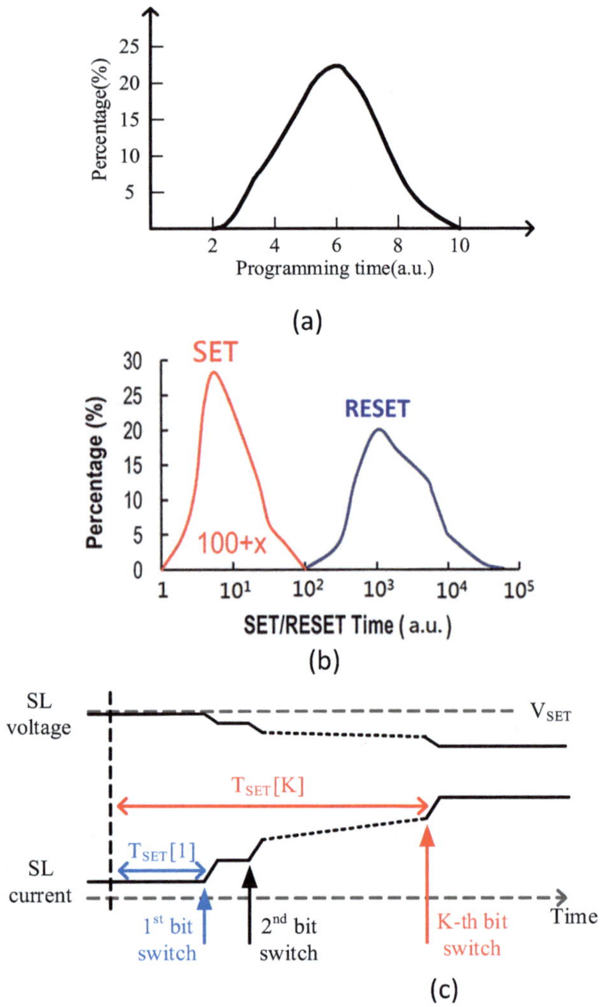


**Fig. 4.26** Program time variation for (**a**) OTP and (**b**) resistive-type emerging NVM devices. (**c**) Current and voltage waveform during program operation

process variation can cause considerable variability in the write durations, such as program time ($T_{PROG}$), SET time ($T_{SET}$), and RESET time ($T_{RESET}$). To cover these tailing cells, the voltage bias for WL and SL must be maintained until the switching completes. For one-time programmable (OTP) and resistive-type memory, the switching of an emerging NVM cell from HRS to LRS causes a considerable DC current ($I_{DC\_CELL}$). As a result, cells with shorter $T_{PROG}/T_{SET}$ experience large $I_{DC\_CELL}$ even after switching to accommodate for the tailing cells. As shown in Fig. 4.26c, this unnecessary $I_{DC\_CELL}$ increases energy consumption, degrades output voltage, and potentially causes write failure in tailing cells.

Several cell-by-cell write termination schemes [55, 68, 69] have been proposed to suppress $I_{DC\_CELL}$ by terminating the program/SET operation as soon as the

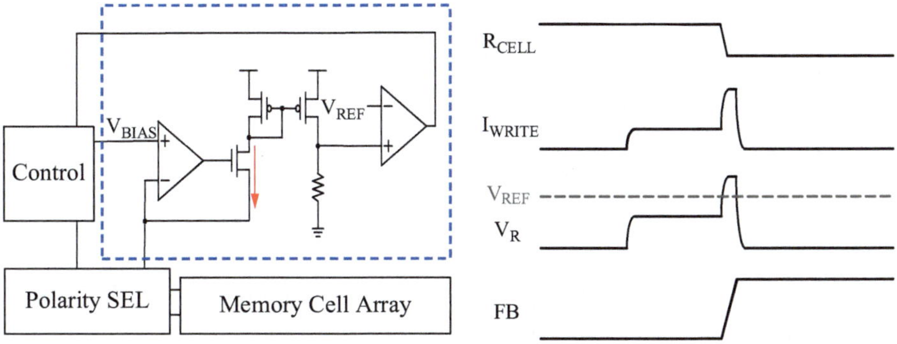


**Fig. 4.27** Circuits and waveforms of the OP-based current-mode write termination (OP-CWT) scheme

switching completes. In the following, we review three write-termination schemes: (1) operational amplified (OP)-based current-mode termination, (2) negative-resistance-based current mode termination, and (3) voltage-mode termination.

Figure 4.27 shows the circuits and waveforms of the OP-based current-mode write termination (OP-CWT) scheme [69]. The current mirror (CM) copies the cell current to a resistive load. Then the OP compares a predefined reference voltage ($V_{REF}$) with the voltage across the resistive load. The voltage bias circuit is disabled when the induced voltage is higher than the $V_{REF}$. This OP-CWT scheme is effective in suppressing $I_{DC\_SET}$; however, the multiple usages of OP and CM incur significant area overhead, slow response time, and lead to considerable DC current consumption.

Figure 4.28 illustrates the negative-resistance-based current mode termination (NR-CWT) scheme reported in [68]. NR-CWT employs a current-mirror-based negative-resistance (−R) circuit to provide feedback for write current control. When writing 1 from LRS to HRS (WR1 = high and WR0 = low), current flows through the cell from SL to BL. When writing 0 from HRS to LRS (WR1 = low and WR0 = high), the current flows from BL to SL. The −R driver reflects on the BL with a current proportional to the cell resistance. When the cell resistance reaches LRS, an increase of cell current pulls down the BL voltage and reduces the current through the current mirror. This further lowers the BL voltage, thereby forming a positive feedback loop. It should be mentioned that the current mirror and positive feedback loop only moderate, rather than terminate, the write current. The current continues to flow through the −R driver.

Figure 4.29 illustrates the voltage-mode write-termination (VWT) scheme for the SET operation [55]. This scheme is implemented using only four transistors to monitor BL voltage for termination. At the beginning of the write operation, the SEL control signal turns on the NRSD transistor to discharge the residue charges on the data-line (DL). During the write operation, the transition from HRS to LRS dramatically drops the voltage across the ReRAM cell, because VWT and ReRAM form a voltage divider network. As a result, the gate voltage of P1 rises dramatically

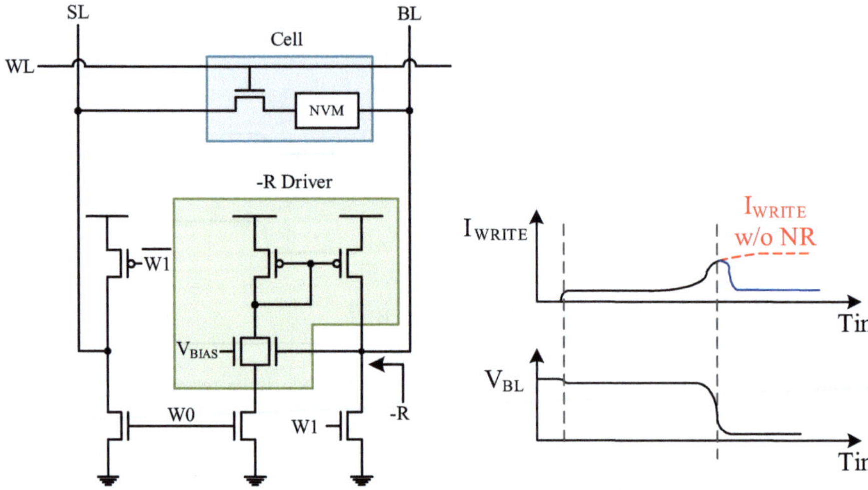


**Fig. 4.28** Circuits and waveforms of the negative-resistance-based current mode termination (NR-CWT) scheme

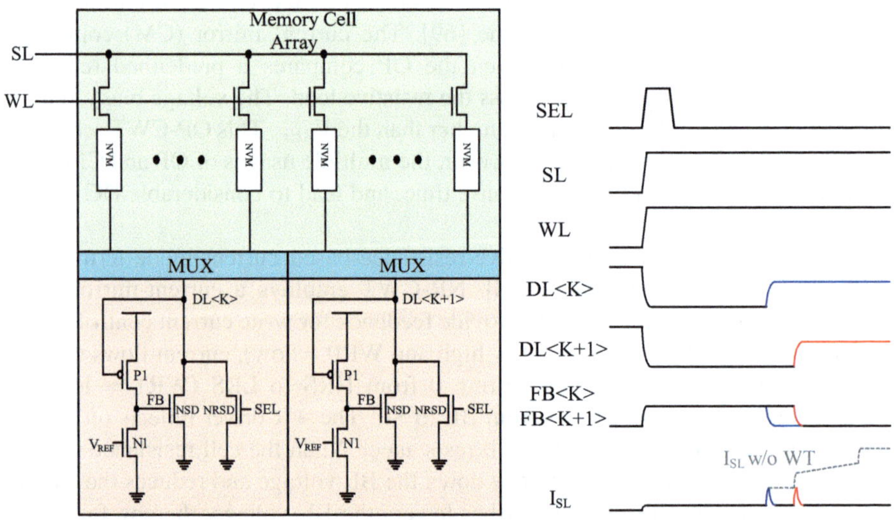


**Fig. 4.29** Circuits and waveforms of the voltage-mode write-termination (VWT) scheme

to turn itself off. Meanwhile, the gate of N1 is biased at $V_{REF}$, it turns off NSD by pulling down its gate terminal. The use of voltage-mode operation and the reuse of $I_{DC\_CELL}$ minimize the area overhead, reduce the power consumption, and improve the response time.

## 4.4 Emerging Resistive NVM for Energy-Efficient Systems

Energy-efficient systems aim to exploit the combined benefits of low-latency volatile memory for power-on computing and low-power nonvolatile memory. However, the expensive write operation and the limited endurance of conventional nonvolatile memory become a bottleneck in the implementation of such low-power systems. Resistive random access memory (ReRAM) is a promising nonvolatile memory candidate for energy-efficient computing because of its low-latency write and low-power operations. This section discusses first the two-macro and one-macro approaches, then the implementation of ReRAM device in nonvolatile SRAM and nonvolatile logic.

### *4.4.1 Two-Macro and One-Macro Approaches*

Effective design of the power-on/off transition is one of the most essential components toward low-power systems. Because critical data must be transferred to NVM during the power-off event and recalled when power returns, the data backup/restore operation must be power-efficient. As illustrated in Fig. 4.30, many low-power systems use the two-macro scheme with volatile memory (SRAM and DRAM) for fast low-voltage access, and nonvolatile flash memory for power-off data storage. The two-macro scheme relaxes the endurance requirement of NVM, because regular accesses and computations will be performed on SRAM/DRAM. If the emerging NVM device has fast random read/write speeds and high endurance, an alternative 3D-IC architecture can be proposed without the need for DRAM. This would eliminate the DRAM self-refresh power during system standby mode, as shown in Fig. 4.30b.

The two-macro approach is not energy-efficient, because the power-off data-back operations consume too much energy. The word-by-word SRAM read/write and the long-latency NVM read/write operations demand long store/restore durations [70–73]. This renders the two-macro approach susceptible to data loss after sudden power failure. Furthermore, the high peak current required by the flash memory write operations can lead to power integrity concerns. In the worst-case scenario, the system may be unable to recover its previous state.

To tackle this challenge, researchers have proposed the concept of nonvolatile logic (nvLogic) and nonvolatile SRAM (nvSRAM) [70–75] to improve the store/restore performances in the traditional two-macro schemes. Figure 4.30 illustrates the chip architecture based on the one-macro solution enabled by nvLogic and nvSRAM. The nonvolatile flip-flop (nvFF) and nvSRAM are new devices that are realized by combining logic devices with NVM devices. This implementation allows parallel data transferring between logic circuits and NVMs, and thus improves the latency and the power efficiency during power-on and power-off. Besides, by integrating NVM cells on top of the logic devices though the BEOL

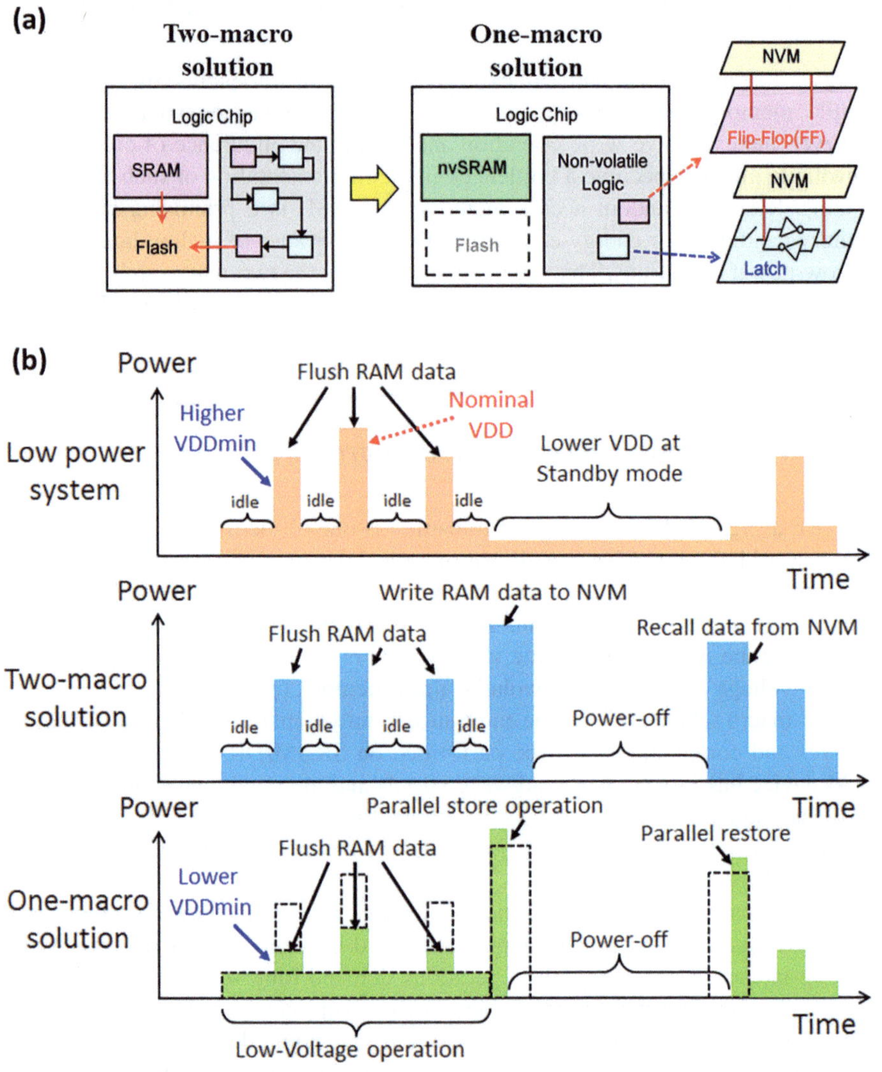


**Fig. 4.30** Conceptual (**a**) illustration and (**b**) power comparison of a low-power system using the two-macro approach and the one-macro approach using nonvolatile memory

process, the approach can provide an alternative low-power architecture with a reliable data-backup scheme with a small area overhead. Figure 4.30 shows the comparison on the performances of the volatile processor based on conventional two-macro solution and the nonvolatile processor (nv-Processor) based on the one-macro solution based on nvLogic [76]. It can be seen that while the volatile processor stops the computation in the frequency power-interrupted scenario, the

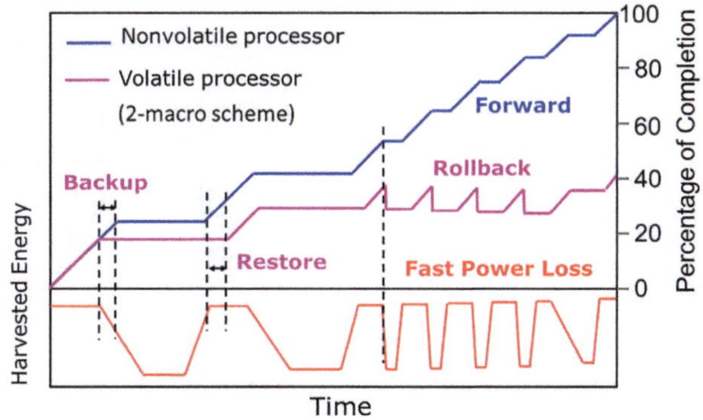


**Fig. 4.31** Comparison on performances of volatile and nonvolatile processors with power interruption [76]

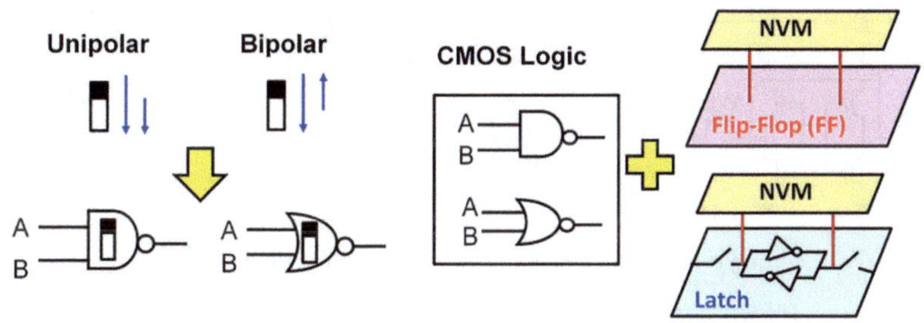


**Fig. 4.32** nvLogic using nonvolatile latch or nonvolatile flip-flop

nv-Processor is capable to proceed with the computation because it requires very low power and latency to store/restore the data during power switching (Fig. 4.31).

## *4.4.2 Nonvolatile Logic and Nonvolatile SRAM*

One challenge of NVM devices preventing them from being directly implemented as computing or accessing units is their limited endurance. Many NVM devices (Flash memory, MRAM, PCRAM, and ReRAM) have many orders of magnitude lesser endurance than volatile memory (SRAM and DRAM) or CMOS logic (latch, flip-flop, and logic-gates). Consequently, endurance-aware designs should perform regular computing/accessing in volatile memory and use nvLogic/nvSRAM only as power-off storages, as shown in Fig. 4.32.

Figure 4.33 illustrates (a) the conventional 6T SRAM cell, along with (b) several reported nonvolatile latch (nvLatch) cells. These nvLatch cells are generally designed with dedicated transistors to isolate the NVM devices during normal mode and provide store/restore functionalities during power-off/on mode. In addition, because the store/restore are bit-to-bit parallel operations, they enable low-latency store/restore operations than the conventional two-macro approaches. We will briefly mention the challenges of these configurations in the following paragraph.

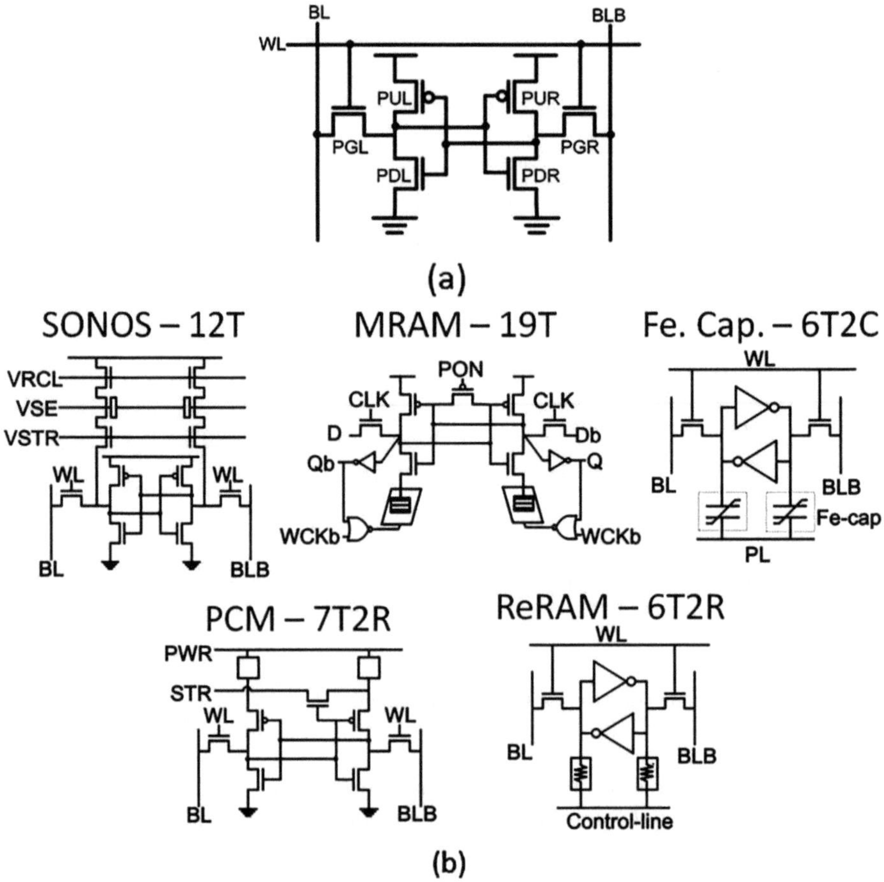


**Fig. 4.33** Comparison between (**a**) conventional 6T latch/SRAM cell and (**b**) reported nonvolatile latch (nvLatch) and nonvolatile SRAM (nvSRAM) cells

The SONOS-12T cell [77] not only occupies a large area, the slow SONOS device write operation causes its store operation to be high voltage and long latency ($T_{STORE}$). Similarly, the MRAM-19T2R latch [78] also suffers from expensive store operation and large area. Compared with the previous two devices, the Fe-capacitor-6T2C cell [79] requires a much smaller area, but it needs an additional ½ $V_{DD}$ voltage bias to access NVM and read SRAM. This necessitates a dedicated voltage regulator that consumes a significant amount of DC current. The PCM-7T2R cell [80] offers a relatively compact area, but the PCM device write characteristics require a large store current and a moderate store time. Furthermore, the PCM-7T2R cell also suffers from cell-$V_{DD}$ ($CV_{DD}$) integrity degradation. Lastly, even though the ReRAM-6T2R cell [81] is the most compacted design by eliminating the NVM switches, it suffers from cell leakage current and SRAM stability issues.

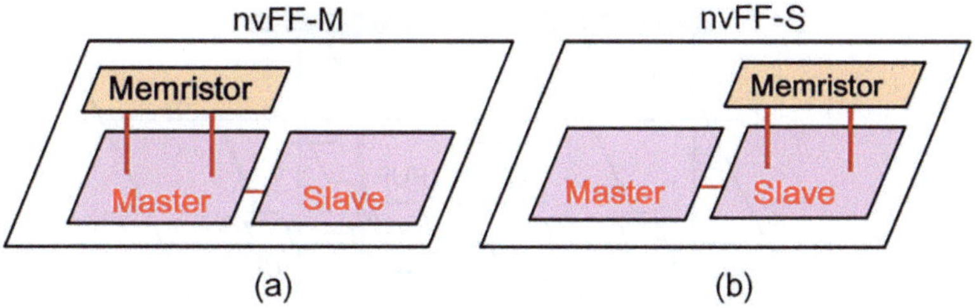


**Fig. 4.34** Nonvolatile flip-flop with memristor in (**a**) master-stage (nvFF-M) and (**b**) slave-stage (nvFF-S)

As previously mentioned, the purpose of additional transistors is to enable isolation and store/restore functionalities. These transistors, however, not only increase the area overhead, but also impose high parasitic loads that increases latency and power consumption of logic gates. Consequently, it is impractical to implement NVM devices to every logic gate. Alternatively, nvLogic gates should only be applied on latches or flip-flops that are on the critical path. Figure 4.34 shows the two nonvolatile flip-flops (nvFF) circuit structures for logic libraries—master-stage nvFF (nvFF-M) and slave-stage nvFF (nvFF-S) [82]. The nvFF-M implements the master-stage using Rnv8T cell [83] and the slave-stage using conventional CMOS circuit, while the nvFF-S employs the exact opposite. These two nvFF configurations can be incorporated in various power-on procedures to accommodate lower VDD, while requiring only 16% of the area overhead using traditional FFs or nvFFs.

### *4.4.3 nvSRAM Example: Rnv8T Cell*

A 3D-stacked resistive nonvolatile 8T2R SRAM cell (Rnv8T) had been proposed [83]. This Rnv8T design achieves low-latency and low-power store operations. The work also suggests a bit-line control-line (BL-CL) sharing scheme to simplify the control of the resistive memory device and a write-assist operation to relax the write constraints. These features allow the Rnv8T to be sized favoring read and suppress read/write failure at a reduced $V_{DD}$. In the following, we first describe the basic operations of the Rnv8T cell, before moving on to discuss cell stability and restore yield. Lastly, we also review some of the recent nvSRAM cell structures.

#### **4.4.3.1 Rnv8T Cell Basic Operations**

The Rnv8T cell structure consists of a typical 6T SRAM cell, 2T ReRAM-switch (RSW), and two resistive memory devices. This design preserves the 6T SRAM advantages, including its low-latency read/write and low-voltage operation. The

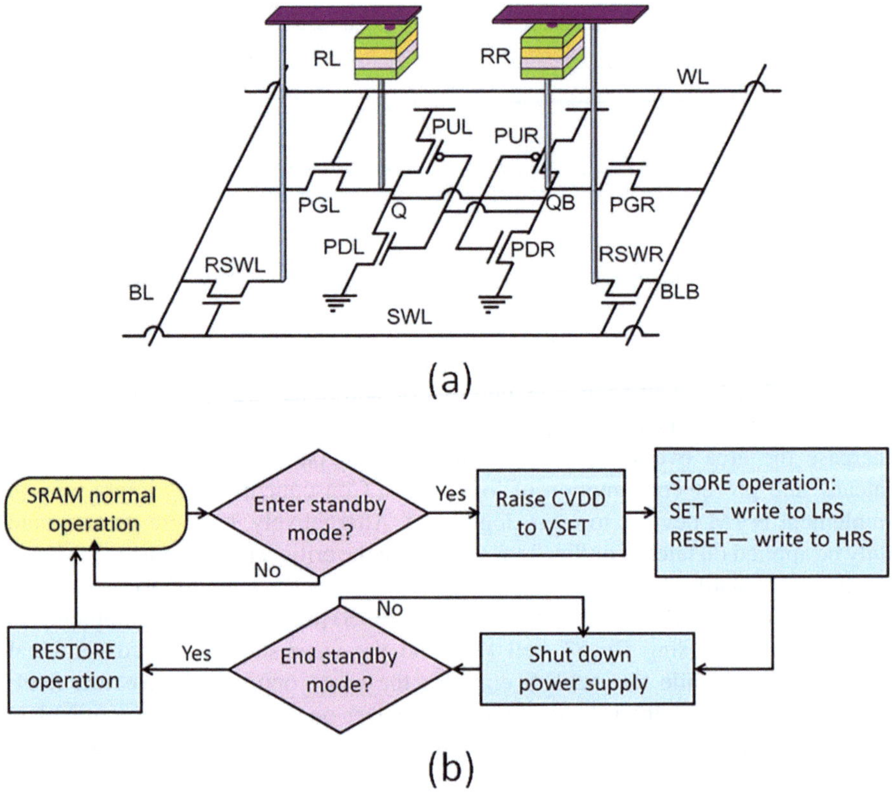


**Fig. 4.35** The Rnv8T (**a**) cell structure and (**b**) operation flowchart

direct connection between ReRAM and SRAM storage nodes (Q and QB) enables nonvolatile storage of both original and complementary data. While other nvSRAM cell designs require a dedicated control line (CL) to carry out the store operations, the Rnv8T scheme shares BLs with the NVM CLs for a more compact cell area. This BL-CL sharing scheme provides SRAM-mode write-assist operation through RSWs. To make the cell area more compact, the two ReRAM devices can be arranged in a 3D configuration on top of the 8T SRAM cell.

Figure 4.35 illustrates the Rnv8T cell structure and operation flowchart. In the SRAM normal mode, the Rnv8T cell executes the typical SRAM read/write operations. During read operations, the switch-line (SWL) is pulled low to turn off RSWs for providing ReRAM device isolation and preventing the SRAM cell disturbance. The differential read in the Rnv8T cell is identical to that in a typical 6T SRAM. When the system switches to standby mode, the Rnv8T macro initiates the store operation by transferring the SRAM data into the ReRAM device. After the data transfer completes, the standby leakage can be completely eliminated by

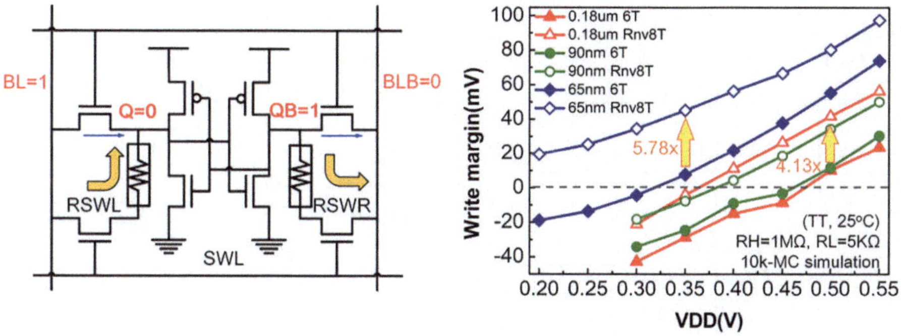


**Fig. 4.36** Write margin under various supply voltage $V_{DD}$ and technology nodes

shutting down the whole macro. During the system wake-up, a restore operation fetches data from ReRAM devices back to the SRAM storage nodes Q and QB.

The write operation initiates by pulling both SWL and WL high. RSWs provide an extra write path and reduce the pulled-up ratio (PR), which is defined as the size ratio of pull-up PMOS to pass-gate NMOS. The BL and BLB data are transferred into the SRAM cell through RSWs and pass gates. As a result, this approach provides a larger write margin (WM) and a faster write speed than that provided by the 6T SRAM cell. In this work, the WM is also referred to as the write trip-point (WTP), which is the voltage between the BL and VSS required for flipping the cell data to the opposite state. To inspect the worst-case WM, we performed Monte Carlo simulation using 10,000 points. The transistor size of both the 6T and the Rnv8T cells are chosen to be the same as a commercial 6T SRAM cell under various technology nodes. Figure 4.36 shows the Rnv8T cell, compared to 6T SRAM cell, the WM is increased by 4.13× at 0.5 V (0.18 μm) and 5.78× at 0.35 V (65 nm).

Figure 4.37a illustrates the Rnv8T cell store operation. In the store mode, data transfer from SRAM to ReRAM requires both SET and RESET operations. For both SET and RESET operations, the SWL is kept high. The SET operation raises both BL and BLB to a high voltage ($V_{SET}$) to switch the RL (if Q = 0) to a low-resistance state (LRS). On the contrary, in RESET operation, the BL and BLB are pulled to ground and the $CV_{DD}$ is biased to $V_{RESET}$. This switches the RR (if QB = 1) to a high-resistance (LRS) state.

Figure 4.37b illustrates the Rnv8T cell restore operation. The purpose of the restore mode is to recall data to the SRAM storage nodes (Q and QB) from the ReRAM devices (RL and RR). First, the rise of SWL turns on RSWs (RSWL and RSWR) to pull both BL and BLB to ground. This clears the remaining charges at the storage nodes and ensures both nodes are at equal states for a later stage comparison. When the $CV_{DD}$ rises, the PMOS transistors charge nodes Q and QB while the RR and RL provide discharge path to ground. The resistance difference between RL and RR generates a voltage differential ($V_{Q\text{-}QB}$) at Q and QB during power-on. The large discharge current of $R_L$ lowers the storage-node voltage rapidly. The $R_H$, on

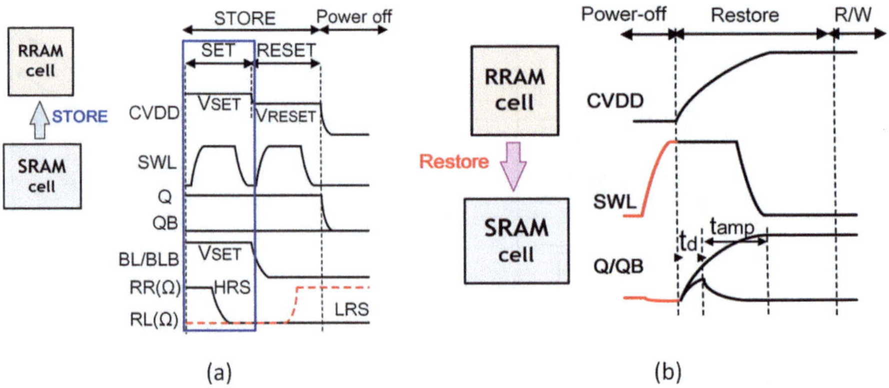


**Fig. 4.37** The (**a**) store and (**b**) restore operation waveforms of the Rnv8T cell

the other hand, provides a small discharge current, the storage-node voltage is kept at a higher level. Finally, the cross-coupled latch amplifies $V_{Q\text{-}QB}$ and restores the data to the SRAM storage nodes.

### 4.4.3.2 Rnv8T Cell Stability

In read operations, the cell stability imposes a minimum $V_{DD}$ ($V_{DDmin}$) limitation. The write-assist feature improves WM and allows trading WM for better read stability by using a read-favored-sizing (RFS) scheme. Using the same amount of area, the Rnv8T transistor sizing can be adjusted to enhance read stability performance through (1) increasing the size ratio of PDL (PDR) over PGL (PGR) to suppress read disturbance, and (2) raising the trip-point of the latch (inv1 and inv2) to prevent data from flipping.

Figure 4.38 compares the read static noise margin (RSNM) [84], WM, and cell $V_{DDmin}$ for Rnv8T cells with and without the RFS scheme. To include process variation into our analysis, we performed a Monte Carlo simulation with 10,000 points. Without the RFS scheme, the $V_{DDmin}$ of the nominal 6T and the Rnv8T cell is limited to 0.55 V. Applying the BL-CL sharing scheme on the Rnv8T cell improves the WM by 4.08× at 0.5 V. It should be mentioned that although WM improved, the $V_{DDmin}$ does not decrease because the RSNM does not benefit from this approach.

After applying the RFS scheme, the WM improvement reduces slightly to 3.32×, whereas the RSNM increases by 11.1× at $V_{DD}$ = 0.6 V. As a result, the Rnv8T cell with the RFS scheme achieved a $V_{DDmin}$ 150 mV lower than that without the RFS scheme. Therefore, the Rnv8T cell with BL-CL sharing and RFS schemes demonstrates superior cell stability and lower SRAM read/write VDDmin than conventional 6T SRAM or other nvSRAM cells.

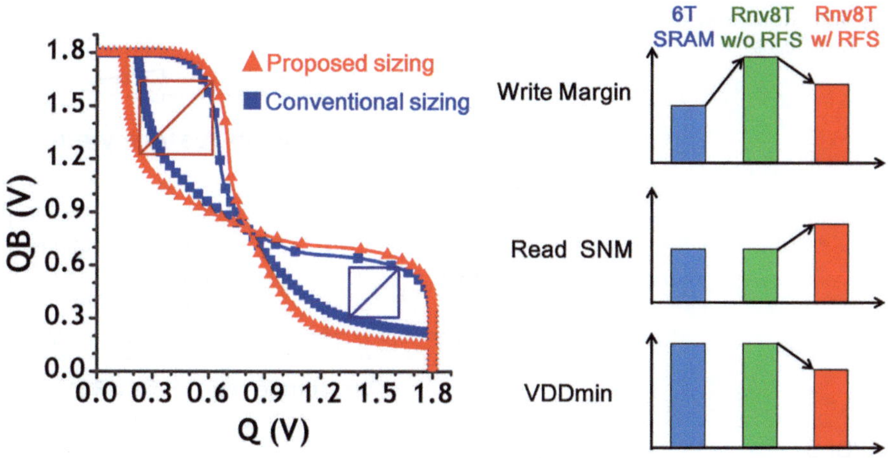


**Fig. 4.38** Trade-off between $V_{DDmin}$, WM, and RSNM in the Rnv8T cell structure

#### 4.4.3.3 Rnv8T Restore Yield

During restore operation, the Rnv8T cell behaves a latch sensing the output from the two ReRAM devices. The sensing margin can be significantly improved by using a differential sensing scheme on two ReRAM devices with opposite states. To include process variation in our analysis, we performed a 40,000-point Monte Carlo simulation. Figure 4.39 shows when the resistance ratio exceeds 3, the restore yield can reach up to 99.6%. In this study, ReRAM devices have a resistive ratio over 10, which ensures a 100% restore yield in the Rnv8T cell. However, because of the read stability limitation, the Rnv8T cell is only capable of achieving $V_{DDmin}$ of 0.45 V for processes with smaller threshold voltage ($V_{TH}$) variation. For nanometer processes, the larger $V_{TH}$ variation causes the $V_{DDmin}$ of Rnv8T cells to degrade. This highlights the need for a new nvSRAM cell structure for low $V_{DD}$ applications, and particularly for nanometer chips.

#### 4.4.3.4 nvSRAM Design Comparisons

A resistive nonvolatile 9T2R SRAM cell (Rnv9T) was reported to solve the read and write failure at low $V_{DD}$ [82]. This Rnv9T cell combines the features of the previously proposed Rnv8T nvSRAM and low-voltage L-shaped 7T SRAM cells [85]. Figure 4.40 shows the circuit structure and the waveform of the low-voltage 7T SRAM cell developed form the Zigzag 8T cell [86].

The Rnv9T cell in Fig. 4.40 uses decoupled 1T read-port and pseudo read word-line (RWL) schemes to eliminate read disturbance. Similar to the Rnv8T cell, the Rnv9T cell also relies on the 2T ReRAM switch for simpler ReRAM control and better WM. Because this structure eliminates read disturbance, the Rnv9T cell

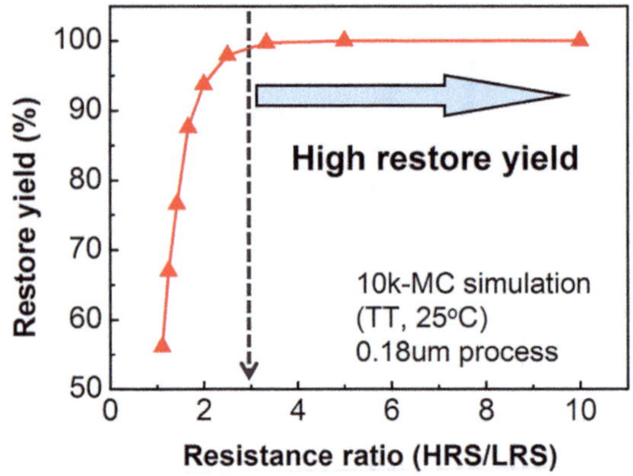


**Fig. 4.39** Restore yield of the Rnv8T cells

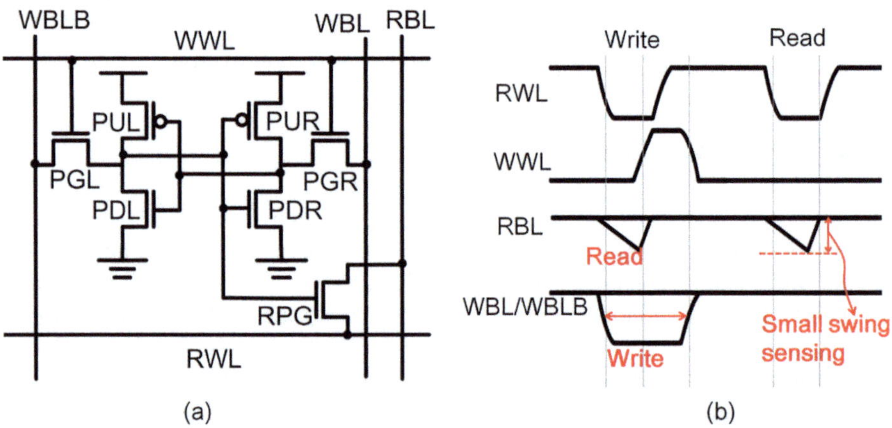


**Fig. 4.40** (**a**) The proposed low-voltage 7T SRAM cell and (**b**) its operation waveform

does not need to rely on the read-favored-sizing (RFS) scheme to improve its read stability. Consequently, this enables a larger WM than that achieved by the Rnv8T. Therefore, with improved read stability and WM, the Rnv9T cell can achieve a lower $V_{DDmin}$ than conventional 6T, low-voltage 7T, and Rnv8T cells.

Figure 4.41a compares the WMs of various nvSRAM designs using a 10,000-point Monte Carlo simulation under various $V_{DD}$'s. The nvSRAM cells were configured as described in previous literatures [78–80, 87], [77, 88], with the same transistor sizing as a conventional 6T SRAM cell. The simulation result shows that the Rnv8T cell can achieve a 3.3× WM improvement, compared to other nvSRAM cells. Figure 4.41b shows that the write time of the Rnv8T cell is also shorter than those of other nvSRAM cells. Finally, Table 4.2 summarizes the performance of different nvSRAM cells. It can be seen that Rnv8T and Rnv9T show superior speed and energy efficiency for store operations compared to other reported cells.

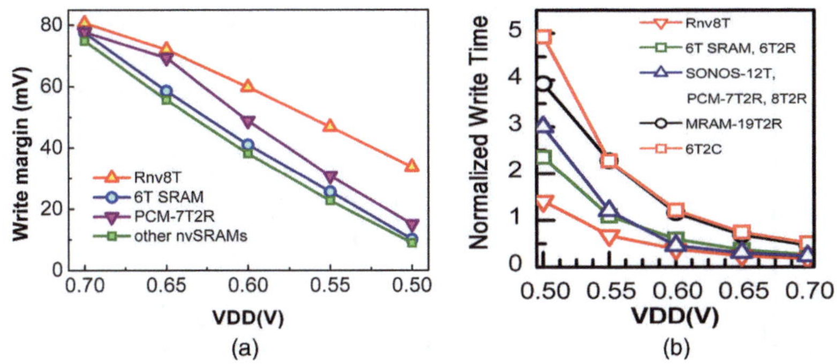


**Fig. 4.41** (**a**) Write margin and (**b**) normalized write time comparison between different nvSRAM cell structures under various $V_{DD}$

**Table 4.2** Performance comparison of various nvSRAM cell structures

| Structure | 9T2R (Rnv9T) | 8T2R (Rnv8T) | 12T | 19T2R | 6T2C | 7T2R | 8T2R | 6T2R |
|---|---|---|---|---|---|---|---|---|
| NV device | ReRAM | ReRAM | SONOS | MRAM | Fe. C | PCM | ReRAM | ReRAM |
| Cell Area | S | S | M | L | S | S | S | S |
| Write speed*1 | + | + | - | - | - | X | - | - |
| VDDmin*1 | ++ | + | X | X | X | X | X | - |
| $T_{STORE}$ | 10ns*2 | 10ns*2 | 4ms | 6ns | 200ns | 200ns | 10ns | 100ns |
| $V_{STORE}$ | 1.8/-1.6 | 1.8/-1.6 | -10/11 | N/A | 3V | 3V | 3V | N/A |
| $I_{STORE}$ | <25uA | <25uA | N/A | 1mA | N/A | 600uA | 100uA | N/A |
| Endurance | $>2\times10^8$ | $>2\times10^8$ | $10^6$ | Inf. | $10^8$ | $10^6$ | N/A | N/A |

*1: compared with 6T SRAM; S: Small M: Medium L: Large
+: Improved −: Degraded X: Unchanged
*2: 10ns =SET+RESET time

### 4.4.4 nvFF Example: SWT1R-nvFF

The low-latency energy-efficient system requires nonvolatile flip-flops (nvFFs) that are capable of achieving parallel data transfer between flip-flops (FFs) and nonvolatile memory (NVM) devices. Conventional two-NVM-based nvFFs face the challenges of excessive store energy ($E_S$) and overwrite stability degradation. To resolve these issues, a SWT1R-nvFF that incorporates a single NVM (1R) with self-write-termination (SWT) had been proposed [89]. This SWT1R-nvFF design reduces $E_S$ by 99% and a 2.7 ns termination latency. The following subsections will describe the nvFFs in detail.

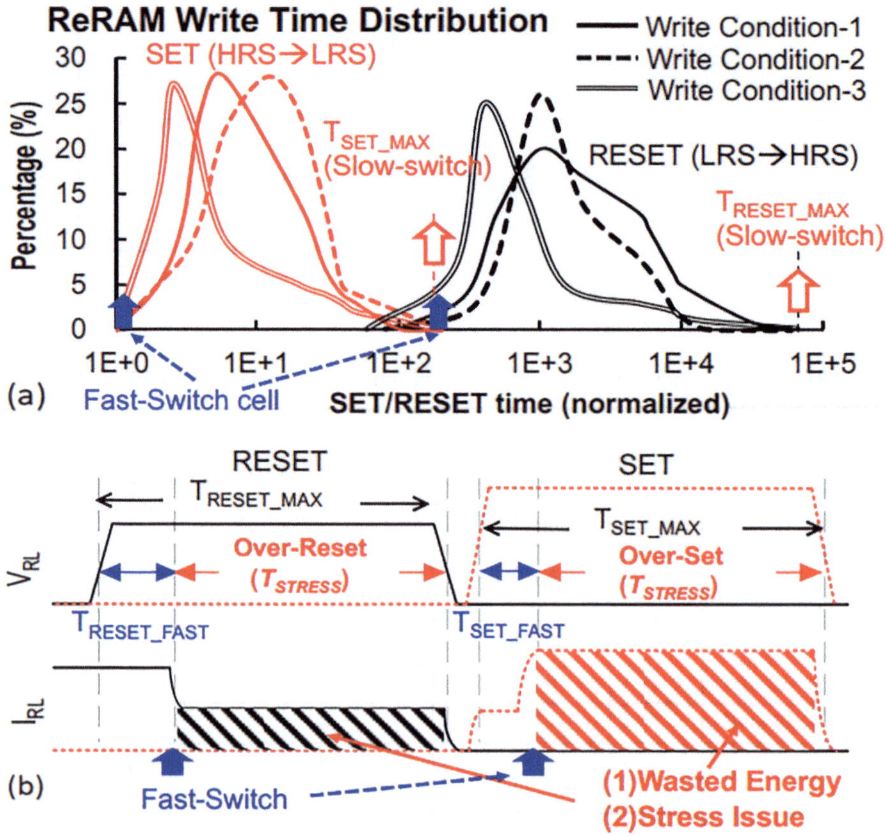


**Fig. 4.42** (**a**) Illustration of the conventional nvFF wide write time distribution challenges, and (**b**) write behavior of a log-process ReRAM

#### 4.4.4.1 SWT1R-nvFF Basic Operations

Figure 4.42 illustrates the wide write time distribution challenges of conventional nvFFs. Conventional NVM-based nvFFs usually employ long write duration to accommodate the slower cells. This approach raises several concerns on energy and reliability. First, the store energy ($E_S$) is high for writing the original and the complementary data to two NVM devices. This excessive waste of $E_S$ is further exacerbated when NVM devices have large variations in SET and RESET duration [90]. In addition, these long write operations may cause cell degradation that leads to endurance and reliability issues.

Figure 4.43 shows the proposed SWT1R-nvFF, it comprises of a master-latch (M-Latch), a dual-mode slave latch (S-Latch), two switches (SW1 and SW2), and an NVM-control unit (NVMCU) with one ReRAM on the left side (RL). There are three modes in this design—(1) regular flip-flop operations, (2) store operations for NVM write, and (3) restore operations for data recall from NVM to FF.

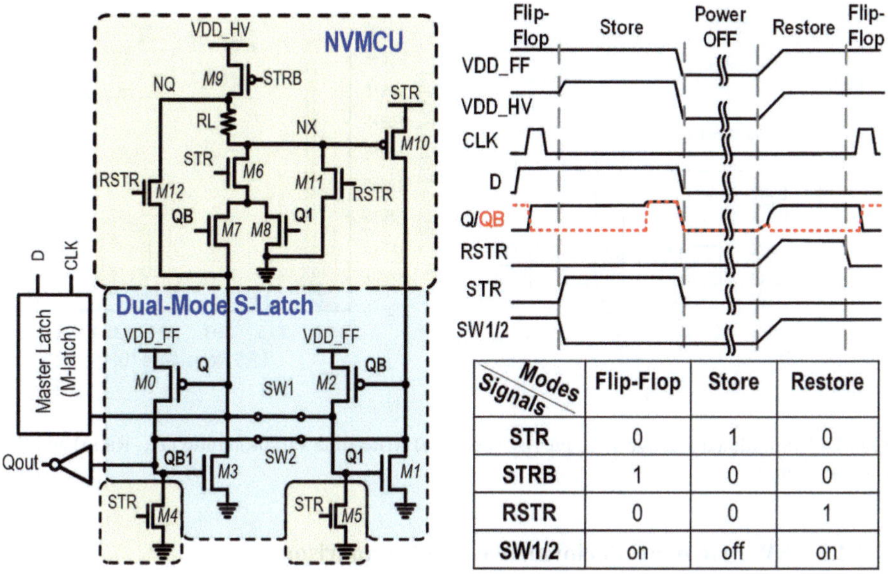


| Signals \ Modes | Flip-Flop | Store | Restore |
|---|---|---|---|
| STR | 0 | 1 | 0 |
| STRB | 1 | 0 | 0 |
| RSTR | 0 | 0 | 1 |
| SW1/2 | on | off | on |

**Fig. 4.43** Structure and operation modes of the proposed SWT1R-nvFF

The flip-flop mode is simple, both switches are on to connect M0-M3 to form a cross-coupled S-Latch. The restore and store signals (RSTR and STR) are kept low to disable M12/M6/M11 from causing voltage and current stress on RL. The device behaves like a typical FF. In the restore mode, RSTR is asserted and the voltage at node NQ ($V_{NQ}$) is determined by the voltage divider network formed by M9-RL-M11. During the rise of VDD, if RL is HRS, the high voltage of $V_{NQ}$ will keep Q at a higher state to restore Q = 1. If RL is LRS, then the low voltage of $V_{NQ}$ will keep Q at a lower state to restore Q = 0.

During store operation, both switches are turned off, while STR is high and STRB is low. In the case of SET program, the path (M9-RL-M6-M7-M3) forms the voltage divider that biases the voltage at node NX ($V_{NX}$) to VDDHV–$V_{SET}$. In the case of a mismatch, $V_{NX}$ is kept low to supply $V_{SET}$ across RL with M3 providing the SET current ($I_{SET}$), while M10 is weakly turned on to maintain QB = 1. When RL switches from HRS to LRS, a large DC-current causes Q to rise. The SET-feedback (SFB) circuit (M0, M3, and M4) then turns off M3 to terminate the SET operation. In the case of RESET program, the write path voltage divider (M9-RL-M6-M8) is activated to bias $V_{NX}$ at VDDHV–$V_{RESET}$. When RL switches from LRS to HRS, $V_{NX}$ is pulled low to turn on M10 and cause QB to rise and Q1 to drop. The RESET-feedback (RFB) circuit (M1, M2, and M5) then turns off M1 and M8 to terminate the RESET operation.

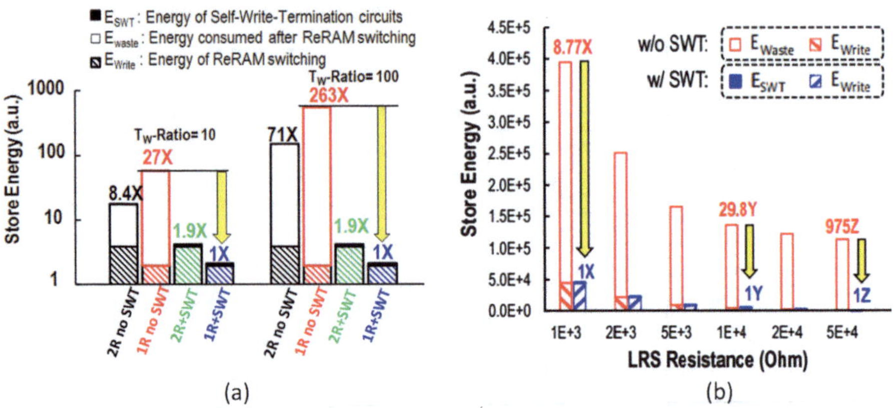


**Fig. 4.44** Simulated average store energy versus (**a**) write-time variation ratio ($T_W$-Ratio) and (**b**) LRS resistance

#### 4.4.4.2 SWT1R-nvFF Performance and Comparison

Figure 4.44a shows that the SWT scheme reduced the store energy ($E_S$) by 27× and store latency ($T_S$) by 10× shorter than the worst-case nvFF. Figure 4.44b presents store energy versus various LRS resistance ($R_{LRS}$) values. During RESET operations, a lower $R_{LRS}$ device consumes large DC current and consumes more power than do devices with a higher $R_{LRS}$. The increase in minimum $R_{LRS}$ enables the SWT scheme to reduce wasted power by 8.77× when the minimum $R_{LRS}$ is 1 kΩ. Lastly, Table 4.3 compares the performances of recent nvFF macros [76, 89, 91–93].

## 4.5 Conclusion

Emerging NVMs can offer potential advantages of low cost, low power, and high speed over conventional NVMs. However, some challenges must be overcome before successfully realizing these memory technologies in practical applications. In this chapter, we first explained the circuit design challenges and reviewed advanced circuit techniques associated with both read and write operations. Then we compared two-macros and one-macro approaches for energy-efficient systems. Lastly, we discussed and compared multiple implementations of nonvolatile logic and nonvolatile SRAM using emerging memory technologies.

**Table 4.3** Comparison of silicon-verified nvFFs

| Publications | [89] | [91] | [92] | [93] | [76] |
|---|---|---|---|---|---|
| Circuits (Slave-Latch Only) | 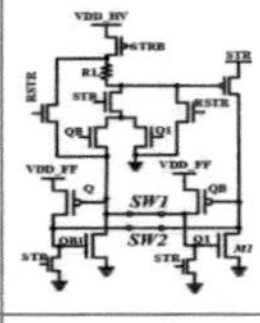 | 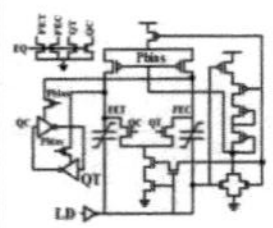 | 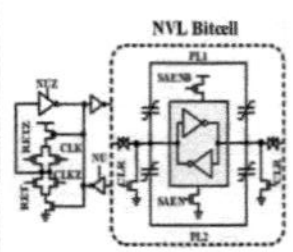 | 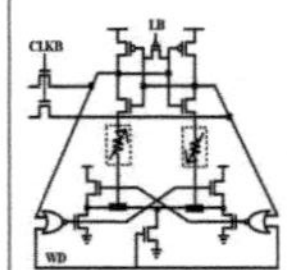 | 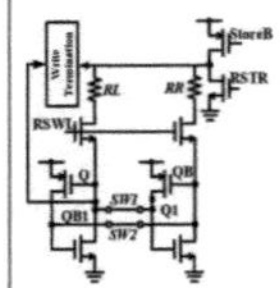 |
| Original NVM device | ReRAM | FeRAM | FeRAM | MTJ | ReRAM |
| CMOS technology | 65 nm | 130 nm | 130 nm | 90 nm | 65 nm |
| Additional devices | 13T + 1NVM (13T + 1R) | 22T + 2NVM (22T + 2R) | 18T + 4NVM (18T + 4R) | 17T + 2NVM (17T + 2R) | 15T + 2NVM (15T + 2R) |
| Self-write termination | Yes | No | No | No | Yes |
| Stress on NVM in flip-flop mode | No | No | No | Current | No |

**Acknowledgements** The authors would like to thank MXIC, TSMC, and ITRI for their supports.

## References

1. R.H. Fowler, L. Nordheim, Electron emission in intense electric fields. Proc R Soc A **119**(781), 173–181 (1928)
2. Y. Choi et al., A 20 nm 1.8 V 8 Gb PRAM with 40 MB/s program bandwidth, in *Proceedings of the IEEE International Solid-State Circuits Conference (ISSCC)* (2012), pp. 46–47
3. H.-S.P. Wong, C. Ahn, J. Cao, H.-Y. Chen, S.W. Fong, Z. Jiang, C. Neumann, S. Qin, J. Sohn, Y. Wu, S. Yu, X. Zheng, H. Li, J.A. Incorvia, S.B. Eryilmaz, K. Okabe, Stanford memory trends, https://nano.stanford.edu/stanford-memory-trends, Accessed 12 Oct 2016
4. R. Atiken, System requirements for memories, in *International Electron Devices Meetings Short Course Material* (2015)
5. B.C. Lee, E. Ipek, O. Mutlu, D. Burger, Architecting phase change memory as a scalable DRAM alternative, in *Proceedings of International Symposium on Computer Architecture (ISCA)* (2009)
6. K. Ikegami et al., Low power and high density STT-MRAM for embedded cache memory using advanced perpendicular MTJ integrations and asymmetric compensation techniques, in *IEEE International Electron Devices Meeting Technical Digest* (2014), pp. 650–653
7. Y. Wu, B. Lee, H.-S.P. Wong, Al2O3-based RRAM using atomic layer deposition (ALD) with 1-μA RESET current. IEEE Electron Devic Lett **31**(12), 1449–1451 (2010)
8. G. Jan et al., Demonstration of fully functional 8Mb perpendicular STT-MRAM chips with sub-5ns writing for non-volatile embedded memories, in *Proceedings of Symposium on VLSI Technology* (2014)
9. D. Ielmini, Emerging memory technologies: ReRAM and PCM, in *International Electron Devices Meetings Short Course Material* (2015)
10. K.J. Lee et al., A 90nm 1.8V 512Mb diode-switch PRAM with 266 MB/s read throughput. IEEE J. Solid State Circuits **43**(1), 150–162 (2008)
11. K. Tsuchida et al., A 64 Mb MRAM with clamped-reference and adequate-reference schemes, in *Proceedings of the IEEE International Solid-State Circuits Conference (ISSCC)* (2010), pp. 258–269
12. S.S. Sheu et al., A 4 Mb embedded SLC resistive-RAM macro with 7.2 ns read-write random-access time and 160ns MLC-access capability, in *Proceedings of the IEEE International Solid-State Circuits Conference (ISSCC)* (2011), pp. 200–201
13. C.-Y. Lee, DRAM life extension challenge and response, in *International Electron Devices Meetings Short Course Material* (2015)
14. M. Stanisavljevic, H. Pozidis, A. Athmanathan, N. Papandreou, T. Mittelholzer, E. Eleftheriou, Demonstration of reliable Triple-Level-Cell (TLC) phase-change memory, in *International Memory Workshop* (2016)
15. D. Worledge et al., Switching distributions and write reliability of perpendicular spin torque MRAM, in *IEEE International Electron Devices Meeting Technical Digest* (2010), pp. 296–299
16. A. Fantini et al., Intrinsic switching variability in HfO2 RRAM, in *IEEE International Memory Workshop (IMW)* (2013)
17. T. Cuong et al., A 7.8 MB/s 64 Gb 4-Bit/cell NAND flash memory on 43 nm CMOS technology, in *Non-volatile Memory Workshop* (2010)
18. P. Zhou, B. Zhao, J. Yang, Y. Zhang, A durable and energy efficient main memory using phase change memory, in *Proceedings of International Symposium on Computer Architecture (ISCA)* (2009)
19. Y. Xie, Introduction, in *Emerging Memory Technologies Design, Architecture And Applications*, (Springer, Berlin, 2014)

20. Y. Xie, *NVSim: A Circuit-Level Performance, Energy, and Area Model for Emerging Non-volatile Memory, a Chapter in Emerging Memory Technologies Design, Architecture and Applications* (Springer, Berlin, 2014)
21. S. Ambrogio et al., Understanding switching variability and random telegraph noise in resistive RAM, in *IEEE International Electron Devices Meeting (IEDM) Technical Digest* (2013), pp. 782–785
22. T.-Y. Liu et al., A 130.7-mm 2-layer 32-Gb ReRAM memory device in 24-nm technology. IEEE J. Solid State Circuits **49**(1), 140–153 (2014)
23. M. J. Breitwisch, "Phase Change Random Access Memory Integration," a Phase Change Materials: Science and Applications, Springer; Berlin, 2009
24. H.-S.P. Wong et al., Phase change memory. Proc. IEEE **98**(12), 2201–2227 (2010)
25. F. Xiong, A.D. Liao, D. Estrade, E. Pop, Low-power switching of phase-change materials with carbon nanotube electrodes. Sci Mag **332**, 568–570 (2011)
26. M. Brightsky, et al., Crystalline-as-deposited ALD phase change material confined PCM cell for high density storage class memory, in *IEEE International Electron Devices Meeting Technical Digest* (2015), pp. 60–63
27. D.H. Im, et al., A unified 7.5 nm dash-type confined cell for high performance PRAM device, in *Proceedings of IEEE International Electron Devices Meeting* (2008)
28. J.J. Nowak et al., Demonstration of ultralow bit error rates for spin-torque magnetic random-access memory with perpendicular magnetic anisotropy. IEEE Magn. Lett. **2**, 3000204 (2011)
29. Everspin, the MRAM Company, https://www.everspin.com/. Accessed 6 Nov 2016
30. T. Devolder, Basics of STT-MRAM, in *International Electron Devices Meetings Short Course Material* (2015)
31. C. Kim et al., A covalent-bonded cross-coupled current-mode sense amplifier for STT-MRAM with 1T1MTJ common source-line structure array, in *IEEE International Solid-State Circuits Conference (ISSCC) Digest of Technical Papers* (2015), pp. 134–136
32. R. Tekemura et al., A 32-Mb SPRAM with 2T1R memory cell, localized bi-directional write driver and '1'/'0' dual-array equalized reference scheme. IEEE J. Solid State Circuits **45**(4), 869–879 (2010)
33. Z.-H. Lin, Y.-H. Wang, Observation of indium ion migration-induced resistive switching in Al/Mg0.5Ca0.5TiO3/ITO. Appl. Phys. Lett. **109**(5), 053507 (2016)
34. J. Park et al., Quantized conductive filament formed by limited Cu source in sub-5nm era, in *IEEE International Electron Devices Meeting Technical Digest* (2011), pp. 64–67
35. C.M. Compagnoni et al., First evidence for injection statistics accuracy limitations in NAND Flash constant-current Fowler-Nordheim programming, in *IEEE International Electron Devices Meeting Technical Digest* (2007), pp. 165–168
36. A. Calderoni, S. Sills, N. Ramaswamy et al., Performance comparison of O-based and Cu-based ReRAM for high-density applications, in *International Memory Workshop* (2014)
37. N. Verma, A.P. Chandrakasan, A 256 kb 65 nm 8T subthreshold SRAM employing sense-amplifier redundancy. IEEE J. Solid State Circuits **43**(1), 141–149 (2008)
38. T.H. Kim, J. Liu, C.H. Kim, A voltage scalable 0.26 V, 64 kb 8T SRAM with Vmin lowering techniques and deep sleep mode. IEEE J. Solid State Circuits **44**(6), 1785–1795 (2009)
39. B. Zhai, S. Hanson, D. Blaauw, et al., A variation-tolerant Sub-200 mV 6-T subthreshold SRAM. IEEE J. Solid State Circuits **43**(10), 2338–2348 (2008)
40. I.J. Chang, J.J. Kim, S.P. Park, et al., A 32 kb 10T sub-threshold SRAM array with bit-interleaving and differential read scheme in 90 nm CMOS. IEEE J. Solid State Circuits **44**(2), 650–658 (2009)
41. Y. Morita, H. Fujiwara, H. Noguchi et al., An area-conscious low-voltage-oriented 8T-SRAM design under DVS environment, in *Symposium on VLSI Circuits Digest Technical Papers* (2007), pp. 256–257
42. J.H. Chen, L.T. Clark, T.H. Chen, An ultra-low-power memory with a subthreshold power supply voltage. IEEE J. Solid State Circuits **41**(10), 2344–2353 (2006)
43. K. Takeuchi, Y. Kameda, S. Fujimura, et al., A 56-nm CMOS 99-mm2 8-Gb multi-level NAND flash memory with 10-MB/s program throughput. IEEE J. Solid State Circuits **42**(1), 219–232 (2007)

44. D.S. Byeon, S.S. Lee, Y.H. Lim et al., An 8 Gb multi-level NAND flash memory with 63 nm STI CMOS process technology, in *IEEE International Solid-State Circuits Conference (ISSCC) Digest Technical Papers*, vol 1 (2005), pp. 46–47
45. S.H. Chang, S.K Lee, S.J. Park et al., A 48 nm 32 Gb 8-level NAND flash memory with 5.5 MB/s program throughput, in *IEEE International Solid-State Circuits Conference (ISSCC) Digest Technical Papers* (2009), pp. 240–241,241a
46. G.G. Marotta, A. Macerola, A. d'Alessandro et al., A 3bit/cell 32 Gb NAND flash memory at 34 nm with 6 MB/s program throughput and with dynamic 2b/cell blocks configuration mode for a program throughput increase up to 13 MB/s, in *IEEE International Solid-State Circuits Conference (ISSCC) Digest Technical Papers* (2010), pp. 444–445
47. K. Kobayashi, T. Nakayama, Y. Miyawaki, et al., A high-speed parallel sensing architecture for multi-megabit flash E2PROMs. IEEE J. Solid State Circuits **25**(1), 79–83 (1990)
48. M. Bauer, R. Alexis, G. Atwood et al., A multilevel-cell 32 Mb flash memory, in *IEEE International Solid-State Circuits Conference (ISSCC) Digest Technical Papers* (1995), pp. 132–133
49. R. Micheloni, L. Crippa, M. Sangalli, et al., The flash memory read path: building blocks and critical aspects. IEEE Proc **91**(4), 537–553 (2003)
50. B.Q. Le, M. Achter, C.G. Chng, et al., Virtual-ground sensing techniques for a 49-ns/200-MHz access time 1.8-V 256-Mb 2-bit-per-cell flash memory. IEEE J. Solid State Circuits **39**(11), 2014–2023 (2004)
51. T. Ogura, M. Hosoda, T. Ogawa, et al., A 1.8-V 256-Mb multilevel cell NOR flash memory with BGO function. IEEE J. Solid State Circuits **41**(11), 2589–2600 (2006)
52. C.C. Chung, H.C. Lin, Y.T. Lin, A multilevel read and verifying scheme for bi-NAND flash memories. IEEE J. Solid State Circuits **42**(5), 1180–1188 (2007)
53. M.F. Chang, S.J. Shen, A process variation tolerant embedded split-gate flash memory using pre-stable current sensing scheme. IEEE J. Solid State Circuits **44**(3), 987–994 (2009)
54. M.F. Chang, C.W. Wu, C.C. Kuo, et al., A low-voltage bulk-drain-driven read scheme for Sub-0.5 V 4 Mb 65 nm logic-process compatible embedded resistive RAM (ReRAM) macro. IEEE J. Solid State Circuits **48**(9), 2250–2259 (2013)
55. M.F. Chang, J.J. Wu, T.F. Chien et al., 19.4 embedded 1Mb ReRAM in 28nm CMOS with 0.27-to-1V read using swing-sample-and-couple sense amplifier and self-boost-write-termination scheme, in *IEEE International Solid-State Circuits Conference (ISSCC) Digest Technical Papers* (2014), pp. 332–333
56. Y.D. Chih, C.H. Wang, C.H. Kuo, *Reference cell circuit for split gate flash memory*, U.S. Patent 6,396,740 (2002)
57. Datasheet, *"sfc 0064_08b9_he" Taiwan Semiconductor Manufacturing Company (TSMC)* (2001)
58. Datasheet, *AF64K8AF25, v1.0 1st Silicon Sdn. Bhd. (X-Fab)* (2005)
59. J. Javanifard, T. Tanadi, H. Giduturi et al., A 45nm self-aligned-contact process 1Gb NOR flash with 5MB/s program speed, in *IEEE International Solid-State Circuits Conference (ISSCC) Digest Technical Papers* (2008), pp. 424–624
60. M.-F. Chang, S.-S. Sheu, K.-F. Lin, et al., A high-speed 7.2-ns read-write random access 4-Mb embedded resistive RAM (ReRAM) macro using process-variation-tolerant current-mode read schemes. IEEE J. Solid State Circuits **48**, 878–891 (2013)
61. M. Jefremow, T. Kern, W. Allers et al., Time-differential sense amplifier for sub-80 mV bitline voltage embedded STT-MRAM in 40 nm CMOS, in *IEEE International Solid-State Circuits Conference (ISSCC) Digest Technical Papers* (2013), pp. 216–217
62. T. Kono, T. Ito, T. Tsuruda, T. Nishiyama, T. Nagasawa, T. Ogawa, Y. Kawashima, H. Hidaka, T. Yamauchi, 40-nm embedded split-gate MONOS (SG-MONOS) flash macros for automotive with 160-MHz random access for code and endurance over 10 M cycles for data at the junction temperature of 170°C. IEEE J. Solid State Circuits **49**, 154–166 (2013)
63. M.F. Chang, Y.F. Lin, Y.C. Liu, et al., An asymmetric-voltage-biased current-mode sensing scheme for fast-read embedded flash macros. IEEE J. Solid State Circuits **50**(9), 2188–2198 (2015)

64. S.N. Wooters, B.H. Calhoun, T.N. Blalock, et al., An energy-efficient subthreshold level converter in 130-nm CMOS. IEEE Trans Circuits Syst II Exp Briefs **57**(4), 290–294 (2010)
65. I.J. Chang, J.J. Kim, K. Kim, et al., Robust level converter for sub-threshold/super-threshold operation:100 mV to 2.5 V. IEEE Trans Very Larg Scale Integr (VLSI) Syst **19**(8), 1429–1437 (2011)
66. S. Lütkemeier, U. Ruckert, et al., A subthreshold to above-threshold level shifter comprising a Wilson current mirror. IEEE Trans Circuits Syst II Exp Briefs **57**(9), 721–724 (2010)
67. M.F. Chang et al., A low-power subthreshold-to-superthreshold level-shifter for sub-0.5V embedded resistive RAM (ReRAM) macro in ultra low-voltage chips, in *2014 IEEE Asia Pacific Conference on Circuits and Systems (APCCAS), Ishigaki* (2014), pp. 695–698
68. D. Halupka, S. Huda, W. Song et al., Negative-resistance read and write schemes for STT-MRAM in 0.13μm CMOS, in *IEEE International Solid-State Circuits Conference (ISSCC) Digest Technical Papers* (2010), pp. 256–257
69. X.Y. Xue, W.X. Jian, J.G. Yang, et al., A 0.13 μm 8 Mb logic-based CuxSiyO ReRAM with self-adaptive operation for yield enhancement and power reduction. IEEE J. Solid State Circuits **48**(5), 1315–1322 (2013)
70. M.-F. Chang et al., Circuit design challenges in embedded memory and resistive RAM (RRAM) for mobile SoC and 3D-IC, in *Proceedings 2011 IEEE Asia South Pacific Design Automation Conference (ASP-DAC)* (2011), pp. 197–203
71. M.-F. Chang et al., Challenges and trends of resistive memory (Memristor) based circuits for 3D-IC applications, in *2011 Symposium on Solid State Devices and Materials (SSDM)* (2011), pp. 1053–1054
72. M.-F. Chang et al., Challenges and trends in low-power 3D die-stacked IC designs using RAM, memristor logic, and resistive memory (ReRAM), in *The IEEE 9th International Conference on ASIC (ASICON)* (2011), pp. 327–330
73. P.-F. Chiu, M.-F. Chang et al., A low store energy, low VDDmin, nonvolatile 8T2R SRAM with 3D stacked RRAM devices for low power mobile applications, in *Symposium on VLSI Circuits Digest Technical Papers* (2010), pp. 229–230
74. H. Ohno et al., Magnetic tunnel junction for nonvolatile CMOS logic, in *IEEE International Electron Devices Meeting (IEDM)* (2010), pp. 9.4.1–9.4.4
75. S.-M. Yoon et al., Phase-change-driven programmable switch for nonvolatile logic applications. IEEE Electron Device Lett **30**(4), 371–373 (2009)
76. Y. Liu et al., A 65nm ReRAM-enabled nonvolatile processor with 6× reduction in restore time and 4× higher clock frequency using adaptive data retention and self-write-termination nonvolatile logic, in *IEEE International Solid-State Circuits Conference (ISSCC) Digest of Technical Papers* (2016), pp. 84–86
77. M. Fliesler et al., A 15ns 4Mb NVSRAM in 0.13u SONOS technology, in *Non-Volatile Semiconductor Memory Workshop (NVSMW)* (2008), pp. 83–86
78. N. Sakimura et al., Nonvolatile magnetic Flip-flop for standby-power-free SoCs. IEEE J. Solid State Circuits **44**(8), 2244–2250 (2009)
79. T. Miwa et al., NV-SRAM: A nonvolatile SRAM with backup ferroelectric capacitors. IEEE J. Solid State Circuits **36**(3), 522–527 (2001)
80. M. Takata et al., Nonvolatile SRAM based on phase change, in *Non-Volatile Semiconductor Memory Workshop (NVSMW)* (2006), pp. 95–96
81. W. Wang et al., Nonvolatile SRAM cell, in *International Electron Device Meeting (IEDM)* (2006), pp. 1–4
82. M.-F. Chang et al., Endurance-aware circuit designs of nonvolatile logic and nonvolatile SRAM using resistive memory (Memristor) device, in *2012 IEEE Asia South Pacific Design Automation Conference (ASP-DAC)* (2012), pp. 329–334
83. P.-F. Chiu, M.-F. Chang, C.-W. Wu, C.-H. Chuang, S.-S. Sheu, Y.-S. Chen, M.-J. Tsai, Low store energy, low VDDmin, 8T2R nonvolatile latch and SRAM with vertical-stacked resistive memory (Memristor) devices for low power mobile applications. IEEE J Solid State Circ **47**(6), 1483–1496 (2012)

84. E. Seevinck, F.J. List, J. Lohstroh, Static-noise margin analysis of MOS SRAM cells. IEEE J. Solid State Circuits **22**(5), 748–754 (1987)
85. L.-F. Chen, *A 7T SRAM Circuit Design for Low Voltage Applications.*, M.S. thesis (Institute of Electronics Engineering, National Tsing Hua University, Hsinchu, 2011)
86. J.-J. Wu et al., A large σVTH/VDD tolerant ZigZag 8T SRAM with area-efficient decoupled differential sensing and fast write-back scheme. IEEE J. Solid State Circuits **46**(4), 815–827 (2011)
87. W. Wang et al., Nonvolatile SRAM cell, in *International Electron Devices Meeting (IEDM) Technical Digest Papers* (2006), pp. 27–30
88. S. Yamamoto et al., Nonvolatile SRAM (NV-SRAM) Using Functional MOSFET Merged with Resistive Switching Devices, in *IEEE Custom Integrated Circuits Conference (CICC) Technical Digest Papers* (2009), pp. 531–534
89. S.P. Lo et al., A ReRAM-based Single-NVM nonvolatile flip-flop with reduced stress-time and write-power against wide distribution in write-time by using self-write-termination scheme for nonvolatile processors in IoT Era, in *IEEE International Electron Devices Meeting (IEDM) Digest Technical Papers* (2016), pp. 420–423
90. M.F. Chang et al., Embedded 1Mb ReRAM in 28nm CMOS with 0.27-to-1V read using swing-sample-and-couple sense amplifier and self-boost-write-termination scheme, in *IEEE International Solid-State Circuits Conference (ISSCC) Digest of Technical Papers* (2014), pp. 332–333
91. M. Qazi, A. Amerasekera, A. P. Chandrakasan, A 3.4 pJ FeRAMenabled D flip-flop in 0.13μm CMOS for nonvolatile processing in digital systems, in *IEEE International Solid-State Circuits Conference (ISSCC) Digest of Technical Papers* (2013), pp. 192–193
92. S.C. Bartling et al., An 8MHz 75μA/MHz zero-leakage non-volatile logic-based Cortex-M0 MCU SoC exhibiting 100% digital state retention at VDD=0V with <400ns wakeup and sleep transitions, in *IEEE International Solid-State Circuits Conference (ISSCC) Digest of Technical Papers* (2013), pp. 432–433
93. N. Sakimura et al., A 90nm 20MHz fully nonvolatile microcontroller for standby-power-critical applications, in *IEEE International Solid-State Circuits Conference (ISSCC) Digest of Technical Papers* (2014), pp. 184–185

# Chapter 5
# The Processing-in-Memory Paradigm: Mechanisms to Enable Adoption

**Saugata Ghose, Kevin Hsieh, Amirali Boroumand, Rachata Ausavarungnirun, and Onur Mutlu**

DRAM, the predominant technology used to build main memory, is a major component of modern computer systems. As the data working set sizes of modern applications grow [37, 48, 79, 225], the need for higher memory capacity and higher memory performance continues to grow as well. However, even though CMOS technology scaling has yet to come to an end, DRAM technology scaling has been unable to keep up with the increasing memory demand from applications [2, 3, 26, 29, 37, 65, 66, 68, 79, 82, 89–91, 110, 114, 116, 117, 120, 125, 129, 137, 138, 143, 151, 152, 160, 226, 233, 239, 240]. For example, if we study the latency and throughput of Double Data Rate (DDR) DRAM over the last 15–20 years, we see that neither has been able to keep up with the growth in application working set size or CPU computational power [26, 28, 107, 114, 125].

A major bottleneck to improving the overall system performance is the high cost of *data movement*. Currently, in order to perform an operation on data that is stored within memory, the CPU must issue a request to the memory controller, which in turn sends a series of commands across an off-chip bus to the DRAM module. The data is then read from the DRAM module, at which point it is returned to the memory controller and typically stored within the CPU cache. Only after the data is placed in the CPU cache can the CPU operate (i.e., perform computation) on the data. The long latency to retrieve data from DRAM is exacerbated by two factors. First, it is difficult to send a large number of requests to memory in parallel, in part because of the narrow bandwidth of the off-chip bus between the memory controller and main memory. Second, despite the time spent on bringing the data into the cache, which is substantial [62, 63], much of the data brought into the caches is

S. Ghose · K. Hsieh · A. Boroumand · R. Ausavarungnirun
Carnegie Mellon University, Pittsburgh, PA, USA

O. Mutlu (✉)
ETH Zürich, Zürich, Switzerland
Carnegie Mellon University, Pittsburgh, PA, USA



R. O. Topaloglu, H.-S. P. Wong (eds.), *Beyond-CMOS Technologies for Next Generation Computer Design*, https://doi.org/10.1007/978-3-319-90385-9_5

*not* reused by the CPU [177, 178], rendering the caching either very inefficient or sometimes even unnecessary. Ultimately, there is significant time and energy wasted on moving data between the CPU and memory, many times with little benefit in return, especially in workloads where caching is not very effective [2, 3].

Recent advances in memory design have unlocked the potential to avoid the unnecessary data movement. In an attempt to improve the scalability of capacity and bandwidth, memory manufacturers have turned to 3D-stacked memories, where multiple layers of memory arrays are stacked on top of each other [119, 132]. These layers are connected together using *through-silicon vias* (TSVs), which provide much greater internal memory bandwidth than the narrow off-chip bus to the CPU. Some prominent examples of these 3D-stacked memory architectures [71, 72, 74, 75, 92, 119] include a *logic layer*, which provides an opportunity to embed general-purpose computational logic *directly within main memory* to take advantage of the high internal bandwidth available.

The idea of performing *processing-in-memory* (PIM), or *near-data processing* (NDP), has been proposed for at least several decades [39, 44, 45, 56, 80, 100, 140, 166, 171, 208, 218]. However, these past efforts were *not* adopted at large scale due to the difficulty of integrating processing elements with DRAM. As a result of the potential enabled by the inclusion of a logic layer in modern memory architectures, various recent works explore a range of PIM architectures for multiple different purposes (e.g., [1–4, 9, 12, 20, 22, 27, 30, 47, 51, 52, 58, 59, 62–64, 67, 68, 81, 93, 95, 96, 98, 118, 123, 124, 131, 133, 149, 163, 172, 176, 195, 196, 200, 202, 203, 205, 221, 243, 246]).

While PIM avoids the need to move data from memory to the CPU for a number of data-intensive functions, it introduces new challenges for system architects and programmers. In particular, PIM processing logic does *not* have quick access to important mechanisms that exist within the CPU, such as address translation and cache coherence, which greatly aid the programmer. Preserving the functionality and efficiency of such mechanisms is essential for PIM, as these mechanisms can (1) preserve the traditional programming models that application developers rely on to productively write programs, and (2) provide significant performance benefits.. As we show in this work (see Sect. 5.4), simply forcing PIM processing logic to send queries to the CPU to accomplish address translation and cache coherence is very inefficient, since the overhead of a query can almost completely *eliminate* the benefits of moving computation to memory. Therefore, it is essential that we provide *PIM-specific* mechanisms that *efficiently* support the functionality of traditional address translation and cache coherence mechanisms and thus provide support for the use of existing programming models to program PIM architectures. Our goal is to design general-purpose address translation and cache coherence mechanisms that can be exploited by any PIM processing logic to provide low-overhead support for common functions, such as pointer chasing in virtual memory and cache coherence.

To this end, we propose two mechanisms to support PIM. The first mechanism, IMPICA, is an in-memory accelerator for pointer chasing, which exploits the high bandwidth available within 3D-stacked memory. IMPICA can traverse a chain of virtual memory pointers within DRAM, *without* having to look up virtual-to-physical address translations in the CPU translation lookaside buffer (TLB) or without using the page walkers within the CPU. The second mechanism, LazyPIM, maintains cache coherence between PIM processing logic and CPU cores *without* sending coherence requests for every memory access. Instead, LazyPIM efficiently provides coherence by having PIM processing logic speculatively acquire coherence permissions, and then later sends compressed *batched* coherence lookups to the CPU to determine whether or not its speculative permission acquisition violated the memory ordering defined by the programming model.

In Sect. 5.1, we cover common design principles of modern PIM architectures. In Sect. 5.2, we discuss key issues that impact the flexibility and adoption of PIM architectures. In Sect. 5.3, we discuss IMPICA, an accelerator that we propose to efficiently support pointer-chasing operations within PIM architectures. In Sect. 5.4, we discuss LazyPIM, a mechanism that we propose to efficiently support cache coherence between PIM processing logic and the CPU cores. In Sect. 5.5, we discuss related work in the area, and in Sect. 5.6 we briefly discuss some future research challenges, with a focus on system-level challenges for the adoption of PIM architectures.

## 5.1 Designing Processing-in-Memory Architectures

Processing-in-memory (PIM) architectures place some form of processing logic (typically accelerators, simple cores, or reconfigurable logic) inside the DRAM subsystem. This *PIM processing logic*, which we also refer to as *PIM cores* or *PIM engines*, interchangeably, can execute portions of applications or entire application kernels, depending on the design of the architecture. In this section, we first discuss how the PIM processing logic is integrated within DRAM modules (Sect. 5.1.1), and then we discuss how applications make use of this PIM processing logic (Sect. 5.1.2).

### *5.1.1 Placing Processing Logic Within the DRAM Subsystem*

Modern PIM architectures rely on implementing processing logic in the DRAM chip itself (e.g., [2–4, 12, 20, 22, 27, 30, 47, 51, 52, 58, 59, 64, 67, 68, 93, 95, 96, 98, 117, 118, 124, 131, 133, 149, 172, 176, 195, 196, 198–200, 202, 203, 221, 243, 246]) or on the DRAM module or the DRAM controller [9, 62, 63, 203]. DRAM consists of multiple arrays of capacitive *cells*, where each cell holds one bit of data. By placing processing logic in close proximity of the cell arrays, PIM architectures are *not* restricted to the limited bandwidth offered by the narrow off-chip bus between the DRAM module and the CPU. Instead, PIM processing logic benefits from the much wider buses that are available within the chip and/or module in modern DRAM architectures.

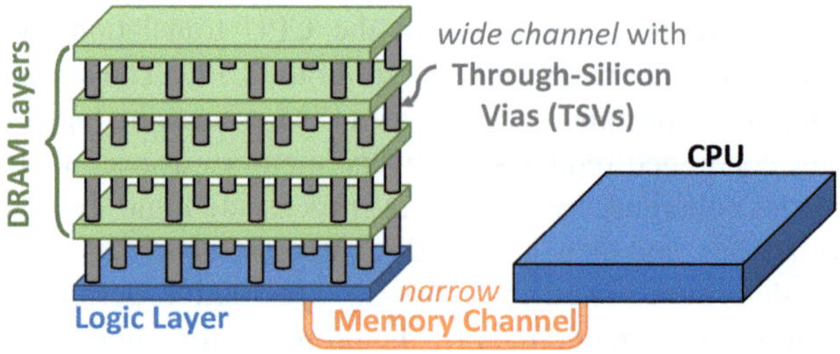


**Fig. 5.1** High-level overview of a 3D-stacked DRAM based architecture

Figure 5.1 shows an overview of a 3D-stacked DRAM based architecture. Examples of 3D-stacked DRAM include High-Bandwidth Memory (HBM) [75, 119] and the Hybrid Memory Cube (HMC) [2, 71, 72]. As the figure shows, a 3D-stacked DRAM consists of multiple layers. 3D-stacked DRAM has a much greater internal data bandwidth than conventional memory, due to its use of *through-silicon vias* (TSVs), which are vertical links that connect the multiple layers of a DRAM stack together [119, 132]. In addition to containing multiple layers of DRAM, a number of 3D-stacked DRAM architectures, such as HBM [75, 119] and HMC [71, 72], include a *logic layer*, typically the bottommost layer, where architects can implement functionality that interacts with both the processor and the DRAM cells [2, 3, 71, 72]. Currently, 3D-stacked DRAM makes limited use of the logic layer (e.g., HMC implements command scheduling logic within the logic layer [71, 72]).

Recent PIM proposals (e.g., [2–4, 12, 20, 22, 30, 47, 51, 52, 58, 59, 64, 67, 68, 93, 95, 96, 98, 118, 131, 133, 149, 163, 172, 176, 196, 221, 243, 246]) add processing logic to the logic layer to exploit the high bandwidth available between the logic layer and the DRAM cell arrays. The proposed PIM processing logic design varies based on the specific architecture, and can range from fixed-function accelerators to simple *in-order* cores, and to reconfigurable logic. The complexity of the processing logic that can be added to the logic layer is currently limited by the manufacturing process technology and thermal design points, which may prevent highly sophisticated processors (e.g., out-of-order processor cores with large caches and sophisticated instruction-level parallelism techniques) from being implemented within the logic layer at this time [43, 172, 243].

### *5.1.2 Using PIM Processing Logic Functionality in Applications*

In order for applications to make use of PIM processing logic that resides within DRAM, each PIM architecture exposes an interface to the CPU. While there currently is no standardization of this interface, most contemporary works on PIM architectures follow similar models for CPU–PIM interactions. PIM processing logic is typically treated as a coprocessor, and executes only when some code (which we refer to as a *PIM kernel*) is launched by the CPU on the PIM processing logic. PIM kernels vary widely in current proposals in terms of granularity. Some works (e.g., [2, 51]) treat an *entire application* thread as a PIM kernel, in order to minimize the amount of synchronization and data sharing that takes place between the main CPU and main compute-capable memory. Many works (e.g., [4, 9, 22, 47, 52, 59, 68, 95, 96, 98, 124, 131, 172, 195, 196, 200, 202, 243]) treat only portions of an application thread (e.g., *functions*) as a PIM kernel, and launch the kernel when a CPU core reaches the PIM kernel call. Yet other works (e.g., [3, 118, 163]) use a much finer granularity for offloading code to PIM: they offload only a *single instruction* as the PIM kernel, which is completed atomically.

An open question for all of these architectures is how a PIM kernel is identified and demarcated, and who is responsible for identification and demarcation. Current works on PIM expect the compiler or programmer to mark sections of the code and/or data that are to be dispatched to the PIM processing logic. When a program reaches a point at which a PIM kernel should be executed, the CPU uses the off-chip memory channel to dispatch the kernel to a free PIM core. The PIM core then executes the kernel, and upon completing the kernel, notifies the CPU using the memory channel. Several works [2, 3, 21, 68] provide a detailed explanation of this process.

Due to the simple nature of PIM processing logic (i.e., that PIM processing logic is expected to be fixed-function accelerators, small in-order general-purpose cores, or simple reconfigurable logic), current PIM architectures do *not replace* the CPU cores with PIM cores; they instead *augment* the existing CPU cores. OS threads continue to run on the CPU cores, and key structures to support application execution, such as large caches, translation lookaside buffers (TLBs), page walkers, and cache coherence hardware, are expected to remain within the CPU. While these decisions minimize the changes that need to be made to write programs for PIM architectures, the decisions introduce new issues that PIM architectures must address to maintain ease of adoption. We discuss two such critical issues in the next section.

## 5.2 Key Issues in Enabling Processing-in-Memory

Pushing some or all of the computation for a program from the CPU to the DRAM introduces new challenges for system architects (as well as programmers) to overcome. In particular, PIM processing logic does *not* have direct access to structures within the CPU that are essential to memory operations, such as address translation and cache coherence hardware. A naive solution is to simply have PIM processing logic access these structures remotely over the off-chip memory channel. Unfortunately, such likely frequent remote accesses can introduce a high performance and energy overhead, and often undermine many, if not all, of the benefits that PIM architectures provide. A second naive solution is to limit the functionality of the PIM processing logic such that it *cannot* perform address translation or cache coherence, and to expose these limitations to programmers. However, this alters the programming model of the system, and can lead to great difficulty for the widespread adoption of PIM as an execution model. In this section, we focus on the address translation and cache coherence challenges, and discuss why naive solutions are not practical. We discuss new PIM-specific solutions that can overcome these challenges in Sects. 5.3 and 5.4.

### *5.2.1 Address Translation*

A large amount of code relies on pointers, which are stored as *virtual* memory addresses. When the application follows a pointer, a core must perform *address translation*, which converts the pointer's stored virtual address into a *physical* address within main memory. If PIM processing logic relies on existing CPU-side address translation mechanisms, any performance gains from performing pointer chasing in memory could easily be nullified, as the processing logic needs to send a long-latency translation request to the CPU via the off-chip channel for each memory access. The translation can sometimes require a page table walk, where the CPU must issue *multiple* memory requests to read the page table, which further increases traffic on the memory channel.

A naive solution is to simply duplicate the TLB and page walker within memory (i.e., within the PIM processing logic). Unfortunately, this is prohibitively difficult or expensive for three reasons: (1) coherence would have to be maintained between the CPU and memory-side TLBs, introducing extra complexity and off-chip requests; (2) the duplicated hardware is very costly in terms of storage overhead and complexity; and (3) a memory module can be used in conjunction with many different processor architectures, which use different page table implementations and formats, and ensuring compatibility between the in-memory TLB/page walker and all of these different designs is difficult.

We explore a tractable solution for PIM address translation as part of our in-memory pointer chasing accelerator, which we discuss in Sect. 5.3.

### *5.2.2 Cache Coherence*

PIM processing logic can modify the data it processes, and this data may also be needed by CPU cores. In a traditional multithreaded execution model that uses shared memory between threads, writes to memory must be coordinated between multiple cores, to ensure that threads do not operate on stale data values. Due to the per-core caches used in CPUs, this requires that when one core writes data to a memory address, cached copies of the data held within the caches of other cores must be updated or invalidated, which is known as *cache coherence*. Cache coherence involves a protocol that is designed to handle write permissions for each core, invalidations and updates, and arbitration when multiple cores request exclusive access to the same memory address. Within a chip multiprocessor (CMP), the per-core caches can perform coherence actions over a shared interconnect. Both snoopy [57, 167] and directory-based [23] coherence mechanisms are employed in existing multiprocessor systems.

Cache coherence is a major system challenge for enabling PIM architectures as general-purpose execution engines. If PIM processing logic is coherent with the processor, the PIM programming model is relatively simple, as it remains similar to conventional shared memory multithreaded programming, which makes PIM architectures easier to adopt in general-purpose systems. Thus, allowing PIM processing logic to maintain such a simple and traditional shared memory programming model can facilitate the widespread adoption of PIM. However, it is impractical for PIM to perform traditional fine-grained cache coherence, as this forces a large number of coherence messages to traverse a narrow off-chip interconnect, potentially undoing the benefits of high-bandwidth and low-latency PIM execution, as we show in Sect. 5.4. Prior works have proposed intermediate solutions that *sidestep* coherence by either requiring the programmer to ensure data coherence or making PIM data non-cacheable in the CPU (e.g., [2–4, 27, 47, 51, 52, 59, 67, 68, 149, 163, 172, 195, 196, 200, 202, 243]). Unfortunately, these solutions either place some restrictions on the programming model or limit the performance and energy gains achievable by a PIM architecture.

In this chapter, we describe a new coherence protocol, which allows PIM processing logic to efficiently perform coherence *without* incurring high overhead or changing the programming model, which we discuss in Sect. 5.4.

## 5.3 IMPICA: An In-Memory Pointer-Chasing Accelerator

Linked data structures such as trees, hash tables, and linked lists are commonly used in many important applications [5, 35, 46, 50, 54, 63, 142, 155, 157, 227, 235]. For example, many databases use B/B$^+$-trees to efficiently index large data sets [46, 54], key-value stores use linked lists to handle collisions in hash tables [50, 142], and graph processing workloads [2, 3, 209] use pointers to represent graph edges. These structures link nodes using pointers, where each node points to at least one other node by storing its address. Traversing the link requires serially accessing consecutive nodes by retrieving the address(es) of the next node(s) from the pointer(s) stored in the current node. This fundamental operation is called *pointer chasing* in linked data structures.

Pointer chasing is currently performed by the CPU cores, as part of an application thread. While this approach eases the integration of pointer chasing into larger programs, pointer chasing can be inefficient within the CPU, as it introduces several sources of performance degradation: (1) dependencies exist between memory requests to the linked nodes, resulting in serialized memory accesses and limiting the available instruction-level and memory-level parallelism [63, 154–156, 187]; (2) irregular allocation or rearrangement of the connected nodes leads to access pattern irregularity [35, 40, 78, 83, 155, 157, 238], causing frequent cache and TLB misses; and (3) link traversals in data structures that diverge at each node (e.g., hash tables, B-trees) frequently go down different paths during different iterations, resulting in little reuse, further limiting cache effectiveness [136]. Due to these inefficiencies, a significant *memory bottleneck* arises when executing pointer chasing operations in the CPU, which stalls on a large number of memory requests that suffer from the long round-trip latency between the CPU and the memory.

Many prior works (e.g., [33–35, 40, 69, 70, 78, 83, 128, 135, 136, 155, 157, 186, 187, 213, 232, 238, 242, 248]) proposed mechanisms to predict and prefetch the next node(s) of a linked data structure early enough to hide the memory latency. Unfortunately, prefetchers for linked data structures suffer from several shortcomings: (1) they usually do *not* provide significant benefit for data structures that diverge at each node [83, 155, 157], due to low prefetcher accuracy and low miss coverage; (2) aggressive prefetchers can consume too much of the limited off-chip memory bandwidth and, as a result, slow down the system [40–42, 62, 78, 109, 113, 204, 216]; and (3) a prefetcher that works well for some pointer-based data structure(s) and access patterns (e.g., a Markov prefetcher designed for mostly-static linked lists [78]) usually does not work efficiently for different data structures and/or access patterns. Thus, it is important to explore new solution directions to alleviate the significant performance and efficiency loss due to pointer chasing.

**Our goal** in this section is to accelerate pointer chasing by *directly minimizing the memory bottleneck* caused by pointer chasing operations. To this end, we propose to perform pointer chasing *inside main memory* by leveraging processing-in-memory (PIM) mechanisms, *avoiding the need to move data to the CPU*. In-memory pointer chasing greatly reduces (1) the latency of the operation, as an address does not need to be brought all the way into the CPU before it can be dereferenced; and (2) the reliance on caching and prefetching in the CPU, which are largely ineffective for pointer chasing over large data structures. In this section, we *describe an in-memory accelerator for chasing pointers* in any linked data structure, called the *In-Memory PoInter Chasing Accelerator* (IMPICA) [67]. IMPICA leverages the low memory access latency at the logic layer of 3D-stacked memory to speed up pointer chasing operations.

We identify *two fundamental challenges that we believe exist for a wide range of in-memory accelerators*, and evaluate them as part of a case study in designing a pointer chasing accelerator in memory. These fundamental challenges are (1) how to achieve high parallelism in the accelerator (in the presence of serial accesses in pointer chasing), and (2) how to effectively perform virtual-to-physical address translation on the memory side without performing costly accesses to the CPU's memory management unit. We call these, respectively, the *parallelism challenge* and the *address translation challenge*.

**The Parallelism Challenge** Parallelism is challenging to exploit in an in-memory accelerator even with the reduced latency and higher bandwidth available within 3D-stacked memory, as the performance of pointer chasing is limited by *dependent sequential accesses*. The serialization problem can be exacerbated when the accelerator traverses multiple streams of links: while traditional out-of-order or multicore CPUs can service memory requests from multiple streams in parallel due to their ability to exploit high levels of instruction- and memory-level parallelism [55, 63, 153, 154, 156, 158, 159, 222], simple accelerators are unable to exploit such parallelism unless they are carefully designed (e.g., [2, 47, 176, 246]).

We observe that accelerator-based pointer chasing is primarily bottlenecked by memory access latency, and that the address generation computation for link traversal takes only a small fraction of the total traversal time, leaving the accelerator idle for a majority of the traversal time. In IMPICA, we exploit this idle time by *decoupling* link address generation from the issuing and servicing of a memory request, which allows the accelerator to generate addresses for one link traversal stream while waiting on the request associated with a different link traversal stream to return from memory. We call this design *address-access decoupling*. Note that this form of decoupling bears resemblance to decoupled access/execute architectures [210–212], and we in fact take inspiration from past works [36, 102, 210–212], except our design is *specialized* for building a pointer chasing accelerator in 3D-stacked memory, and this paper solves specific challenges within the context of pointer chasing acceleration.

**The Address Translation Challenge** An in-memory pointer chasing accelerator must be able to perform address translation, as each pointer in a linked data structure node stores the *virtual* address of the next node, even though main memory is *physically* addressed. To determine the next address in the pointer chasing sequence, the accelerator must resolve the virtual-to-physical address mapping. As we discuss in Sect. 5.2.1, relying on existing CPU-side address translation mechanisms or duplicating the TLB and page walker within DRAM are impractical solutions.

We observe that traditional address translation techniques do *not* need to be employed for pointer chasing, as link traversals are (1) limited to linked data structures, and (2) touch only certain data structures in memory. We exploit this in IMPICA by allocating data structures accessed by IMPICA into contiguous *regions* within the virtual memory space, and designing a new address translation mechanism, the *region-based page table*, which is optimized for in-memory acceleration. Our approach provides translation within memory at low latency and low cost, while minimizing the cost of maintaining TLB coherence.

**Evaluation** By solving both key challenges, IMPICA provides significant performance and energy benefits for pointer chasing operations and applications that use such operations. First we examine three microbenchmarks, each of which performs pointer chasing on a widely used data structure (linked list, hash table, B-tree), and find that IMPICA improves their performance by 92%, 29%, and 18%, respectively, on a quad-core system over a state-of-the-art baseline. Second, we evaluate IMPICA on a real database workload, DBx1000 [241], on a quad-core system, and show that IMPICA increases *overall* database transaction throughput by 16% and reduces transaction latency by 13%. Third, IMPICA reduces *overall* system energy, by 41%, 23%, and 10% for the three microbenchmarks and by 6% for DBx1000. These benefits come at a very small hardware cost: our evaluations show that IMPICA comprises only 7.6% of the area of a small embedded core (the ARM Cortex-A57 [7]).

Our IMPICA proposal, originally published in the ICCD 2016 conference [67], makes the following major contributions:

- This is the first work to propose an in-memory accelerator for pointer chasing. Our proposal, IMPICA, accelerates linked data structure traversal by chasing pointers inside the logic layer of 3D-stacked memory, thereby eliminating inefficient, high-latency serialized data transfers between the CPU and main memory.
- We identify two fundamental challenges in designing an efficient in-memory pointer chasing accelerator (Sect. 5.3.2). These challenges can greatly hamper performance if the accelerator is not designed *carefully* to overcome them. First, multiple streams of link traversal can unnecessarily get serialized at the accelerator, thereby degrading performance (the *parallelism challenge*). Second, an in-memory accelerator needs to perform virtual-to-physical address translation for each pointer, but this critical functionality does *not* exist on the memory side (the *address translation challenge*).
- IMPICA solves the *parallelism challenge* by decoupling link address generation from memory accesses, and utilizes the idle time during memory accesses to service *multiple* pointer chasing streams simultaneously. We call this approach *address-access decoupling* (Sect. 5.3.3.1).
- IMPICA solves the *address translation challenge* by allocating data structures it accesses into contiguous virtual memory regions, and using an optimized and low-cost *region-based page table* structure for address translation (Sect. 5.3.3.3).
- We evaluate IMPICA extensively using both microbenchmarks and a real database workload. Our results (Sect. 5.3.6) show that IMPICA improves both system performance and energy efficiency for all of these workloads, while requiring only very modest hardware overhead in the logic layer of 3D-stacked DRAM.

## *5.3.1 Motivation*

To motivate the need for a pointer chasing accelerator, we first examine the usage of pointer chasing in contemporary workloads. We then discuss opportunities for acceleration within 3D-stacked memory.

### 5.3.1.1 Pointer Chasing in Modern Workloads

Pointers are ubiquitous in fundamental data structures such as linked lists, trees, and hash tables, where the nodes of the data structure are linked together by storing the addresses (i.e., pointers) of neighboring nodes. Pointers make it easy to dynamically add/delete nodes in these data structures, but link traversal is often serialized, as the address of the next node can be known only after the current node is fetched. The serialized link traversal is commonly referred to as *pointer chasing*.

Due to the flexibility of insertion/deletion, pointer-based data structures and link traversal algorithms are essential building blocks in programming, and they enable a very wide range of workloads. For instance, at least seven different types of modern data-intensive applications rely *heavily* on linked data structures: (1) **databases and file systems** use B/$B^+$-trees for indexing tables or metadata [46, 53, 54, 183]; (2) **in-memory caching** applications based on key-value stores, such as Memcached [50] and Masstree [142], use linked lists to resolve hash table collisions and tree-like $B^+$-trees as their main data structures; (3) **graph processing workloads** use pointers to represent the edges that connect the vertex data structures together [2, 209]; (4) **garbage collectors** in high level languages typically maintain reference relations using trees [76, 77, 227]; (5) **3D video games** use binary space partitioning trees to determine the objects that need to be rendered [76, 164]; (6) **dynamic routing tables** used by networks employ balanced search trees for high-performance IP address lookups [224]; and (7) **hash table based DNA read mappers** that store and find potential locations of a read in a reference genome index [5, 6, 96, 98, 115, 235, 236].

While linked data structures are widely used in many modern applications, chasing pointers is very inefficient in general-purpose processors. There are three major reasons behind the inefficiency. First, the inherent serialization that occurs when accessing consecutive nodes limits the available instruction-level and memory-level parallelism [78, 128, 136, 153–159, 186, 187]. As a result, out-of-order execution provides only limited performance benefit when chasing pointers [155–158]. Second, as nodes can be inserted and removed dynamically, they can get allocated to different regions of memory. The irregular memory allocation causes pointer chasing to exhibit irregular access patterns, which lead to frequent cache and TLB misses [40, 78, 83, 155, 238]. Third, for data structures that diverge at each node, such as B-trees, link traversals often go down different paths during different iterations, as the inputs to the traversal function change. As a result, lower-level nodes that were recently referenced during a link traversal are unlikely to be reused in subsequent traversals, limiting the effectiveness of many caching policies [99, 128, 136], such as LRU replacement.

To quantify the performance impact of chasing pointers in real-world workloads, we profile two popular applications that heavily depend on linked data structures, using a state-of-the-art Intel Xeon system[1]: (1) *Memcached* [50], an in-memory caching system, using a real Twitter dataset [48] as its input; and (2) *DBx1000* [241], an in-memory database system, using the TPC-C benchmark [223] as its input. We profile the pointer chasing code within the application separately from other parts

[1]We use the Intel® VTune™ profiling tool on a machine with a Xeon® W3550 processor (3 GHz, 8-core, 8 MB LLC) [73] and 18 GB memory. We profile each application for 10 min after it reaches steady state.

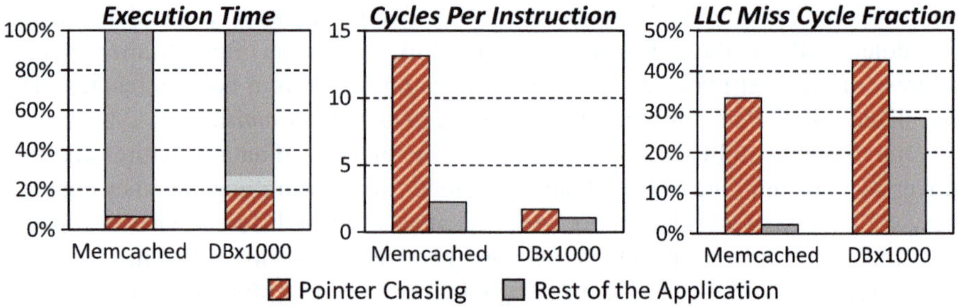


**Fig. 5.2** Profiling results of pointer chasing portions of code vs. the rest of the application code in Memcached and DBx1000. Figure adapted from [67]

of the application code. Figure 5.2 shows how pointer chasing compares to the rest of the application in terms of execution time, cycles per instruction (CPI), and the ratio of last-level cache (LLC) miss cycles to the total cycles.

We make three major observations from Fig. 5.2. First, both Memcached and DBx1000 spend a significant fraction of their total execution time (7% and 19%, respectively) on pointer chasing, as a result of dependent cache misses [63, 155, 157, 187]. Though these percentages might sound small, real software often does *not* have a single type of operation that consumes this significant a fraction of the total time. Second, we find that pointer chasing is significantly more inefficient than the rest of the application, as it requires much higher cycles per instruction (6× in Memcached, and 1.6× in DBx1000). Third, pointer chasing is largely memory-bound, as it exhibits much higher cache miss rates than the rest of the application and as a result spends a much larger fraction of cycles waiting for LLC misses (16× in Memcached, and 1.5× in DBx1000). From these observations, we conclude that (1) pointer chasing consumes a significant fraction of execution time in two important sophisticated applications, (2) pointer chasing operations are bound by memory, and (3) executing pointer chasing code in a modern general-purpose processor is very inefficient and thus can lead to a large performance overhead. Other works made similar observations for different workloads [35, 63, 155, 157, 187].

Prior works (e.g., [33–35, 40, 69, 70, 78, 83, 128, 135, 136, 155, 157, 186, 187, 213, 232, 238, 242, 248]) proposed specialized prefetchers that predict and prefetch the next node of a linked data structure to hide memory latency. While prefetching can mitigate part of the memory latency problem, it has three major shortcomings. First, the efficiency of a prefetcher degrades significantly when the traversal of linked data structures diverges into multiple paths and the access order is irregular [83, 155, 157]. Second, prefetchers can sometimes slow down the entire system due to contention caused by inaccurate as well as accurate prefetch requests [40–42, 62, 78, 109, 113, 204, 216]. Third, these specialized hardware prefetchers are usually designed for specific data structure implementations, and tend to be very inefficient when dealing with other data structures. For example, a Markov prefetcher [78] can potentially be very effective for static linked lists, but it becomes very inefficient for trees with dynamic access patterns. It is difficult to design a prefetcher that is efficient and effective for *all* types of linked data structures. **Our goal** in this work is to improve the performance of pointer chasing applications *without* relying on prefetchers, regardless of the types and access patterns of linked data structures used in an application.

#### 5.3.1.2 Accelerating Pointer Chasing in 3D-Stacked Memory

We propose to improve the performance of pointer chasing by leveraging processing-in-memory (PIM) to alleviate the memory bottleneck. Instead of sequentially fetching *each node* from memory and sending it to the CPU when an application is looking for a particular node, PIM-based pointer chasing consists of (1) traversing the linked data structures *in memory*, and (2) returning only the final node found to the CPU.

Unlike prior works that proposed general architectural models for in-memory computation by embedding logic in main memory [1, 3, 9, 20, 27, 47, 51, 52, 59, 62–64, 68, 81, 123, 131, 149, 163, 172, 195, 196, 200, 202, 203, 205, 221, 243], we propose to design a *specialized In-Memory PoInter Chasing Accelerator* (IMPICA) that exploits the logic layer of 3D-stacked memory [71, 72, 74, 75, 119]. 3D die-stacked memory achieves low latency (and high bandwidth) by stacking memory dies on top of a logic die, and interconnecting the layers using through-silicon vias (TSVs). Figure 5.3 shows a binary tree traversal using IMPICA, compared to a traditional architecture where the CPU traverses the binary tree. The traversal sequentially accesses the nodes from the root to a particular node (e.g., **H**→**E**→**A** in Fig. 5.3a). In a traditional architecture (Fig. 5.3b), these serialized accesses to the nodes miss in the caches and three memory requests are sent to memory serially across a high-latency off-chip channel. In contrast, IMPICA traverses the tree inside the logic layer of 3D-stacked memory, and as Fig. 5.3c shows, only the final node (**A**) is sent from the memory to the host CPU in response to the traversal request. Doing the traversal in memory minimizes both traversal latency (as queuing delays in the on-chip interconnect and the CPU-to-memory bus are eliminated) and off-chip bandwidth consumption, as shown in Fig. 5.3c.

Our accelerator architecture has three major advantages. First, it improves performance and reduces memory bandwidth consumption by eliminating the round trips required for memory accesses over the CPU-to-memory interconnects. Second, it frees the CPU to execute other work than linked data structure traversal, thereby increasing system throughput. Third, it minimizes the cache contention caused by pointer chasing operations.

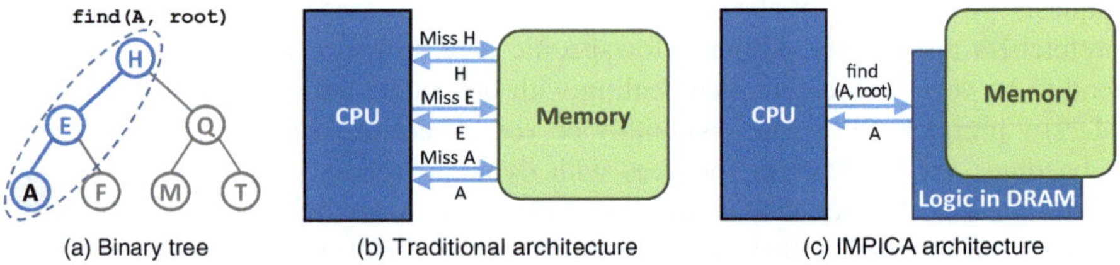


**Fig. 5.3** Pointer chasing (**a**) in a traditional architecture (**b**) and in IMPICA with 3D-stacked memory (**c**). Figure adapted from [67]

## *5.3.2 Design Challenges*

We identify and describe two new challenges that are crucial to the performance and functionality of our new pointer chasing accelerator in memory: (1) the *parallelism challenge*, and (2) the *address translation challenge*. Section 5.3.3 describes our IMPICA architecture, which centers around two key ideas that solve these two challenges.

### 5.3.2.1 Challenge 1: Parallelism in the Accelerator

A pointer chasing accelerator supporting a multicore system needs to handle *multiple* link traversals (from different cores) in parallel at low cost. A simple accelerator that can handle only one request at a time (which we call a *non-parallel accelerator*) would serialize the requests and could potentially be slower than using multiple CPU cores to perform the multiple traversals. As depicted in Fig. 5.4a, while a non-parallel accelerator speeds up each *individual* pointer chasing operation done by one of the CPU cores due to its shorter memory latency, the accelerator is slower *overall* for two pointer chasing operations, as *multiple cores* can operate in *parallel* on independent pointer chasing operations.

To overcome this deficiency, an in-memory accelerator needs to exploit parallelism when it services requests. However, the accelerator must do this *at low cost* and *low complexity*, due to its placement within the logic layer of 3D-stacked memory, where complex logic, such as out of order execution circuitry, is currently not feasible. The straightforward solution of adding multiple accelerators to service independent pointer chasing operations (e.g., [99]) does not scale well, and also can lead to excessive energy dissipation (and, thus, potentially thermal violations) and die area usage in the logic layer.

A key observation we make is that pointer chasing operations are bottlenecked by memory stalls, as shown in Fig. 5.2. In our evaluation, the memory access time is 10–15× the computation time (see Sect. 5.3.5 for our methodology). As a result, the accelerator spends a significant amount of time waiting for memory, causing its compute resources to sit idle. This makes typical in-order or out-of-order

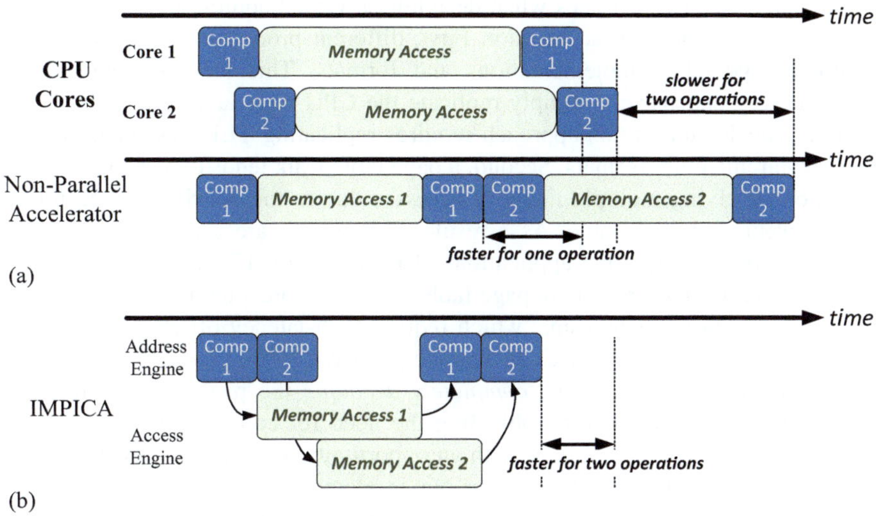


**Fig. 5.4** Execution time of two independent pointer chasing operations, broken down into address computation time (*Comp*) and memory access time. Figure adapted from [67]. (**a**) Pointer chasing on two CPU cores vs. one non-parallel accelerator. (**b**) Pointer chasing using our IMPICA proposal

execution engines *inefficient* for an in-memory pointer-chasing accelerator. If we utilize the hardware resources in a more efficient manner, we can enable parallelism by handling *multiple* pointer chasing operations *within a single accelerator*.

Based on our observation, we *decouple* address generation from memory accesses in IMPICA using two engines (the address engine and the access engine), allowing the accelerator to generate addresses from one pointer chasing operation while it *concurrently* performs memory accesses for a different pointer chasing operation (as shown in Fig. 5.4b). We describe the details of our decoupled accelerator design in Sect. 5.3.3.

#### 5.3.2.2 Challenge 2: Virtual Address Translation

A second challenge arises when pointer chasing is moved out of the CPU cores, which are equipped with facilities for address translation. Within the program data structures, each pointer is stored as a virtual address, and requires *translation* to a physical address before its memory access can be performed. This is a challenging task for an in-memory accelerator, which has no easy access to the virtual address translation engine that sits in the CPU core. While sequential array operations could potentially be constrained to work within page boundaries or directly in physical memory, indirect memory accesses that come with pointer-based data structures require some support for virtual memory address translation, as they might touch many parts of the virtual address space.

There are two major issues when designing a virtual address translation mechanism for an in-memory accelerator. First, different processor architectures have different page table implementations and formats. This lack of compatibility makes it very expensive to simply replicate the CPU page table walker in the in-memory accelerator as this approach requires replicating TLBs and page walkers for many architecture formats. Second, a page table walk tends to be a high-latency operation involving multiple memory accesses due to the heavily layered format of a conventional page table. As a result, TLB misses are a major performance bottleneck in data-intensive applications [10, 11, 13, 15–17, 139, 173, 175, 215]. If the accelerator requires many page table walks that are supported by the CPU's address translation mechanisms, which require high-latency off-chip accesses for the accelerator, its performance can degrade greatly.

To address these issues, we *completely decouple* the page table of IMPICA from that of the CPU, thereby obviating the need for compatibility between the two page tables. This presents us with an opportunity to develop a new page table design that is much more efficient for our in-memory accelerator. We make two key observations about the behavior of a pointer chasing accelerator. First, the accelerator operates only on certain data structures that can be mapped to *contiguous regions* in the virtual address space, which we refer to as *IMPICA regions*. As a result, it is possible to map contiguous IMPICA regions with a *smaller, region-based* page table without needing to duplicate the page table mappings for the *entire* address space. Second, we observe that if we need to map *only* IMPICA regions, we can collapse the hierarchy present in conventional page tables, which allows us to limit the hardware and storage overhead of the IMPICA page table. We describe the IMPICA page table in detail in Sect. 5.3.3.3.

## 5.3.3 IMPICA Architecture

We propose a new in-memory accelerator, IMPICA, that addresses the two design challenges that face in-memory accelerators for pointer chasing. The IMPICA architecture consists of a single specialized core designed to decouple address generation from memory accesses. Our approach, which we call *address-access decoupling*, allows us to *efficiently* overcome the parallelism challenge (Sect. 5.3.3.1). The IMPICA core uses a novel *region-based page table* design to perform efficient address translation locally in the accelerator, which allows us to overcome the address translation challenge (Sect. 5.3.3.3).

### 5.3.3.1 IMPICA Core Architecture

Our IMPICA core uses what we call address-access decoupling, where we separate the core into two parts: (1) an *address engine*, which generates the address specified by the pointer; and (2) an *access engine*, which performs memory access operations

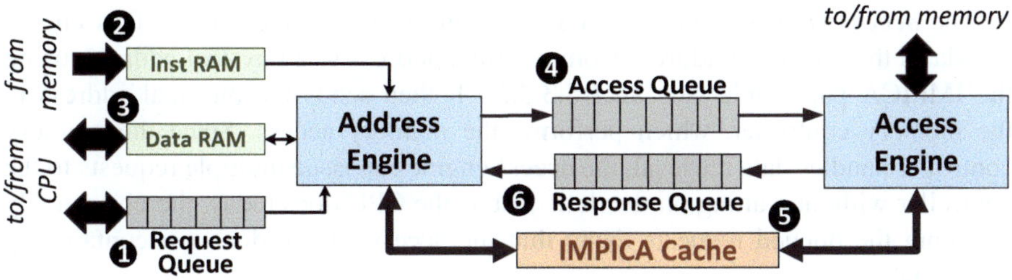


**Fig. 5.5** IMPICA core architecture. Figure adapted from [67]

using addresses generated by the address engine. The key advantage of this design is that the address engine supports fast context switching between multiple pointer chasing operations, allowing it to utilize the idle time during memory access(es) to compute addresses from a different pointer chasing operation. As Fig. 5.4b shows, an IMPICA core can process multiple pointer chasing operations faster than multiple cores because it has the ability to overlap address generation with memory accesses.

Our address-access decoupling has similarities to, and is in fact inspired by, the decoupled access-execute (DAE) architecture [210–212], with two key differences. First, the goal of DAE is to exploit instruction-level parallelism (ILP) within a *single* thread, whereas our goal is to exploit thread-level parallelism (TLP) across pointer chasing operations from *multiple* threads. Second, unlike DAE, the decoupling in IMPICA does not require any programmer or compiler effort. Our approach is much simpler than both general-purpose DAE and out-of-order execution [169, 170, 222], as it can switch between different independent execution streams, without the need for dependency checking.

Figure 5.5 shows the architecture of the IMPICA core. The host CPU initializes a pointer chasing operation by moving its code to main memory, and then enqueuing the request in the *request queue* (❶ in Fig. 5.5). Section 5.3.4.1 describes the details of the CPU interface.

The *address engine* services the enqueued request by loading the pointer chasing code into its *instruction RAM* (❷). This engine contains all of IMPICA's functional units, and executes the code in its instruction RAM while using its *data RAM* (❸) as a stack. All instructions that do not involve memory accesses, such as ALU operations and control flow, are performed by the address engine. The number of pointer chasing operations that can be processed in parallel is limited by the size of the stack in the data RAM.

When the address engine encounters a memory instruction, it enqueues the address (along with the data RAM stack pointer) into the *access queue* (❹), and then performs a *context switch* to an independent stream. For the switch, the engine pushes the hardware context (i.e., architectural registers and the program counter) into the data RAM stack. When this is done, the address engine can work on a different pointer chasing operation.

The *access engine* services requests waiting in the access queue. This engine translates the enqueued address from a virtual address to a physical address, using the IMPICA page table (see Sect. 5.3.3.3). It then sends the physical address to the memory controller, which performs the memory access. Since the memory controller handles data retrieval, the access engine can issue multiple requests to the controller without waiting on the data, just as the CPU does today, thereby quickly servicing the queued requests. Note that the access engine does *not* contain any functional units.

When the access engine receives data back from the memory controller, it stores this data in the *IMPICA cache* (❺), a small cache that contains data destined for the address engine. The access queue entry corresponding to the returned data is moved from the access queue to the *response queue* (❻).

The address engine monitors the response queue. When a response queue entry is ready, the address engine reads it, and uses the stack pointer to access and reload the registers and PC that were pushed onto the data RAM stack. It then resumes execution for the pointer chasing operation, continuing until it encounters the next memory instruction.

#### 5.3.3.2 IMPICA Cache

IMPICA uses a cache to deliver data fetched by the access engine to the address engine. The cache employs three features that cater to pointer-chasing applications. First, it uses *cache line locking* to guarantee that data is not displaced from the cache until the address engine processes the data. Cache line locking is achieved using a *lock bit* in the tag that is set when the cache line is inserted, and is cleared only after the address engine processes the associated entry in the response queue. If all of the cache lines in a set are locked, the access engine stalls until one of the cache lines becomes unlocked. Second, when a traversal is completed, the IMPICA cache *immediately* evicts cache lines fetched by that pointer-chasing operation. A *request ID* associated with the tag is used to determine if a cache line belongs to a completed task. Third, the IMPICA cache prioritizes nodes that closer to the root of the data structure in the cache, by leveraging the observation that pointer-based structures traverse multiple paths and usually do *not* re-reference the leaf nodes. To achieve this, the cache sets a **root bit** in the tag if a cache line is fetched by the first few memory accesses of a pointer-chasing operation. Figure 5.6 shows the structure of the IMPICA cache, including cache line metadata.

#### 5.3.3.3 IMPICA Page Table

IMPICA uses a *region-based* page table (RPT) design optimized for in-memory pointer chasing, leveraging the continuous ranges of accesses (*IMPICA regions*) discussed in Sect. 5.3.2.2. Figure 5.7 shows the structure of the RPT in IMPICA. The RPT is split into three levels: (1) a first-level *region table*, which needs to

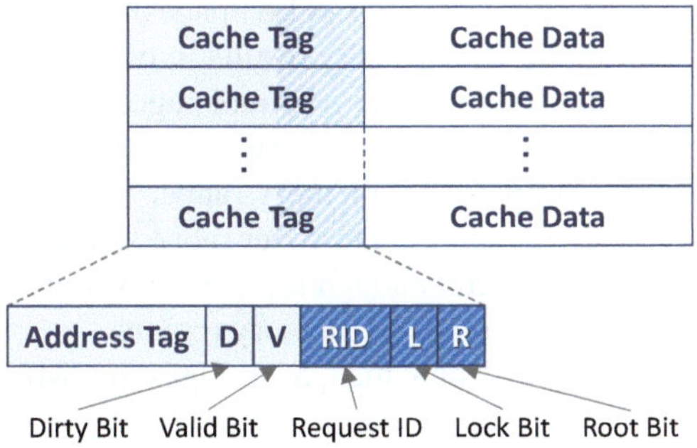


**Fig. 5.6** Structure of the IMPICA cache

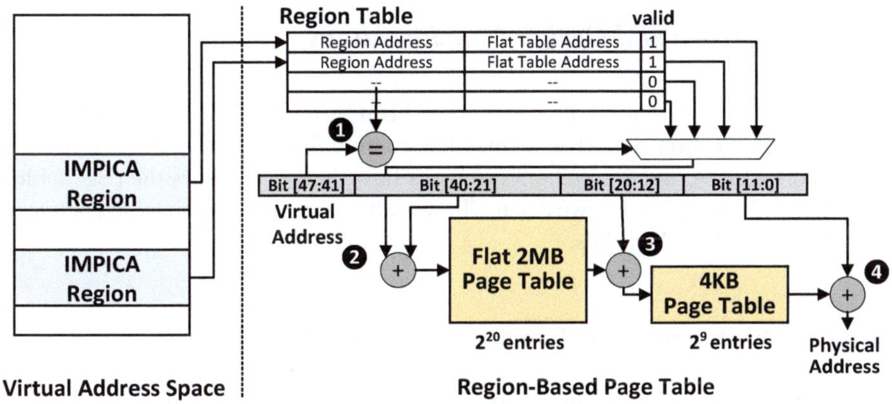


**Fig. 5.7** IMPICA virtual memory architecture. Figure adapted from [67]

map only a small number of the contiguously allocated IMPICA regions; (2) a second-level *flat page table* for each region with a larger (e.g., 2 MB) page size; and (3) third-level *small page tables* that use conventional small (e.g., 4 kB) pages. In the example in Fig. 5.7, when a 48-bit virtual memory address arrives for translation, bits 47–41 of the address are used to index the region table (❶ in Fig. 5.7) to find the corresponding flat page table. Bits 40–21 are used to index the flat page table (❷), providing the location of the small page table, which is indexed using bits 20–12 (❸). The entry in the small page table provides the physical page number of the page, and bits 11–0 specify the offset within the physical page (❹).

The RPT is optimized to take advantage of the properties of pointer chasing. The region table is almost always cached in the IMPICA cache, as the total number of IMPICA regions is small, requiring small storage (e.g., a 4-entry region table needs only 68 B of cache space). We employ a flat table with large (e.g., 2 MB) pages at the second level in order to reduce the number of page misses, though this requires more memory capacity than the conventional 4-level page table structure. As the number of regions touched by the accelerator is limited, this additional capacity overhead remains constrained. Our page table can optionally use traditional smaller page sizes to maximize memory management flexibility. The OS can freely choose large

(2 MB) pages or small (4 kB) pages at the last level. Thanks to this design, a page walk in the RPT usually results in only two misses, one for the flat page table and the other for the last-level small page table. This represents a 2× improvement over a conventional four-level page table, while our flattened page table still provides coverage for a 2 TB memory range. The size of the IMPICA region is configurable and can be increased to cover more virtual address space. We believe that our RPT design is general enough for use in a variety of in-memory accelerators that operate on a specific range of memory regions.

We discuss how the OS manages the IMPICA RPT in Sect. 5.3.4.3.

## 5.3.4 *Interface and Design Considerations*

In this section, we discuss how we expose IMPICA to the CPU and the operating system (OS). Section 5.3.4.1 describes the communication interface between the CPU and IMPICA. Section 5.3.4.3 discusses how the OS manages the page tables in IMPICA. In Sect. 5.3.4.4, we discuss how cache coherence is maintained between the CPU and IMPICA caches.

### 5.3.4.1 CPU Interface and Communication Model

We use a packet-based interface between the CPU and IMPICA. Instead of communicating individual operations or operands, the packet-based interface buffers requests and sends them in a burst to minimize the communication overhead. Executing a function in IMPICA consists of four steps on the interface. (1) The CPU sends to memory a packet comprising the function call and parameters. (2) This packet is written to a specific location in memory, which is memory-mapped to the *data RAM* in IMPICA and triggers IMPICA execution. (3) IMPICA loads the specific function into the *instruction RAM* with appropriate parameters, by reading the values from predefined memory locations. (4) Once IMPICA finishes the function execution, it writes the return value back to the memory-mapped locations in the *data RAM*. The CPU periodically polls these locations and receives the IMPICA output. Note that the IMPICA interface is similar to the interface proposed for the Hybrid Memory Cube (HMC) [2, 71, 72].

### 5.3.4.2 IMPICA Programming Model

The programming model for IMPICA is similar to the CPU programming model. An IMPICA program can be written in C with a new API that handles passing the parameters and returning the results to the IMPICA accelerator. Figure 5.8a shows the pseudocode for a B-tree traversal in the CPU, and Fig. 5.8b shows the equivalent pseudocode for IMPICA. We observe that the code fragments are very similar,

```
bt_node *find_leaf(bt_node *root, uint64_t key)
{
    bt_node *c = root;
    uint64_t = i;

    while (!c->is_leaf) {
        for (i = 0; i < c->num_keys; i++) {
            if (key < c->keys[i])
                break;
        }
        c = c->pointers[i];
    }
    return c;
}
```

(a) Conventional B-tree traversal pseudocode

❶ Get parameters with special API

❷ Write results with special API

```
void pica_find_leaf()
    bt_node *c = __param(ROOT);
    uint64_t key = __param(KEY);
    uint64_t i;

    while (!c->is_leaf) {
        for (i = 0; i < c->num_keys; i++) {
            if (key < c->keys[i])
                break;
        }
        c = c->pointers[i];
    }
    __param(RESULT) = c;
}
```

(b) B-tree traversal pseudocode in IMPICA

**Fig. 5.8** B-tree traversal pseudocode demonstrating the differences between the (**a**) conventional and (**b**) IMPICA programming models

differing in only two places. First, the parameters passed in the function call of the CPU code are accessed with the `__param` API call in IMPICA (❶ in Fig. 5.8). The `__param` API call ensures that the program explicitly reads the parameters from the predefined memory-mapped locations of the data RAM. Second, instead of using the `return` statement, IMPICA uses the same `__param` API call to write the return value to a specific memory location (❷). This API call makes sure that the CPU can receive the output through the IMPICA interface.

#### 5.3.4.3 Page Table Management

In order for the RPT to identify IMPICA regions, the regions must be tagged by the application. For this, the application uses a special API to allocate pointer-based data structures. This API allocates memory to a contiguous virtual address space. To ensure that all API allocations are contiguous, the OS reserves a portion of the unused virtual address space for IMPICA, and always allocates memory for IMPICA regions from this portion. The use of such a special API requires minimal changes to applications, and it allows the system to provide more efficient virtual address translation. This also allows us to ensure that when multiple memory stacks are present within the system, the OS can allocate all IMPICA regions belonging to a single application (along with the associated IMPICA page table) into one memory stack, thereby avoiding the need for the accelerator to communicate with a remote memory stack.

The OS maintains coherence between the IMPICA RPT and the CPU page table. When memory is allocated in the IMPICA region, the OS allocates the IMPICA page table. The OS also shoots down TLB entries in IMPICA if the CPU performs any updates to IMPICA regions. While this makes the OS page fault handler more complex, the additional complexity does not cause a noticeable performance impact, as page faults occur rarely and take a long time to service in the CPU.

#### 5.3.4.4 Cache Coherence

Coherence must be maintained between the CPU and IMPICA caches, and with memory, to avoid using stale data and thus ensure correct execution. We maintain coherence by executing *every* function that operates on the IMPICA regions in the accelerator. This solution guarantees that no data is shared between the CPU and IMPICA, and that IMPICA always works on up-to-date data. Other PIM coherence solutions (e.g., LazyPIM in Sect. 5.4, or those proposed by prior works [3, 52]) can also be used to allow the CPU to update the linked data structures, but we choose not to employ these solutions in our evaluation, as our workloads do *not* perform any such updates.

#### 5.3.4.5 Handling Multiple Memory Stacks

Many systems need to employ multiple memory stacks to have enough memory capacity, as the current die-stacking technology can integrate only a limited number of DRAM dies into a single memory stack [145]. In systems that use multiple memory stacks, the efficiency of an in-memory accelerator such as IMPICA could be significantly degraded whenever the data that the accelerator accesses is placed on different memory stacks. Without any modifications, IMPICA would have to go through the off-chip memory channels to access the data, which would effectively eliminate the benefits of in-memory computation.

**Table 5.1** Major simulation parameters used for IMPICA evaluations

| | |
|---|---|
| *Baseline main processor (CPU)* | |
| ISA | ARMv8 (64-bits) |
| Core configuration | 4 OoO cores, 2 GHz, 8-wide issue, 128-entry ROB |
| Operating system | 64-bit Linux from Linaro [127] |
| L1 I/D cache | 32 kB/2-way each, 2-cycle |
| L2 cache | 1 MB/8-way, shared, 20-cycle |
| *Baseline main memory parameters* | |
| Memory configuration | DDR3-1600, 8 banks/device, FR-FCFS scheduler [182, 249] |
| DRAM bus bandwidth | 12.8 GB/s for CPU, 51.2 GB/s for IMPICA |
| *IMPICA accelerator* | |
| Accelerator core | 500 MHz, 16 entries for each queue |
| Cache[a] | 32 kB/2-way |
| Address translator | 32 TLB entries with region-based page table |
| RAM | 16 kB data RAM and 16 kB instruction RAM |

[a]Based on our experiments on a real Intel Xeon machine, we find that this is large enough to satisfactorily represent the behavior of 1,000,000 transactions

Fortunately, this challenge can be tackled with our proposed modifications to the operating system (OS) in Sect. 5.3.4.3. As we can identify the memory regions that IMPICA needs to access, the OS can easily map *all IMPICA regions* of an application into the same memory stack. In addition, the OS can allocate all IMPICA page tables into the same memory stack. This ensures that an IMPICA accelerator can access all of that data that it needs from within the memory stack that it resides in without incurring any additional hardware cost or latency overhead.

### *5.3.5 Evaluation Methodology for IMPICA*

We use the gem5 [18] full-system simulator with DRAMSim2 [185] to evaluate our proposed design. We choose the 64-bit ARMv8 architecture, the accuracy of which has been validated against real hardware [60]. We conservatively model the internal memory bandwidth of the memory stack to be 4× that of the external bandwidth, similar to the configuration used in prior works [47, 243]. Our simulation parameters are summarized in Table 5.1. Our source code is available openly at our research group's GitHub site [188, 190]. This distribution includes the source code of our microbenchmarks as well.

#### 5.3.5.1 Workloads

We use three data-intensive microbenchmarks, which are essential building blocks in a wide range of workloads, to evaluate the native performance of pointer-chasing operations: linked lists, hash tables, and B-trees. We also evaluate the performance improvement in a real data-intensive workload, measuring the transaction latency and transaction throughput of DBx1000 [241], an in-memory OLTP database. We modify all four workloads to offload each pointer chasing request to IMPICA. To minimize communication overhead, we map the IMPICA registers to user mode address space, thereby avoiding the need for costly kernel code intervention.

**Linked List** We use the linked list traversal microbenchmark [247] derived from the *health* workload in the Olden benchmark suite [184]. The parameters are configured to approximate the performance of the *health* workload. We measure the performance of the linked list traversal after 30,000 iterations.

**Hash Table** We create a microbenchmark from the hash table implementation of *Memcached* [50]. The hash table in Memcached resolves hash collisions using chaining via linked lists. When there are more than $1.5n$ items in a table of $n$ buckets, it doubles the number of buckets. We follow this rule by inserting $1.5 \times 2^{20}$ random keys into a hash table with $2^{20}$ buckets. We run evaluations for 100,000 random key look-ups.

**B-Tree** We use the B-tree implementation of DBx1000 for our B-tree microbenchmark. It is a 16-way B-tree that uses a 64-bit integer as the key of each node. We randomly generate 3,000,000 keys and insert them into the B-tree. After the insertions, we measure the performance of the B-tree traversal with 100,000 random keys. This is the most time-consuming operation in the database index lookup.

**DBx1000** We run DBx1000 [241] with the TPC-C benchmark [223]. We set up the TPC-C tables with 2000 customers and 100,000 items. For each run, we spawn four threads and bind them to four different CPUs to achieve maximum throughput. We run each thread for a warm-up period for the duration of 2000 transactions, and then record the software and hardware statistics for the next 5000 transactions per thread,[2] which takes 300–500 million CPU cycles.

#### 5.3.5.2 Die Area and Energy Estimation

We estimate the die area of the IMPICA processing logic at the 40 nm process node based on recently published work [134]. We include the most important components: processor cores, L1/L2 caches, and the memory controller. We use the area of ARM Cortex-A57 [7, 49], a small embedded processor, for the baseline main CPU. We *conservatively* estimate the die area of IMPICA using the area of the Cortex-R4 [8], an 8-stage dual issue RISC processor with 32 kB I/D caches. We believe the actual area of an optimized IMPICA design can be much smaller. Table 5.2 lists the area estimate of each component.

[2]We sweep the size of the IMPICA cache from 32 to 128 kB, and find that it has negligible effect on our results.

**Table 5.2** Die area estimates using a 40 nm process for IMPICA evaluations

| | |
|---|---|
| Baseline CPU core (Cortex-A57) | 5.85 $mm^2$ per core |
| L2 cache | 5 $mm^2$ per MB |
| Memory controller | 10 $mm^2$ |
| Complete baseline chip | 38.4 $mm^2$ |
| IMPICA Core (including 32 kB I/D caches) | 0.45 $mm^2$ (1.2% of the baseline chip area) |

IMPICA comprises only 7.6% the area of a single baseline main CPU core, or only 1.2% the total area of the baseline chip (which includes four CPU cores, 1 MB L2 cache, and one memory controller). Note that we conservatively model IMPICA as a RISC core. A much more specialized engine can be designed for IMPICA to solely execute pointer chasing code. Doing so would reduce the area and energy overheads of IMPICA greatly, but can reduce the generality of the pointer chasing access patterns that IMPICA can accelerate. We leave such optimizations, evaluations, and analyses for future work.

We use McPAT [122] to estimate the energy consumption of the CPU, caches, memory controllers, and IMPICA. We conservatively use the configuration of the Cortex-R4 to estimate the energy consumed by IMPICA. We use DRAMSim2 [185] to analyze DRAM energy.

## *5.3.6 Evaluation of IMPICA*

We first evaluate the effect of IMPICA on system performance, using both our microbenchmarks (Sect. 5.3.6.1) and the DBx1000 database (Sect. 5.3.6.2). We investigate the impact of different IMPICA page table designs in Sect. 5.3.6.3, and examine system energy consumption in Sect. 5.3.6.4. We compare a system containing IMPICA to an accelerator-free baseline that includes an additional 128 kB of L2 cache (which is equivalent to the area of IMPICA) to ensure area-equivalence across evaluated systems.

### 5.3.6.1 Microbenchmark Performance

Figure 5.9 shows the speedup of IMPICA and the baseline with extra 128 kB of L2 cache over the baseline for each microbenchmark. IMPICA achieves significant speedups across all three data structures—1.92× for the linked list, 1.29× for the hash table, and 1.18× for the B-tree. In contrast, the extra 128 kB of L2 cache provides very small speedup (1.03×, 1.01×, and 1.02×, respectively). We conclude that IMPICA is much more effective than the area-equivalent additional L2 cache for pointer chasing operations.

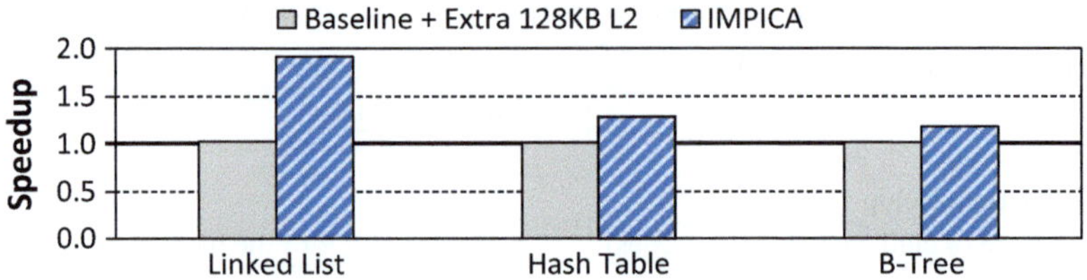


**Fig. 5.9** Microbenchmark performance with IMPICA. Figure adapted from [67]

To provide insight into why IMPICA improves performance, we present total (i.e., combined CPU and IMPICA) TLB misses per kilo instructions (MPKI), cache miss latency, and total memory bandwidth usage for these microbenchmarks in Fig. 5.10. We make three observations.

First, a major factor contributing to the performance improvement is the reduction in TLB misses. The TLB MPKI in Fig. 5.10a depicts the total (i.e., combined CPU and IMPICA) TLB misses in both the baseline system and IMPICA. The pointer chasing operations have low locality and pollute the CPU TLB. This leads to a higher overall TLB miss rate in the application. With IMPICA, the pointer chasing operations are offloaded to the accelerator. This reduces the pollution and contention at the CPU TLB, reducing the overall number of TLB misses. The linked list has a significantly higher TLB MPKI than the other data structures because linked list traversal requires far fewer instructions in an iteration. It simply accesses the next pointer, while a hash table or a B-tree traversal needs to compare the keys in the node to determine the next step.

Second, we observe a significant reduction in last-level cache miss latency with IMPICA. Figure 5.10b compares the average cache miss latency between the baseline last-level cache and the IMPICA cache. On average, the cache miss latency of IMPICA is only 60–70% of the baseline cache miss latency. This is because IMPICA leverages the faster and wider TSVs in 3D-stacked memory as opposed to the narrow, high-latency DRAM interface used by the CPU.

Third, as Fig. 5.10c shows, IMPICA effectively utilizes the internal memory bandwidth of 3D-stacked memory, which is cheap and abundant. There are two reasons for high bandwidth utilization: (1) IMPICA runs much faster than the baseline so it generates more traffic within the same amount time; and (2) IMPICA always accesses memory at a larger granularity, retrieving each full node in a linked data structure with a single memory request, while a CPU issues multiple requests for each node as it can fetch only one cache line at a time. The CPU can avoid using some of its limited memory bandwidth by skipping some fields in the data structure that are not needed for the current loop iteration. For example, some keys and pointers in a B-tree node can be skipped whenever a match is found. In contrast, IMPICA utilizes the wide internal bandwidth of 3D-stacked memory to retrieve a full node on each access.

We conclude that IMPICA is effective at significantly improving the performance of important linked data structures.

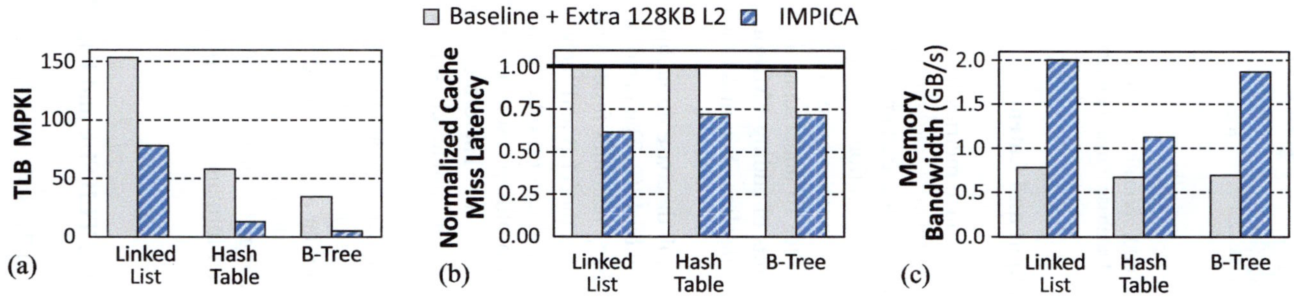


**Fig. 5.10** Key architectural statistics for the evaluated microbenchmarks. Figure adapted from [67]. (**a**) Total TLB misses per kilo-instruction (MPKI). (**b**) Average last-level cache miss latency. (**c**) Total DRAM memory bandwidth utilization

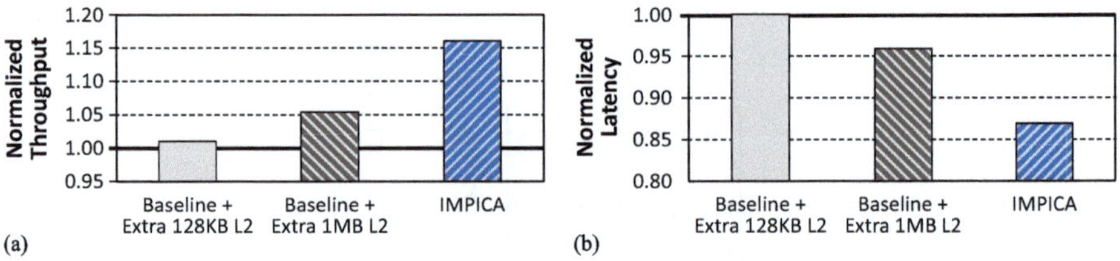


**Fig. 5.11** Performance results for DBx1000, normalized to the baseline. Figure adapted from [67]. (**a**) Database transaction throughput. (**b**) Database transaction latency

#### 5.3.6.2 Real Database Throughput and Latency

Figure 5.11 presents two key performance metrics for our evaluation of DBx1000: *database throughput* and *database latency*. *Database throughput* represents how many transactions are completed within a certain period, while *database latency* is the average time to complete a transaction. We normalize the results of three configurations to the baseline. As mentioned earlier, the die area increase of IMPICA is similar to a 128 kB cache. To understand the effect of additional LLC space better, we also show the results of adding 1 MB of cache, which takes about 8× the area of IMPICA, to the baseline. We make two observations from our analysis of DBx1000.

First, IMPICA improves the overall database throughput by 16% and reduces the average database transaction latency by 13%. The performance improvement is due to three reasons: (1) database indexing becomes faster with IMPICA, (2) offloading database indexing to IMPICA reduces the TLB and cache contention due to pointer chasing in the CPU, and (3) the CPU can do other useful tasks in parallel while waiting for IMPICA. Note that our profiling results in Fig. 5.2 show that DBx1000 spends 19% of its time on pointer chasing. Therefore, a 16% overall improvement is very close to the upper bound that *any* pointer chasing accelerator can achieve for this database.

Second, IMPICA yields much higher database throughput than simply providing additional cache capacity. IMPICA improves the database throughput by 16%, while an extra 128 kB of cache (with a similar area overhead as IMPICA) does so by only 2%, and an extra 1 MB of cache (8× the area of IMPICA) by only 5%.

We conclude that by accelerating the fundamental pointer chasing operation, IMPICA efficiently improves the performance of a sophisticated real data-intensive workload.

#### 5.3.6.3 Sensitivity to the IMPICA TLB Size and Page Table Design

To understand the effect of different TLB sizes and page table designs in IMPICA, we evaluate the speedup in the amount of time spent on address translation for IMPICA when different IMPICA TLB sizes (32 and 64 entries) and accelerator page

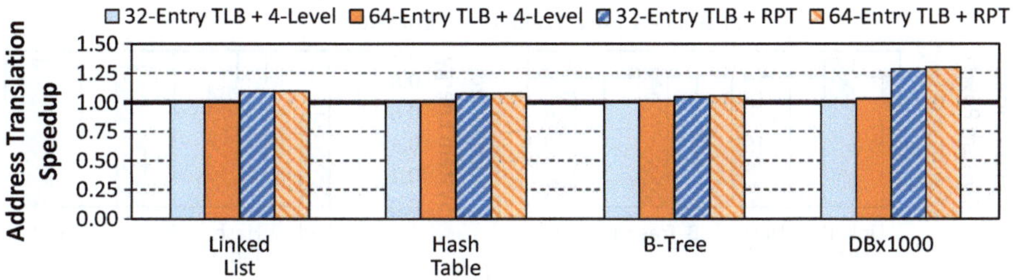


**Fig. 5.12** Speedup of address translation with different TLB sizes and page table designs. Figure adapted from [67]

table structures (the baseline 4-level page table; and the region-based page table, or RPT) are used inside the accelerator. Figure 5.12 shows the speedup in address translation time relative to IMPICA with a 32-entry TLB and the conventional 4-level page table. Two observations are in order.

First, the performance of IMPICA is largely unaffected from small changes in the IMPICA TLB size. Doubling the IMPICA TLB entries from 32 to 64 barely improves the address translation time. This observation reflects the irregular nature of pointer chasing. Second, the benefit of the RPT is much more significant in a sophisticated workload (DBx1000) than in microbenchmarks. This is because the working set size of the microbenchmarks is much smaller than that of the database system. When the working set is small, the operating system needs only a small number of page table entries in the first and second levels of a traditional page table. These entries are used frequently, so they stay in the IMPICA cache much longer, reducing the address translation overhead. This caching benefit goes away with a larger working set, which would require a significantly larger TLB and IMPICA cache to reap locality benefits. The benefit of RPT is more significant in such a case because RPT does not rely on this caching effect. Its region table is *always* small irrespective of the workload working set size and it has fewer page table levels. Thus, we conclude that RPT is a much more efficient and high-performance page table design for our IMPICA accelerator than conventional page table design.

#### 5.3.6.4 Energy Efficiency

Figure 5.13 shows the system power and system energy consumption for the microbenchmarks and DBx1000. We observe that the overall system *power* increases by 5.6% on average, due to the addition of IMPICA and higher utilization of internal memory bandwidth. However, as IMPICA significantly reduces the execution time of the evaluated workloads, the overall system *energy* consumption reduces by 41%, 24%, and 10% for the microbenchmarks, and by 6% for DBx1000. We conclude that IMPICA is an energy-efficient accelerator for pointer chasing.

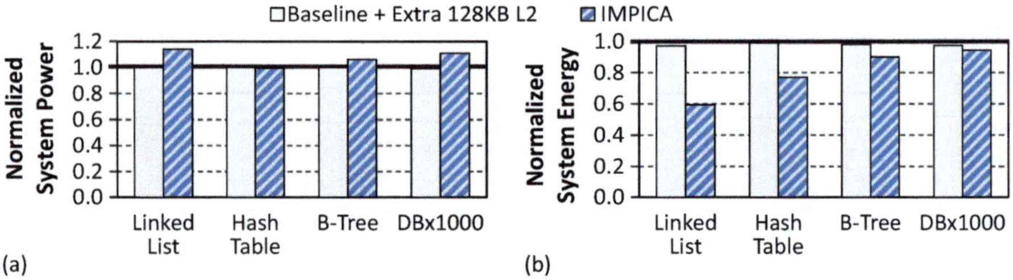


**Fig. 5.13** Effect of IMPICA on system power (**a**) and system energy consumption (**b**). Figure (**b**) adapted from [67]

### *5.3.7 Summary of IMPICA*

We introduce the design and evaluation of an *in-memory accelerator*, called IMPICA, for performing pointer chasing operations in 3D-stacked memory. We identify two major challenges in the design of such an in-memory accelerator: (1) the *parallelism challenge* and (2) the *address translation challenge*. We provide new solutions to these two challenges: (1) *address-access decoupling* solves the parallelism challenge by decoupling the address generation from memory accesses in pointer chasing operations and exploiting the idle time during memory accesses to execute multiple pointer chasing operations in parallel, and (2) the *region-based page table* in 3D-stacked memory solves the address translation challenge by tracking only those limited set of virtual memory regions that are accessed by pointer chasing operations. Our evaluations show that for both commonly-used linked data structures and a real database application, IMPICA significantly improves both performance and energy efficiency. We conclude that IMPICA is an efficient and effective accelerator design for pointer chasing. We also believe that the two challenges we identify (parallelism and address translation) exist in various forms in other in-memory accelerators (e.g., for graph processing), and, therefore, our solutions to these challenges can be adapted for use by a broad class of (in-memory) accelerators. We believe ample future work potential exists on examining other solutions for these two challenges as well as our solutions for them within the context of other in-memory accelerators, such as those described in [2, 22, 68, 96, 98, 195, 196, 200, 202]. We also believe that examining solutions like IMPICA for other, non-in-memory accelerators is a promising direction to examine.

## 5.4 LazyPIM: An Efficient Cache Coherence Mechanism for Processing-in-Memory

As discussed in Sect. 5.2.2, cache coherence is a major challenge for PIM architectures, as traditional coherence cannot be performed along the off-chip memory channel without potentially undoing the benefits of high-bandwidth and low-energy

PIM execution. To work around the limitations presented by cache coherence, most prior works assume a limited amount of sharing between the PIM kernels and the processor threads of an application. Thus, they sidestep coherence by employing solutions that restrict PIM to execute on non-cacheable data (e.g., [2, 47, 51, 149, 243]) or force processor cores to flush or not access any data that could *potentially* be used by PIM (e.g., [3, 4, 27, 47, 52, 59, 67, 68, 163, 172, 195, 196, 200, 202]). In fact, the IMPICA accelerator design, described in Sect. 5.3, falls into the latter category.

To understand the trade-offs that can occur by sidestepping coherence, we analyze several data-intensive applications. We make two *key observations* based on our analysis: (1) some portions of the applications are better suited for execution in processor threads, and these portions often concurrently access the same region of data as the PIM kernels, leading to *significant data sharing*; and (2) poor handling of coherence eliminates a significant portion of the performance benefits of PIM. As a result, we find that a good coherence mechanism is *required* to ensure the correct execution of the program while maintaining the benefits of PIM (see Sect. 5.4.2). **Our goal** in this section is to describe a cache coherence mechanism for PIM architectures that *logically behaves* like traditional coherence, but retains all of the benefits of PIM.

To this end, we propose *LazyPIM*, a new cache coherence mechanism that efficiently batches coherence messages sent by the PIM processing logic. During PIM kernel execution, a PIM core *speculatively* assumes that it has acquired coherence permissions without sending a coherence message, and maintains all data updates speculatively in its cache. Only when the kernel finishes execution, the processor receives compressed information from the PIM core, and checks if any coherence conflicts occurred. If a conflict exists (see Sect. 5.4.3), the dirty cache lines in the processor are flushed, and the PIM core rolls back and re-executes the kernel. Our execution model *for PIM processing logic* is similar to *chunk-based execution* [24] (i.e., each *batch* of consecutive instructions executes atomically), which prior work has harnessed for various purposes [24, 38, 61, 161, 174, 192]. Unlike past works, however, the processor in LazyPIM executes conventionally and *never rolls back*, which can make it easier to enable PIM.

We make the following key contributions in this section:

- We propose a new hardware coherence mechanism for PIM. Our approach (1) reduces the off-chip traffic between the PIM cores and the processor, (2) avoids the costly overheads of prior approaches to provide coherence for PIM, and (3) retains the same logical coherence behavior as architectures without PIM to keep programming simple.
- LazyPIM improves average performance by 49.1% (coming within 5.5% of an ideal PIM mechanism), and reduces off-chip traffic by 58.8%, over the best prior coherence approach.

### *5.4.1 Baseline PIM Architecture*

In our evaluation, we assume that the compute units inside memory consist of simple *in-order* cores. These PIM cores, which are ISA-compatible with the out-of-order processor cores, are much weaker in terms of performance, as they lack large caches and sophisticated ILP techniques, but are more practical to implement within the DRAM logic layer, as we discussed earlier in Sect. 5.1.1. Each PIM core has private L1 I/D caches, which are kept coherent using a MESI directory [23, 167] within the DRAM logic layer. A second directory in the processor acts as the main coherence point for the system, interfacing with both the processor cache and the PIM coherence directory. Like prior PIM works [2–4, 47, 51, 68, 172], we assume that direct segments [13] are used for PIM data, and that PIM kernels operate only on physical addresses.

### *5.4.2 Motivation for Coherence Support in PIM*

Applications benefit the most from PIM execution when their memory-intensive parts, which often exhibit poor locality and contribute to a large portion of execution time, are dispatched to PIM processing logic. On the other hand, compute-intensive parts or those parts that exhibit high locality *must remain on the processor cores* to maximize performance [3, 68].

Prior work mostly assumes that there is only a limited amount of sharing between the PIM kernels and the processor. However, *this is not the case* for many important applications, such as graph and database workloads. For example, in multithreaded graph frameworks, each thread performs a graph algorithm (e.g., connected components, PageRank) on a shared graph [2, 209, 237]. We study a number of these algorithms [209], and find that (1) only certain portions of each algorithm are well suited for PIM, and (2) the PIM kernels and processor threads access the shared graph and intermediate data structures concurrently. Another example is modern in-memory databases that support Hybrid Transactional/Analytical Processing (HTAP) workloads [144, 193, 203, 219]. The analytical portions of these databases are well suited for PIM execution [99, 148, 234]. In contrast, even though transactional queries access the *same* data, they perform better if they are executed on the main processor (i.e., the CPU), as they are short-lived and latency sensitive, accessing only a few rows each. Thus, concurrent accesses from both PIM kernels (analytics) and processor threads (transactions) are inevitable.

The shared data needs to remain coherent between the processor and PIM cores. Traditional, or *fine-grained*, coherence protocols (e.g., MESI [23, 167]) have several qualities well suited for pointer-intensive data structures, such as those in graph workloads and databases. Fine-grained coherence allows the processor or PIM to acquire permissions for only the pieces of data that are *actually accessed*. In addition, fine-grained coherence can ease programmer effort when developing

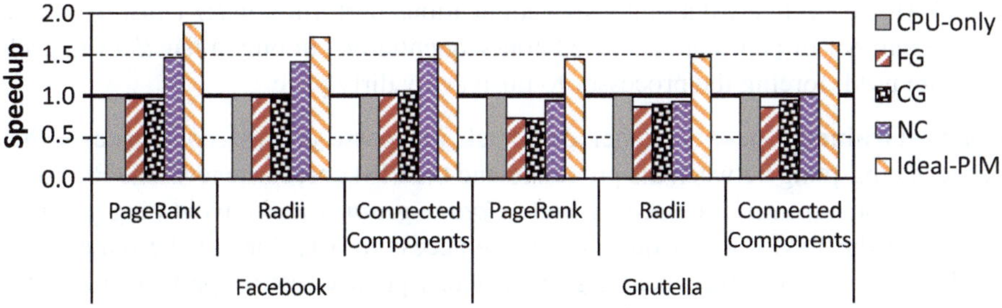


**Fig. 5.14** PIM speedup with 16 threads, normalized to CPU-only, with three different and ideal coherence mechanisms. Figure adapted from [20]

PIM applications, as multithreaded programs already use this programming model. Unfortunately, if a PIM core participates in traditional coherence, it would have to send a message for *every cache miss* to the processor over a narrow shared interconnect (we call this type of interconnect traffic as *PIM coherence traffic*).

We study four mechanisms to evaluate how coherence protocols impact PIM: (1) *CPU-only*, a baseline where PIM is disabled; (2) *FG*, fine-grained coherence, which is the MESI protocol, variants of which are employed in many state-of-the-art systems; (3) *CG*, coarse-grained lock based coherence, where PIM cores gain *exclusive* access to all PIM data during PIM kernel execution; and (4) *NC*, non-cacheable, where the PIM data is not cacheable in the CPU. We describe CG and NC in more detail below. Figure 5.14 shows the speedup of PIM with these four mechanisms for certain graph workloads, normalized to CPU-only.[3] To illustrate the impact of inefficient coherence mechanisms, we also show the performance of an *ideal* mechanism where there is no performance penalty for coherence (*Ideal-PIM*). As shown in Fig. 5.14, employing PIM with a state-of-the-art fine-grained coherence (*FG*) mechanism *always* performs worse than CPU-only execution.

To reduce the impact of PIM coherence traffic, there are three general alternatives to fine-grained coherence for PIM execution: (1) coarse-grained coherence, (2) coarse-grained locks, and (3) making PIM data non-cacheable in the processor. We briefly examine these alternatives.

**Coarse-Grained Coherence** One approach to reduce PIM coherence traffic is to maintain a single coherence entry for *all* of the PIM data. Unfortunately, this can still incur high overheads, as the processor must flush *all* of the dirty cache lines within the PIM data region *every time* the PIM core acquires permissions, *even if the PIM kernel may not access most of the data.* For example, with just four processor threads, the number of cache lines flushed for PageRank is 227× the number of lines *actually required by the PIM kernel.*[3] Coherence at a smaller granularity, such as page-granularity [52], does not cause flushes for pages not accessed by the PIM kernel. However, many data-intensive applications perform *pointer chasing*, where a large number of pages are accessed non-sequentially, but only a *few lines* in each page are used, forcing the processor to flush *every* dirty page.

**Coarse-Grained Locks** Another drawback of coarse-grained coherence is that data can ping-pong between the processor and the PIM cores whenever the PIM data region is concurrently accessed by both. *Coarse-grained locks* avoid ping-ponging by having the PIM cores acquire *exclusive* access to a region for the duration of the PIM kernel. However, coarse-grained locks greatly restrict performance. Our application study shows that PIM kernels and processor threads often work in parallel on the same data region, and coarse-grained locks frequently cause thread serialization. PIM with coarse-grained locks (*CG* in Fig. 5.14) performs 8.4% *worse*, on average, than CPU-only execution. We conclude that using coarse-grained locks is not suitable for many important applications for PIM execution.

**Non-Cacheable PIM Data** Another approach sidesteps coherence by marking the PIM data region as *non-cacheable* in the processor [2, 47, 51, 149, 243], so that DRAM always contains up-to-date data. For applications where PIM data is almost exclusively accessed by the PIM processing logic, this incurs little penalty, but for many applications, the processor also accesses PIM data often. For our graph applications with a representative input (arXiV) (see footnote 3), the processor cores generate 42.6% of the total number of accesses to PIM data. With so many processor accesses, making PIM data non-cacheable results in a high performance and bandwidth overhead. As shown in Fig. 5.14, though marking PIM data as non-cacheable (*NC*) sometimes performs better than CPU-only, it still loses up to 62.7% (on average, 39.9%) of the improvement of Ideal-PIM. Therefore, while this approach avoids the overhead of coarse-grained mechanisms, it is a poor fit for applications that rely on processor involvement, and thus restricts the applications where PIM is effective.

We conclude that prior approaches to PIM coherence eliminate a significant portion of the benefits of PIM when data sharing occurs, due to their high coherence overheads. In fact, they sometimes cause PIM execution to *consistently degrade performance*. Thus, an *efficient* alternative to fine-grained coherence is necessary to retain PIM benefits across a wide range of applications.

### *5.4.3 LazyPIM Mechanism for Efficient PIM Coherence*

Our goal is to design a coherence mechanism that maintains the logical behavior of traditional coherence while retaining the large performance benefits of PIM. To

[3]See Sect. 5.4.5 for our experimental evaluation methodology.

this end, we propose *LazyPIM*, a new coherence mechanism that lets PIM kernels *speculatively* assume that they have the required permissions from the coherence protocol, *without* actually sending off-chip messages to the main (processor) coherence directory during execution. Instead, coherence states are updated only *after* the PIM kernel completes, at which point the PIM core transmits a single batched coherence message (i.e., a compressed *signature* containing *all* addresses that the PIM kernel read from or wrote to) back to the processor coherence directory. The directory checks to see whether any *conflicts* occurred. If a conflict exists, the PIM kernel *rolls back* its changes, conflicting cache lines are written back by the processor to DRAM, and the kernel re-executes. If no conflicts exist, speculative data within the PIM core is *committed*, and the processor coherence directory is updated to reflect the data held by the PIM core. Note that in LazyPIM, the processor *always* executes *non-speculatively*, which ensures minimal changes to the processor design, thereby likely enabling easier adoption of PIM.

LazyPIM avoids the pitfalls of the coherence mechanisms discussed in Sect. 5.4.2 (FG, CG, NC). With its compressed signatures, LazyPIM causes much less PIM coherence traffic than traditional fine-grained coherence. Unlike coarse-grained coherence and coarse-grained locks, LazyPIM checks coherence *only after* it completes PIM execution, avoiding the need to unnecessarily flush a large amount of data. Unlike non-cacheable, LazyPIM allows processor threads to cache the data used by PIM kernels within the processor cores as well, avoiding the need for the processor to perform a large number of off-chip accesses that can hurt performance greatly. LazyPIM also allows for efficient concurrent execution of processor threads and PIM kernels: by executing speculatively, the PIM cores do *not* invoke coherence requests during concurrent execution, avoiding data ping-ponging between the PIM cores and the processor.

**Conflicts** In LazyPIM, a PIM kernel *speculatively* assumes during execution that it has coherence permissions on a cache line, without checking the processor coherence directory. In the meantime, the processor continues to execute *non-speculatively*. To resolve PIM kernel speculation, LazyPIM provides *coarse-grained atomicity*, where all PIM memory updates are treated as if they *all* occur *at the moment that a PIM kernel finishes execution*. (We explain how LazyPIM enables this in Sect. 5.4.4.) If, before the PIM kernel finishes, the processor updates a cache line that the PIM kernel read during its execution, a *conflict* occurs. LazyPIM detects and handles all potential conflicts once the PIM kernel finishes executing.

Figure 5.15 shows an example timeline where a PIM kernel is launched on PIM core PIM0 while execution continues on processor cores CPU0 and CPU1. Due to the use of coarse-grained atomicity, PIM kernel execution behaves as if *the entire kernel's memory accesses* take place at the moment coherence is checked (i.e., at the end of kernel execution), *regardless of the actual time at which the kernel's accesses are performed*. Therefore, for *every* cache line read by PIM0, if CPU0 or CPU1 modifies the line before the coherence check occurs, PIM0 unknowingly uses stale data, leading to incorrect execution. Figure 5.15 shows two examples of this: (1) CPU0's write to line C *during* kernel execution; and (2) CPU0's write to

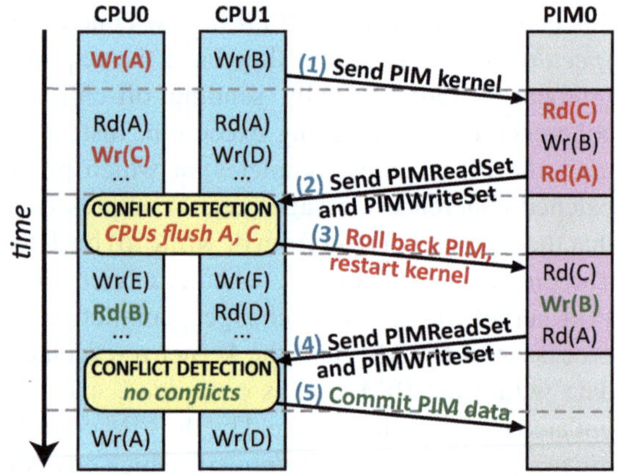


**Fig. 5.15** Example timeline of LazyPIM coherence sequence. Figure reproduced from [20]

line A *before* kernel execution, which was not written back to DRAM. To detect such conflicts, we record the addresses of processor writes and PIM kernel reads into two signatures, and then check to see if any addresses in them match (i.e., conflict) *after* the PIM kernel finishes (see Sect. 5.4.4.2).

If the PIM kernel writes to a cache line that is subsequently read by the processor before the kernel finishes (e.g., the second write by PIM0 to line B in Fig. 5.15), this is *not* a conflict. With coarse-grained atomicity, any read by the processor during PIM execution is ordered *before* the PIM kernel's write. LazyPIM ensures that the processor cannot read the PIM kernel's writes, by marking the PIM kernel writes as speculative inside the PIM processing logic until the kernel finishes (see Sect. 5.4.4.2). This is also the case when the processor and a PIM kernel write to the *same* cache line. Note that this ordering does not violate consistency models, such as sequential consistency.[4]

If the PIM kernel writes to a cache line that is subsequently *written to* by the processor before the kernel finishes, this is *not* a conflict. With coarse-grained atomicity, any write by the processor during PIM kernel execution is ordered before the PIM core's write since the PIM core write effectively takes place *after the PIM kernel finishes*. When the two writes modify different words in the same cache line, LazyPIM uses a per-word dirty bit mask in the PIM L1 cache to merge the writes, similar to prior work [108]. Note that the dirty bit mask is only in the PIM L1 cache; processor caches remain unchanged.

More details on the operation of the LazyPIM coherence mechanism are provided in our arXiv paper [21].

[4]A thorough treatment of memory consistency [106] is outside the scope of this work. Our goal is to deal with the coherence problem in PIM, not handle consistency issues.

## *5.4.4 Architectural Support for LazyPIM*

### 5.4.4.1 LazyPIM Programming Model

We provide a simple interface to port applications to LazyPIM. We show the implementation of a simple LazyPIM kernel within a program in Code Example 5.1. The programmer selects the portion(s) of the code to execute on PIM cores, using two macros (`#PIM_begin` and `#PIM_end`). The compiler converts the macros into instructions that we add to the ISA, which *trigger* and *end* PIM kernel execution. LazyPIM also needs to know which parts of the allocated data *might* be accessed by the PIM cores, which we refer to as the *PIM data region*.[5] We assume that either the programmer or the compiler can annotate all of the PIM data region using compiler directives or a PIM memory allocation API (`@PIM`). This information is saved in the page table using per-page flag bits, via communication to the system software using the system call interface.

Code Example 5.1 shows a portion of the compute function used by PageRank, as modified for execution with LazyPIM. All of our modifications are shown in bold. In this example, we want to execute only the `edgeMap()` function (Line 13) on the PIM cores. To ensure that LazyPIM tracks all data accessed during the `edgeMap()` call, we mark all of this data using `@PIM`, including any objects passed by value (e.g., `GA` on Line 1), any objects allocated in the function (e.g., those on Lines 4–6), and any objects allocated during functions that are executed on the PIM cores (e.g., the `PR_F<vertex>` object on Line 13). To tell the compiler that we want to execute only `edgeMap()` on the PIM cores, we surround it with the `#PIM_begin` and `#PIM_end` compiler directives on Lines 11 and 15, respectively. No other modifications are needed to execute our example code with LazyPIM.

### 5.4.4.2 Speculative Execution

When an application reaches a *PIM kernel trigger* instruction, the processor dispatches the kernel's starting PC to a free PIM core. The PIM core *checkpoints* the starting PC and registers, and starts executing the kernel. The kernel *speculatively* assumes that it has coherence permissions for *every* line it accesses, without *actually* checking the processor directory. We add a one-bit flag to each line in the PIM core cache, to mark all data updates as speculative. If a speculative line is selected for eviction, the core rolls back to the starting PC and discards the updates.

LazyPIM tracks three sets of addresses during PIM kernel execution. These are recorded into three *signatures*, as shown in Fig. 5.16: (1) the *CPUWriteSet* (all *CPU writes* to the PIM data region), (2) the *PIMReadSet* (all *PIM reads*), and

[5] The programmer should be conservative in identifying PIM data regions, and should not miss *any possible data* that may be touched by a PIM core. If any data *not marked* as PIM data is accessed by the PIM core, the program can produce incorrect results.

```
PageRankCompute(@PIM Graph GA) { // GA is accessed by PIM cores in edgeMap()
  const int n = GA.n;                   // not accessed by PIM cores
  const double damping = 0.85, epsilon = 0.0000001; // not accessed by PIM cores
  @PIM double* p_curr, p_next;   // accessed by PIM cores in edgeMap()
  @PIM bool* frontier;                  // accessed by PIM cores in edgeMap()
  @PIM vertexSubset Frontier(n, n, frontier); // accessed by PIM in edgeMap()
  double L1_norm;                       // not accessed by PIM cores
  long iter = 0;                        // not accessed by PIM cores
  ...
  while(iter++ < maxIters) {
    #PIM_begin
      // only the edgeMap() function is offloaded to the PIM cores
      vertexSubset output =
          edgeMap(GA, Frontier, @PIM PR_F<vertex>(p_curr, p_next, GA.V), 0);
      // PR_F<vertex> object allocated during edgeMap() call, needs annotation
    #PIM_end
    vertexMap(Frontier, PR_Vertex_F(p_curr, p_next, damping, n));  // run on CPU

    // compute L1-norm between p_curr and p_next
    L1_norm = fabs(p_curr - p_next);   // run on CPU
    if(L1_norm < epsilon) break;       // run on CPU
    ...
  }
  Frontier.del();
}
```

**Listing 5.1** Example PIM program implementation. Modifications for PIM execution are shown in bold

(3) the *PIMWriteSet* (all *PIM writes*). When the kernel starts, the dirty lines in the processor cache containing PIM data are recorded in the CPUWriteSet, by scanning the tag store (potentially using a Dirty-Block Index [201]). The processor uses the page table flag bits from Sect. 5.4.4.1 to identify which writes need to be added to the CPUWriteSet during kernel execution. The PIMReadSet and PIMWriteSet are updated for *every* read and write performed by the PIM kernel. When the kernel finishes execution, the three signatures are used to resolve speculation (see Sect. 5.4.4.3)

The signatures use parallel Bloom filters [19], which employ simple Boolean logic to hash multiple addresses into a single (256B) fixed-length register. If the speculative coherence requests were sent back to the processor without any sort of compression at the end of PIM kernel execution, the coherence messages would still

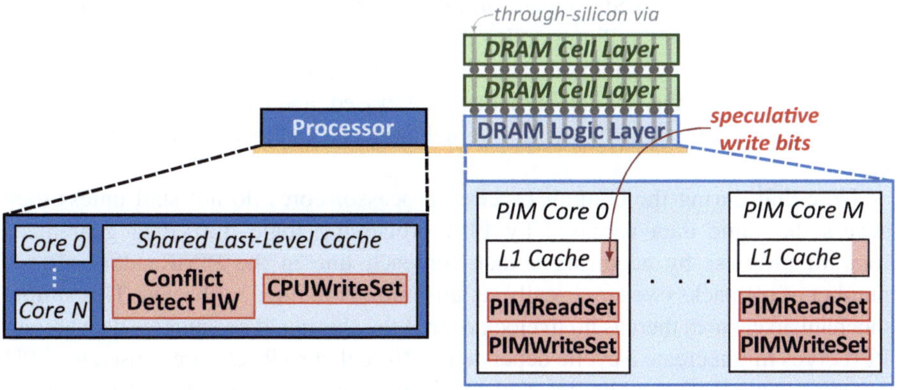


**Fig. 5.16** High-level additions (in bold) to PIM architecture to support LazyPIM. Figure adapted from [20]

consume a large amount of off-chip traffic, nullifying most of the benefits of the speculation. Bloom filters allow LazyPIM to compress these coherence messages into a much smaller size, while guaranteeing that there are no false negatives [19] (i.e., no coherence messages are lost during compression). The addresses of *all* data accessed speculatively by the PIM cores can be extracted and compared from the Bloom filter [19, 24]. The hashing introduces a limited number of false positives, with the false positive rate increasing as we store more addresses in a single fixed-length Bloom filter. In our evaluated system, each signature is 256B long, and can store up to 607 addresses without exceeding a 20.0% false positive rate (with *no* false negatives). To store more addresses, we use multiple filters to guarantee an upper bound on the false positive rate.

#### 5.4.4.3 Handling Conflicts

As Fig. 5.15 shows, we need to detect conflicts that occur during PIM kernel execution. In LazyPIM, when the kernel finishes executing, both the PIMReadSet and PIMWriteSet are sent back to the processor.

If no matches are detected between the PIMReadSet and the CPUWriteSet (i.e., no conflicts have occurred), PIM kernel *commit* starts. Any addresses (including false positives) in the PIMWriteSet are invalidated from the processor cache. A message is then sent to the PIM core, allowing it to write its speculative cache lines back to DRAM. During the commit process, all coherence directory entries for the PIM data region are locked to ensure atomicity of commit. Finally, all signatures are erased.

If an overlap is found between the PIMReadSet and the CPUWriteSet, a conflict may have occurred. At this point, only the dirty lines in the processor that match in the PIMReadSet are flushed back to DRAM. During this flush, all PIM data directory entries are locked to ensure atomicity. Once the flush completes, a message is sent to the PIM core, telling it to invalidate all speculative cache lines, and to *roll back* the PC to the checkpointed value. Now that all possibly conflicting cache lines are written back to DRAM, all signatures are erased, and the PIM core *restarts* the kernel. After re-execution of the PIM kernel finishes, conflict detection is performed again.

Note that during the commit process, processor cores do not stall unless they access the same data accessed by PIM processing logic. LazyPIM guarantees forward progress by acquiring a lock for each line in the PIMReadSet after a number of rollbacks (we empirically set this number to three rollbacks). This simple mechanism ensures there is no livelock even if the sharing of speculative data among PIM cores might create a cyclic dependency. Note that rollbacks are caused by CPU accesses to conflicting addresses, and not by the sharing of speculative data between PIM cores. As a result, once we lock conflicting addresses following three rollbacks, the PIM cores will not roll back again as there will be no conflicts, guaranteeing forward progress.

#### 5.4.4.4 Hardware Overhead

LazyPIM's overhead consists mainly of (1) 1 bit per page (0.003% of DRAM capacity) and 1 bit per TLB entry for the page table flag bits (Sect. 5.4.4.1); (2) a 0.2% increase in PIM core L1 size to mark speculative data (Sect. 5.4.4.2); (3) a 1.6% increase in PIM core L1 size for the dirty bit mask (Sect. 5.4.3); and (4) in the worst case, 12 kB for the signatures per PIM core (Sect. 5.4.4.2). This overhead can be greatly optimized (as part of future work): for PIM kernels that need multiple signatures, we could instead divide the kernel into smaller chunks where each chunk's addresses fit in a single signature, lowering signature overhead to 512B. We leave a detailed evaluation of LazyPIM hardware overhead optimization to future work. Some ideas related to this and a detailed hardware overhead analysis are presented in our arXiv paper [21].

### *5.4.5 Methodology for LazyPIM Evaluation*

We study two types of data-intensive applications: graph workloads and databases. We use three Ligra [209] graph applications (PageRank, Radii, Connected Components), with input graphs constructed from real-world network datasets [217]: Facebook, arXiV High Energy Physics Theory, and Gnutella25 (peer-to-peer). We also use an in-house prototype of a modern in-memory database (IMDB) [144, 193, 203, 219] that supports HTAP workloads. Our transactional workload consists of 200K transactions, each randomly performing reads or writes on a few randomly chosen tuples. Our analytical workload consists of 256 analytical queries that use the select and join operations on randomly-chosen tables and columns.

PIM kernels are selected from these applications with the help of OProfile [165]. We conservatively select candidate PIM kernels, choosing portions of functions

**Table 5.3** Evaluated system configuration for LazyPIM evaluation

| | |
|---|---|
| *Main processor (CPU)* | |
| ISA | x86-64 |
| Core configuration | 4–16 cores, 2 GHz, 8-wide issue |
| Operating system | 64-bit Linux from Linaro [127] |
| L1 I/D cache | 64 kB per core, private, 4-way associative, 64B blocks, 2-cycle lookup |
| L2 cache | 2 MB, shared, 8-way associative, 64B blocks, 20-cycle lookup |
| Cache coherence | MESI directory [23, 167] |
| *PIM cores* | |
| ISA | x86-64 |
| Core configuration | 4–16 cores, 2 GHz, 1-wide issue |
| L1 I/D cache | 64 kB per core, private, 4-way associative, 64B blocks, 2-cycle lookup |
| Cache coherence | MESI directory [23, 167] |
| *Main memory parameters* | |
| Memory configuration | HMC 2.0 [72], one 4 GB cube, 16 vaults per cube, 16 banks per vault, |
| | FR-FCFS scheduler [182, 249] |

where the application (1) spends the majority of its cycles, and (2) generates the majority of its last-level cache misses. From these candidates, we pick kernels that we believe minimize the coherence overhead for each evaluated mechanism, by minimizing data sharing between the processor and PIM processing logic. We modify each application to ship the selected PIM kernels to the PIM cores. We manually annotate the PIM data set.

For our evaluations, we modify the gem5 simulator [18]. We use the x86-64 architecture in full-system mode, and use DRAMSim2 [185] to perform detailed timing simulation of DRAM. Table 5.3 shows our system configuration.

### 5.4.6 Evaluation of LazyPIM

We first analyze the off-chip traffic reduction of LazyPIM. This off-chip reduction leads to bandwidth and energy savings. We then analyze LazyPIM's effect on system performance. We show system performance results normalized to a processor-only baseline (*CPU-only*, as defined in Sect. 5.4.2), and compare LazyPIM's performance with using fine-grained coherence (*FG*), coarse-grained locks (*CG*), or non-cacheable data (*NC*) for PIM data.

#### 5.4.6.1 Off-Chip Memory Traffic

Figure 5.17a shows the normalized off-chip memory traffic of the PIM coherence mechanisms for a 16-core architecture (with 16 processor cores and 16 PIM cores) Fig. 5.17b shows the normalized off-chip memory traffic as the number of threads

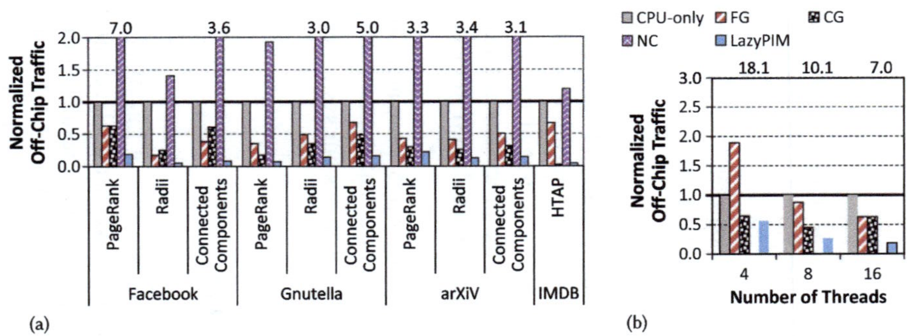


**Fig. 5.17** Effect of LazyPIM on off-chip memory traffic, normalized to CPU-only. Figure adapted from [20]. (**a**) 16-Thread off-chip memory traffic. (**b**) Off-chip memory traffic sensitivity to thread count for PageRank

increases, for PageRank using the Facebook graph. LazyPIM significantly reduces the *overall* off-chip traffic (up to 81.2% over CPU-only, 70.1% over FG, 70.2% over CG, and 97.3% over NC), and scales better with thread count. LazyPIM reduces off-chip memory traffic by 58.8%, on average, over CG, the best prior approach in terms of off-chip traffic.

CG has greater traffic than LazyPIM, the majority of which is due to having to flush dirty cache lines before each PIM kernel invocation. Due to false sharing, the number of flushes scales *superlinearly* with thread count (not shown), increasing 13.1× from 4 to 16 threads. LazyPIM avoids this cost with speculation, as *only* the *necessary* flushes are performed *after* the PIM kernel finishes execution. As a result, it reduces the flush count (e.g., by 94.0% for 16-thread PageRank using Facebook), and thus lowers overall off-chip memory traffic (by 50.3% for our example).

NC suffers from the fact that *all* processor accesses to PIM data must go to DRAM, increasing average off-chip memory traffic by 3.3x over CPU-only. NC off-chip memory traffic also scales poorly with thread count, as more processor threads generate a greater number of accesses. In contrast, LazyPIM allows processor cores to cache PIM data, by enabling coherence efficiently.

#### 5.4.6.2 Performance

Figure 5.18a shows the performance improvement for 16 threads. Without any coherence overhead, Ideal-PIM significantly outperforms CPU-only across *all* applications, showing PIM's potential on these workloads. Poor handling of coherence by FG, CG, and NC leads to drastic performance losses compared to Ideal-PIM, indicating that an efficient coherence mechanism is essential for PIM performance. For example, in some cases, NC and CG actually perform *worse* than CPU-only, and for PageRank running on the Gnutella graph, all prior mechanisms degrade performance. In contrast, LazyPIM consistently retains most of Ideal-PIM's benefits for all applications, coming within 5.5% on average. LazyPIM outperforms all of the other approaches, improving over the best-performing prior approach (NC) by 49.1%, on average.

Figure 5.18b shows the performance of PageRank using Gnutella as we increase the thread count. LazyPIM comes within 5.5% of Ideal-PIM, which has no coherence overhead (as defined in Sect. 5.4.2), and improves performance by 73.2% over FG, 47.0% over CG, 29.4% over NC, and 39.4% over CPU-only, on average. With NC, the processor threads incur a large penalty for accessing DRAM frequently. CG suffers greatly due to (1) flushing dirty cache lines, and (2) blocking all processor threads that access PIM data during execution. In fact, processor threads are blocked for up to 73.1% of the total execution time with CG. With more threads, the negative effects of blocking worsen CG's performance. FG also loses a significant portion of Ideal-PIM's improvements, as it sends a large amount of off-chip messages. Note that NC, despite its high off-chip traffic, performs better than CG and FG, as it neither blocks processor cores nor slows down PIM execution.

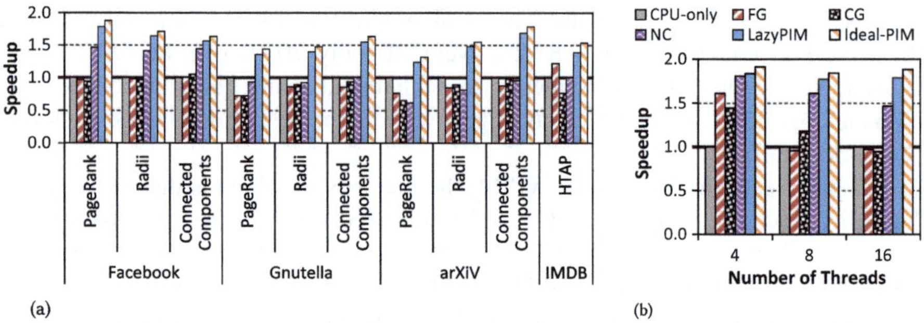


**Fig. 5.18** Speedup of cache coherence mechanisms, normalized to CPU-only. Figure adapted from [20]. (**a**) Speedup for all applications with 16 threads. (**b**) Speedup sensitivity to thread count for Gnutella

One reason for the difference in performance between LazyPIM and Ideal-PIM is the number of conflicts that are detected at the end of PIM kernel execution. As we discuss in Sect. 5.4.4.3, any detected conflict causes a rollback, where the PIM kernel must be re-executed. We study the number of commits that contain conflicts for two representative 16-thread workloads: Components using the Enron graph, and HTAP-128 (results not shown). If we study an idealized version of full kernel commit, where no false positives exist, we find that a relatively high percentage of commits contain conflicts (47.1% for Components and 21.3% for HTAP). Using realistic signatures for full kernel commit, which includes the impact of false positives, the conflict rate increases to 67.8% for Components and 37.8% for HTAP. Despite the high number of commits that induce rollback, the overall performance impact of rollback is low, as LazyPIM comes within 5.5% of the performance of Ideal-PIM. We find that for all of our applications, a kernel never rolls back more than once, limiting the performance impact of conflicts. We can further improve the performance of LazyPIM by optimizing the commit process to reduce the rollback overhead, which we explore in our arXiv paper [21].

### *5.4.7 Summary of LazyPIM*

We propose LazyPIM, a new cache coherence mechanism for PIM architectures. Prior approaches to PIM coherence generate very high off-chip traffic for important data-intensive applications. LazyPIM avoids this by avoiding coherence lookups *during* PIM kernel execution. The key idea is to use compressed coherence *signatures* to batch the lookups and verify correctness *after* the kernel completes. As a result of the more efficient approach to coherence employed by LazyPIM, applications that performed poorly under prior approaches to PIM coherence can now take advantage of the benefits of PIM execution. LazyPIM improves average performance by 49.1% (coming within 5.5% of an ideal PIM mechanism), and reduces off-chip traffic by 58.8%, over the best prior approach to PIM coherence while retaining the conventional multithreaded programming model.

## 5.5 Related Work

We briefly survey related work in processing-in-memory, accelerator design, mechanisms for handling pointer chasing, and techniques for pointer chasing.

**Early Processing-in-Memory (PIM) Proposals** Early proposals for PIM architectures had limited to no adoption, as the proposed logic integration was too costly and did not solve many of the obstacles facing the adoption of PIM. The earliest such proposals date from the 1970s, where small processing elements are combined with small amounts of RAM to provide a distributed array of memories that perform computation [208, 218]. Some of the other early works, such as EXECUBE [100], Terasys [56], IRAM [171], and Computational RAM [44, 45], add logic within DRAM to perform vector operations. Yet other early works, such as FlexRAM [80], DIVA [39], Smart Memories [140], and Active Pages [166], propose more versatile substrates that tightly integrate logic and reconfigurability within DRAM itself to increase flexibility and the available compute power.

**Processing in 3D-Stacked Memory** With the advent of 3D-stacked memories, we have seen a resurgence of PIM proposals [133, 199]. Recent PIM proposals add compute units within the logic layer to exploit the high bandwidth available. These works primarily focus on the design of the underlying logic that is placed within memory, and in many cases propose special-purpose PIM architectures that cater only to a limited set of applications. These works include accelerators for MapReduce [176], matrix multiplication [246], data reorganization [4], graph processing [2, 163], databases [12], in-memory analytics [52], genome sequencing [96, 98], data-intensive processing [58], consumer device workloads [22], and machine learning workloads [30, 93, 118]. Some works propose more generic architectures by adding PIM-enabled instructions [3], GPGPUs [68, 172, 243], single-instruction multiple-data (SIMD) processing units [149], or reconfigurable hardware [47, 51, 59] to memory.

**Processing Using Memory** A number of recent works have examined how to perform memory operations directly within the memory array itself, which we refer to as *processing using memory* [199]. These works take advantage of inherent architectural properties of memory devices to perform operations in bulk. While such works can significantly improve computational efficiency within memory, they still suffer from many of the same programmability and adoption challenges that PIM architectures face, such as the address translation and cache coherence challenges that we focus on in this chapter. Mechanisms for processing using memory can perform a variety of functions, such as bulk copy and data initialization for DRAM [27, 28, 197, 200], bulk bitwise operations for DRAM [124, 195, 196, 202] and phase-change memory (PCM) [123], and simple arithmetic operations for SRAM [1, 81] and memristors [103–105, 121, 205].

**Processing in the DRAM Module or Memory Controller** Several works have examined how to embed processing functionality near memory, but not within the DRAM chip itself. Such an approach can reduce the cost of PIM manufacturing, as the DRAM chip does not need to be modified or specialized for any particular functionality. However, these works (1) are often unable to take advantage of the high internal bandwidth of 3D-stacked DRAM, which reduces the efficiency of PIM execution, and (2) may still suffer from many of the same challenges faced by architectures that embed logic within the DRAM chip. Examples of this work include Chameleon [9], which proposes a method of integrating logic within the DRAM module but outside of the chip to reduce manufacturing costs, Gather-Scatter DRAM [203], which embeds logic within the memory controller to remap a single memory request across multiple rows and columns within DRAM, and work by Hashemi et al. [62, 63] to embed logic in the memory controller that accelerates dependent cache misses and performs runahead execution [153, 154, 156, 158].

**Addressing Challenges to PIM Adoption** Recent work has examined design challenges for systems with PIM support that can affect PIM adoption. A number of these works improve PIM programmability, such as LazyPIM [20, 21], which provides efficient cache coherence support for PIM (as we described in detail in Sect. 5.4) the study by Sura et al. [221], which optimizes how programs access PIM data, and work by Liu et al. [131], which designs PIM-specific concurrent data structures to improve PIM performance. Other works tackle hardware-level design challenges, including IMPICA [67], which introduces in-memory support for address translation and pointer chasing (as we described in detail in Sect. 5.3), work by Hassan et al. [64] to optimize the 3D-stacked DRAM architecture for PIM, and work by Kim et al. [95] to enable the distribution of PIM data across multiple memory stacks.

**Coherence for PIM Architectures** In order to avoid the overheads of fine-grained coherence, many prior works on PIM architectures design their systems in such a way that they do not need to utilize traditional coherence protocols. Instead, these works use one of two alternatives. Some works restrict PIM processing logic to execute on only non-cacheable data (e.g., [2, 47, 51, 149, 243]), which forces cores within the CPU to read PIM data directly from DRAM. Other works use coarse-grained coherence or coarse-grained locks, which force processor cores to not access any data that could *potentially* be used by the PIM processing logic, or to flush this data back to DRAM before the PIM kernel begins executing (e.g., [3, 4, 27, 47, 52, 59, 67, 67, 68, 163, 172, 195, 196, 200, 202]). Both of these approaches generate significant coherence overhead, as discussed in Sect. 5.4.2. Unlike these approaches, LazyPIM (Sect. 5.4) places no restriction on the way in which processor cores and PIM processing logic can access data. Instead, LazyPIM uses PIM-side coherence speculation and efficient coherence message compression to provide cache coherence, which avoids the communication overheads associated with traditional coherence protocols.

**Accelerators in CPUs** There have been various CPU-side accelerator proposals for database systems (e.g., [32, 99, 230, 231]) and key-value stores [126]. Widx [99] is a database indexing accelerator that uses a set of custom RISC cores in the CPU to accelerate hash index lookups. While a hash table is one of our data structures of interest, IMPICA (Sect. 5.3) differs from Widx in three major ways. First, it is an *in-memory* (as opposed to CPU-side) accelerator, which poses very different design challenges. Second, we solve the address translation challenge for in-memory accelerators, while Widx uses the CPU address translation structures. Third, we enable parallelism within a single accelerator core, while Widx achieves parallelism by replicating several RISC cores.

**Prefetching for Linked Data Structures** Many works propose mechanisms to prefetch data in linked data structures to hide memory latency. These proposals are hardware-based (e.g., [34, 35, 69, 70, 78, 155, 157, 186, 242]), software-based (e.g., [128, 136, 187, 232, 238]), pre-execution-based (e.g., [33, 135, 213, 248]), or software/hardware-cooperative (e.g., [40, 83, 186]) mechanisms. These approaches have two major drawbacks. First, they usually rely on predictable traversal sequences to prefetch accurately. As a result, many of these mechanisms can become very inefficient if the linked data structure is complex or when access patterns are less regular. Second, the pointer chasing or prefetching is performed at the CPU cores or at the memory controller, which likely leads to pollution of the CPU caches and TLBs by these irregular memory accesses.

## 5.6 Other System-Level Challenges for PIM Adoption

IMPICA (Sect. 5.3) and LazyPIM (Sect. 5.4) demonstrate the need for and gains that can be achieved by designing system-level solutions that are applicable across a wide variety of PIM architectures. In order for PIM to achieve widespread adoption, we believe there are a number of other system-level challenges that must be addressed. In this section, we discuss six research directions that aim towards solving these challenges: (1) the PIM programming model, (2) data mapping, (3) runtime scheduling support for PIM, (4) the granularity of PIM scheduling, (5) evaluation infrastructures and benchmark suites for PIM, and (6) applying PIM to emerging memory technologies.

**PIM Programming Model** Programmers need a well-defined interface to incorporate PIM functionality into their applications. Determining the programming model for how a programmer should invoke and interact with PIM processing logic is an open research direction. Using a set of special instructions allows for very fine-grained control of when PIM processing logic is invoked, which can potentially result in a significant performance improvement. However, this approach can potentially introduce overheads while taking advantage of PIM, due to the need to frequently exchange information between PIM processing logic and the CPU. Hence, there is a need for researchers to investigate how to integrate PIM instructions with other compiler-based methods or library calls that can support PIM integration, and how these approaches can ease the burden on the programmer. For example, one of our recent works [68] examines compiler-based mechanisms to decide what portions of code should be offloaded to PIM processing logic in a GPU-based system. Another recent work [172] examines system-level techniques that decide which GPU application kernels are suitable for PIM execution.

**Data Mapping** Determining the ideal memory mapping for data used by PIM processing logic is another important research direction. To maximize the benefits of PIM, data that needs to be read from or written to by a single PIM kernel instance should be mapped to the same memory stack. Hence, it is important to examine both static and adaptive data mapping mechanisms to intelligently map (or remap) data. Even with such data mapping mechanisms, it is beneficial to provide low-cost and low-overhead data migration mechanisms to facilitate easier PIM execution, in case the data mapping needs to be adapted to execution and access patterns at runtime. One of our recent works provides a mechanism that provides programmer-transparent data mapping support for PIM [68]. Future work can focus on developing new data mapping mechanisms, as well as designing systems that can take advantage of these new data mapping mechanisms.

**PIM Runtime Scheduling Support** At least four key runtime issues in PIM are to decide (1) when to enable PIM execution, (2) what to execute near data, (3) how to map data to multiple (hybrid) memory modules such that PIM execution is viable and effective, and (4) how to effectively share/partition PIM mechanisms/accelerators at runtime across multiple threads/cores to maximize performance and energy efficiency. It is possible to build on our recent works that employ locality prediction [3] and combined compiler and dynamic code identification and scheduling in GPU-based systems [68, 172]. Several key research questions that should be investigated include:

- What are simple mechanisms to enable and disable PIM execution? How can PIM execution be throttled for highest performance gains? How should data locations and access patterns affect where/whether PIM execution should occur?
- Which parts of the application code should be executed on PIM? What are simple mechanisms to identify such code?
- What are scheduling mechanisms to share PIM accelerators between multiple requesting cores to maximize PIM's benefits?

**Granularity of PIM Scheduling** To enable the widespread adoption of PIM, we must understand the ideal granularity at which PIM operations can be scheduled without sacrificing PIM execution's efficiency and limiting changes to the shared memory programming model. Two key issues for scheduling code for PIM execution are (1) how large each part of the code should be (i.e., the granularity of PIM execution), and (2) the frequency at which code executing on a PIM engine should synchronize with code executing on the CPU cores (i.e., the granularity of PIM synchronization).

The optimal granularity of PIM execution remains an open question. For example, is it best to offload only a single instruction to the PIM processing logic? Should PIM kernels consist of a set of instructions, and if so, how large is each set? Do we limit PIM execution to work only on entire functions, entire threads, or even entire applications? If we offload too short a piece of code, the benefits of executing the code near memory may be unable to overcome the overhead of invoking PIM execution (e.g., communicating registers or data, taking checkpoints).

Once code begins to execute on PIM processing logic, there may be times where the code needs to synchronize with code executing on the CPU. For example, many shared memory applications employ locks, barriers, or memory fences to coordinate access to data and ensure correct execution. PIM system architects must determine (1) whether code executing on PIM should allow the support of such synchronization operations; and (2) if they do allow such operations, how to perform them efficiently. Without an efficient mechanism for synchronization, PIM processing logic may need to communicate frequently with the CPU when synchronization takes place, which can introduce overheads and undermine the benefits of PIM execution. Research on PIM synchronization can build upon our prior work, where we limit PIM execution to atomic instructions to avoid the need for synchronization [3], or where provide support within LazyPIM to perform synchronization during PIM kernel execution [21].

**PIM Evaluation Infrastructures and Benchmark Suites** To ease adoption, it is critical that we accurately assess the benefits of PIM. Accurate assessment for PIM requires (1) a set of real-world memory-intensive applications that have the potential to benefit significantly when executed near memory, and (2) a simulation/evaluation infrastructure that allows architects and system designers to precisely analyze the benefits and overhead of adding PIM processing logic to memory and executing code on this processing logic.

In order to identify what processing logic should be introduced near memory, and to know what properties are ideal for PIM kernels, we must begin by developing a real-world benchmark suite of applications that can potentially benefit from PIM. While many data-intensive applications, such as pointer chasing and bulk memory copy, can potentially benefit from PIM, it is crucial to examine important candidate applications for PIM execution, and for researchers to agree on a common set of these candidate applications to focus the efforts of the community. We believe that these applications should come from a number of popular and emerging domains. Examples of potential domains include data-parallel applications, neural networks, machine learning, graph processing, data analytics, search/filtering, mobile workloads, bioinformatics, Hadoop/Spark programs, and in-memory data stores. Many of these applications have large data sets and can benefit from high memory bandwidth and low memory latency benefits provided by PIM mechanisms. As an example, in our prior work, we have started analyzing mechanisms for accelerating graph processing [2, 3]; pointer chasing [62, 67]; databases [20, 21, 67, 203]; consumer workloads [22], including web browsing, video encoding/decoding, and machine learning; and GPGPU workloads [68, 172].

Once we have established a set of applications to explore, it is essential for researchers to develop an extensive and flexible application profiling and simulation infrastructure and mechanisms that can (1) identify parts of these applications for which PIM execution can be beneficial; and (2) simulate in-memory acceleration. A systematic process for identifying potential PIM kernels within an application can not only ease the burden for performing PIM research, but could also inspire tools that programmers and compilers can use to automate the process of offloading portions of existing applications to PIM processing logic. Once we have identified potential PIM kernels, we need a simulator to accurately model the energy and performance consumed by PIM hardware structures, available memory bandwidth, and communication overhead when we execute the kernels near memory. Highly-flexible memory simulators (e.g., Ramulator [92, 189], SoftMC [66, 191]) can be combined with full-system simulation infrastructures (e.g., gem5 [18]) to provide a robust environment that can evaluate how various PIM architectures affect the *entire compute stack*, and can allow designers to identify memory characteristics (e.g., internal bandwidth, trade-off between number of PIM engines and memory capacity) that affect the efficiency of PIM execution.

**Applicability to Emerging Memory Technologies** As DRAM scalability issues are becoming more difficult to work around [2, 3, 26, 29, 37, 65, 66, 68, 79, 82, 89–91, 110, 114, 116, 117, 120, 125, 129, 137, 138, 143, 151, 152, 160, 226, 233, 239, 240], there has been a growing amount of work on emerging non-volatile memory technologies to replace DRAM. Examples of these emerging memory technologies include *phase-change memory* (PCM) [110–112, 179, 228, 240, 245], *spin-transfer torque magnetic RAM* (STT-MRAM) [101, 162], *metal-oxide resistive RAM* (RRAM) [229], and *memristors* [31, 220]. These memories have the potential to offer much greater memory capacity and high internal memory bandwidth. Processing-in-memory techniques can take advantage of this potential, by exploiting the high available internal memory bandwidth, and by making use of the underlying memory device behavior, to perform computation.

PIM can be especially useful in single-level store settings [14, 88, 146, 181, 206, 207, 244], where multiple memory and storage technologies (including emerging memory technologies) are presented to the system as a single monolithic memory, which can be accessed quickly and at high volume by applications. By performing some of the computation in memory, PIM can take advantage of the high bandwidth and capacity available within a single-level store without being bottlenecked by the limited off-chip bandwidth between the various memory and system software components of the single-level store and the CPU.

Given the worsening DRAM scaling issues, and the limited bandwidth available between memory and the CPU, we believe that there is a growing need to investigate PIM processing logic that is designed for emerging memory technologies. We believe that many PIM techniques can be applicable in these technologies. Already, several prior works propose to exploit memory device behavior to perform processing using memory, where the memory consists of PCM [123] or memristors [103–

105, 121, 205]. Future research should explore how PIM can take advantage of emerging memory technologies in other ways, and how PIM can work effectively in single-level stores.

## 5.7 Conclusion

Circuit and device technology scaling for main memory, built predominantly with DRAM, is already showing signs of coming to an end, with three major issues emerging. First, the reliability and data retention capability of DRAM have been decreasing, as shown by various error characterization and analysis studies [25, 66, 82, 84–87, 97, 107, 129, 130, 141, 147, 150, 152, 168, 180, 194, 214], and new failure mechanisms have been slipping into devices in the field (e.g., Rowhammer [89, 91, 94, 152]). Second, main memory performance improvements have not grown as rapidly as logic performance improvements have for several years now, resulting in significant performance bottlenecks [2, 26, 28, 107, 114, 116, 117, 120, 125, 143, 151, 160, 197, 233]. Third, the increasing application demand for memory places even greater pressure on the main memory system in terms of both performance and energy efficiency [2, 3, 22, 26, 29, 37, 68, 79, 82, 89–91, 110, 114, 116, 117, 120, 125, 129, 137, 138, 143, 151, 152, 160, 226, 233, 239, 240]. To solve these issues, there is an increasing need for architectural and system-level approaches [151, 152, 160].

A major hindrance to memory performance and energy efficiency is the high cost of moving data between the CPU and memory. Currently, this cost must be paid *every time* an application needs to perform an operation on data that is stored within memory. The recent advent of 3D-stacked memory architectures, which contain a layer dedicated for logic within the same stack as memory layers, open new possibilities to reduce unnecessary data movement by allowing architects to shift some computation into memory. Processing-in-memory (PIM), or near-data processing, allows architects to introduce simple PIM processing logic (which can be specialized acceleration logic, general-purpose cores, or reconfigurable logic) into the logic layer of the memory, where the PIM processing logic has access to the high internal bandwidth and low memory access latency that exist within 3D-stacked memory. As a result, PIM architectures can reduce costly data movement over the memory channel, lower memory access latency, and thereby also reduce energy consumption.

A number of challenges exist in enabling PIM at the system level, such that PIM can be adopted easily in many system designs. In this work, we examine two such key design issues, which we believe require efficient and elegant solutions to enable widespread adoption of PIM in real systems. First, because applications store memory references as virtual addresses, PIM processing logic needs to perform *address translation* to determine the physical addresses of these references during execution. However, PIM processing logic does not have an efficient way of accessing to the translation lookaside buffer or the page table walkers that reside in the CPU. Second, because PIM processing logic can often access the same data structures that are being accessed and modified by the CPU, a system that incorporates PIM cores needs to support cache coherence between the CPU and PIM cores to ensure that all of the cores are using the correct version of the data. Naive solutions to overcome the address translation and cache coherence challenges either place significant restrictions on the types of computation that can be performed by PIM processing logic, which can break the existing multithreaded programming model and prevent the widespread adoption of PIM, or force PIM processing logic to communicate with the CPU frequently, which can undo the benefits of moving computation to memory. Using key observations about the behavior of address translation and cache coherence for several memory-intensive applications, we propose two solutions that (1) provide general purpose support for translation and coherence in PIM architectures, (2) maintain the conventional multithreaded programming model, and (3) do not incur high communication overheads. The first solution, IMPICA, provides an efficient in-memory accelerator for pointer chasing that can perform efficient address translation from within memory. The second solution, LazyPIM, provides an efficient cache coherence protocol that does not restrict how PIM processing logic and the CPU share data, by using speculation and coherence message compression to minimize the overhead of PIM coherence requests.

We hope that our solutions to the address translation and cache coherence challenges can ease the adoption of PIM-based architectures, by easing both the design and programmability of such systems. We also hope that the challenges and ideas discussed in this chapter can inspire other researchers to develop other novel solutions that can ease the adoption of PIM architectures.

**Acknowledgements** We thank all of the members of the SAFARI Research Group, and our collaborators at Carnegie Mellon, ETH Zürich, and other universities, who have contributed to the various works we describe in this chapter. Thanks also goes to our research group's industrial sponsors over the past 9 years, especially Google, Huawei, Intel, Microsoft, NVIDIA, Samsung, Seagate, and VMware. This work was also partially supported by the Intel Science and Technology Center for Cloud Computing, the Semiconductor Research Corporation, the Data Storage Systems Center at Carnegie Mellon University, and NSF grants 1212962, 1320531, and 1409723.

## References

1. S. Aga, S. Jeloka, A. Subramaniyan, S. Narayanasamy, D. Blaauw, R. Das, Compute caches, in *HPCA* (2017)
2. J. Ahn, S. Hong, S. Yoo, O. Mutlu, K. Choi, A scalable processing-in-memory accelerator for parallel graph processing, in *ISCA* (2015)
3. J. Ahn, S. Yoo, O. Mutlu, K. Choi, PIM-enabled instructions: a low-overhead, locality-aware processing-in-memory architecture, in *ISCA* (2015)
4. B. Akin, F. Franchetti, J.C. Hoe, Data reorganization in memory using 3D-stacked DRAM, in *ISCA* (2015)

5. C. Alkan et al., Personalized copy number and segmental duplication maps using next-generation sequencing. Nat. Genet. **41**, 1061 (2009)
6. M. Alser, H. Hassan, H. Xin, O. Ergin, O. Mutlu, C. Alkan, GateKeeper: a new hardware architecture for accelerating pre-alignment in DNA short read mapping. Bioinformatics **33**, 3355–3363 (2017)
7. ARM Holdings, ARM Cortex-A57. http://www.arm.com/products/processors/cortex-a/cortex-a57-processor.php
8. ARM Holdings, ARM Cortex-R4. http://www.arm.com/products/processors/cortex-r/cortex-r4.php
9. H. Asghari-Moghaddam, Y.H. Son, J.H. Ahn, N.S. Kim, Chameleon: versatile and practical near-DRAM acceleration architecture for large memory systems, in *MICRO* (2016)
10. R. Ausavarungnirun, J. Landgraf, V. Miller, S. Ghose, J. Gandhi, C.J. Rossbach, O. Mutlu, Mosaic: a GPU memory manager with application-transparent support for multiple page sizes, in *MICRO* (2017)
11. R. Ausavarungnirun, V. Miller, J. Landgraf, S. Ghose, J. Gandhi, A. Jog, C.J. Rossbach, O. Mutlu, MASK: redesigning the GPU memory hierarchy to support multi-application concurrency, in *ASPLOS* (2018)
12. O.O. Babarinsa, S. Idreos, JAFAR: near-data processing for databases, in *SIGMOD* (2015)
13. A. Basu, J. Gandhi, J. Chang, M.D. Hill, M.M. Swift, Efficient virtual memory for big memory servers, in *ISCA* (2013)
14. A. Bensoussan, C.T. Clingen, R.C. Daley, The Multics virtual memory: concepts and design, in *CACM* (1972)
15. A. Bhattacharjee, Large-reach memory management unit caches, in *MICRO* (2013)
16. A. Bhattacharjee, M. Martonosi, Inter-core cooperative TLB for chip multiprocessors, in *ASPLOS* (2010)
17. A. Bhattacharjee, D. Lustig, M. Martonosi, Shared last-level TLBs for chip multiprocessors, in *HPCA* (2011)
18. N. Binkert, B. Beckman, A. Saidi, G. Black, A. Basu, The gem5 Simulator, in *CAN* (2011)
19. B.H. Bloom, Space/time trade-offs in hash coding with allowable errors. Commun. ACM **13**, 422–426 (1970)
20. A. Boroumand, S. Ghose, M. Patel, H. Hassan, B. Lucia, K. Hsieh, K.T. Malladi, H. Zheng, O. Mutlu, LazyPIM: an efficient cache coherence mechanism for processing-in-memory, in *CAL* (2016)
21. A. Boroumand, S. Ghose, M. Patel, H. Hassan, B. Lucia, N. Hajinazar, K. Hsieh, K.T. Malladi, H. Zheng, O. Mutlu, LazyPIM: efficient support for cache coherence in processing-in-memory architectures (2017). arXiv:1706.03162 [cs:AR]
22. A. Boroumand, S. Ghose, Y. Kim, R. Ausavarungnirun, E. Shiu, R. Thakur, D. Kim, A. Kuusela, A. Knies, P. Ranganathan, O. Mutlu, Google workloads for consumer devices: mitigating data movement bottlenecks, in *ASPLOS* (2018)
23. L.M. Censier, P. Feutrier, A new solution to coherence problems in multicache systems, in *IEEE TC* (1978)
24. L. Ceze, J. Tuck, P. Montesinos, J. Torrellas, BulkSC: bulk enforcement of sequential consistency, in *ISCA* (2007)
25. K.K. Chang, D. Lee, Z. Chishti, A.R. Alameldeen, C. Wilkerson, Y. Kim, O. Mutlu, Improving DRAM performance by parallelizing refreshes with accesses, in *HPCA* (2014)
26. K.K. Chang, A. Kashyap, H. Hassan, S. Ghose, K. Hsieh, D. Lee, T. Li, G. Pekhimenko, S. Khan, O. Mutlu, Understanding latency variation in modern DRAM chips: experimental characterization, analysis, and optimization, in *SIGMETRICS* (2016)
27. K.K. Chang, P.J. Nair, D. Lee, S. Ghose, M.K. Qureshi, O. Mutlu, Low-cost inter-linked subarrays (LISA): enabling fast inter-subarray data movement in DRAM, in *HPCA* (2016)
28. K.K. Chang, Understanding and improving the latency of DRAM-based memory systems. Ph.D. dissertation, Carnegie Mellon University, 2017

29. K.K. Chang, A.G. Yağlıkçı, S. Ghose, A. Agrawal, N. Chatterjee, A. Kashyap, D. Lee, M. O'Connor, H. Hassan, O. Mutlu, Understanding reduced-voltage operation in modern DRAM devices: experimental characterization, analysis, and mechanisms, in *SIGMETRICS* (2017)
30. P. Chi, S. Li, C. Xu, T. Zhang, J. Zhao, Y. Liu, Y. Wang, Y. Xie, PRIME: a novel processing-in-memory architecture for neural network computation in ReRAM-based main memory, in *ISCA* (2016)
31. L. Chua, Memristor—the missing circuit element, in *IEEE TCT* (1971)
32. E.S. Chung, J.D. Davis, J. Lee, LINQits: big data on little clients, in *ISCA* (2013)
33. J.D. Collins, H. Wang, D.M. Tullsen, C.J. Hughes, Y. Lee, D.M. Lavery, J.P. Shen, Speculative precomputation: long-range prefetching of delinquent loads, in *ISCA* (2001)
34. J.D. Collins, S. Sair, B. Calder, D.M. Tullsen, Pointer cache assisted prefetching, in *MICRO* (2002)
35. R. Cooksey, S. Jourdan, D. Grunwald, A stateless, content-directed data prefetching mechanism, in *ASPLOS* (2002)
36. N.C. Crago, S.J. Patel, OUTRIDER: efficient memory latency tolerance with decoupled strands, in *ISCA* (2011)
37. J. Dean, L.A. Barroso, The tail at scale, in *CACM* (2013)
38. J. Devietti, B. Lucia, L. Ceze, M. Oskin, DMP: deterministic shared memory multiprocessing, in *ASPLOS* (2009)
39. J. Draper, J. Chame, M. Hall, C. Steele, T. Barrett, J. LaCoss, J. Granacki, J. Shin, C. Chen, C.W. Kang, I. Kim, G. Daglikoca, The architecture of the DIVA processing-in-memory chip, in *SC* (2002)
40. E. Ebrahimi, O. Mutlu, Y. Patt, Techniques for bandwidth-efficient prefetching of linked data structures in hybrid prefetching systems, in *HPCA* (2009)
41. E. Ebrahimi, O. Mutlu, C.J. Lee, Y.N. Patt, Coordinated control of multiple prefetchers in multi-core systems, in *MICRO* (2009)
42. E. Ebrahimi, C.J. Lee, O. Mutlu, Y.N. Patt, Prefetch-aware shared resource management for multi-core systems, in *ISCA* (2011)
43. Y. Eckert, N. Jayasena, G.H. Loh, Thermal feasibility of die-stacked processing in memory, in *WoNDP* (2014)
44. D.G. Elliott, W.M. Snelgrove, M. Stumm, Computational RAM: a memory-SIMD hybrid and its application to DSP, in *CICC* (1992)
45. D. Elliott, M. Stumm, W.M. Snelgrove, C. Cojocaru, R. McKenzie, Computational RAM: implementing processors in memory, in *IEEE Design & Test* (1999)
46. R. Elmasri, *Fundamentals of Database Systems* (Pearson, Boston, 2007)
47. A. Farmahini-Farahani, J.H. Ahn, K. Morrow, N.S. Kim, NDA: near-DRAM acceleration architecture leveraging commodity DRAM devices and standard memory modules, in *HPCA* (2015)
48. M. Ferdman, A. Adileh, O. Kocberber, S. Volos, M. Alisafaee, D. Jevdjic, C. Kaynak, A.D. Popescu, A. Ailamaki, B. Falsafi, Clearing the clouds: a study of emerging scale-out workloads on modern hardware, in *ASPLOS* (2012)
49. M. Filippo, Technology preview: ARM next generation processing, in *ARM TechCon* (2012)
50. B. Fitzpatrick, Distributed caching with memcached. Linux J. **2004**, 5 (2004)
51. M. Gao, C. Kozyrakis, HRL: efficient and flexible reconfigurable logic for near-data processing, in *HPCA* (2016)
52. M. Gao, G. Ayers, C. Kozyrakis, Practical near-data processing for in-memory analytics frameworks, in *PACT* (2015)
53. S. Ghemawat, H. Gobioff, S.-T. Leung, The Google file system, in *SOSP* (2003)
54. D. Giampaolo, *Practical File System Design with the BE File System* (Morgan Kaufmann Publishers Inc., San Francisco, 1998)
55. A. Glew, MLP yes! ILP no!, in *ASPLOS WACI* (1998)
56. M. Gokhale, B. Holmes, K. Iobst, Processing in memory: the Terasys massively parallel PIM array. IEEE Comput. **28**, 23–31 (1995)

57. J.R. Goodman, Using cache memory to reduce processor-memory traffic, in *ISCA* (1983)
58. B. Gu, A.S. Yoon, D.-H. Bae, I. Jo, J. Lee, J. Yoon, J.-U. Kang, M. Kwon, C. Yoon, S. Cho, J. Jeong, D. Chang, Biscuit: a framework for near-data processing of big data workloads, in *ISCA* (2016)
59. Q. Guo, N. Alachiotis, B. Akin, F. Sadi, G. Xu, T.M. Low, L. Pileggi, J.C. Hoe, F. Franchetti, 3D-stacked memory-side acceleration: accelerator and system design, in *WoNDP* (2014)
60. A. Gutierrez, J. Pusdesris, R.G. Dreslinski, T. Mudge, C. Sudanthi, C.D. Emmons, M. Hayenga, N. Paver, Sources of error in full-system simulation, in *ISPASS* (2014)
61. L. Hammond, V. Wong, M. Chen, B.D. Carlstrom, J.D. Davis, B. Hertzberg, M.K. Prabhu, H. Wijaya, C. Kozyrakis, K. Olukotun, Transactional memory coherence and consistency, in *ISCA* (2004)
62. M. Hashemi, O. Mutlu, Y.N. Patt, Continuous runahead: transparent hardware acceleration for memory intensive workloads, in *MICRO* (2016)
63. M. Hashemi, Khubaib, E. Ebrahimi, O. Mutlu, Y.N. Patt, Accelerating dependent cache misses with an enhanced memory controller, in *ISCA* (2016)
64. S.M. Hassan, S. Yalamanchili, S. Mukhopadhyay, Near data processing: impact and optimization of 3D memory system architecture on the uncore, in *MEMSYS* (2015)
65. H. Hassan, G. Pekhimenko, N. Vijaykumar, V. Seshadri, D. Lee, O. Ergin, O. Mutlu, ChargeCache: reducing DRAM latency by exploiting row access locality, in *HPCA* (2016)
66. H. Hassan, N. Vijaykumar, S. Khan, S. Ghose, K. Chang, G. Pekhimenko, D. Lee, O. Ergin, O. Mutlu, SoftMC: a flexible and practical open-source infrastructure for enabling experimental DRAM studies, in *HPCA* (2017)
67. K. Hsieh, S. Khan, N. Vijaykumar, K.K. Chang, A. Boroumand, S. Ghose, O. Mutlu, Accelerating pointer chasing in 3D-stacked memory: challenges, mechanisms, evaluation, in *ICCD* (2016)
68. K. Hsieh, E. Ebrahimi, G. Kim, N. Chatterjee, M. O'Conner, N. Vijaykumar, O. Mutlu, S. Keckler, Transparent offloading and mapping (TOM): enabling programmer-transparent near-data processing in GPU systems, in *ISCA* (2016)
69. Z. Hu, M. Martonosi, S. Kaxiras, TCP: tag correlating prefetchers, in *HPCA* (2003)
70. C.J. Hughes, S.V. Adve, Memory-side prefetching for linked data structures for processor-in-memory systems, in *JPDC* (2005)
71. Hybrid Memory Cube Consortium, HMC Specification 1.1 (2013)
72. Hybrid Memory Cube Consortium, HMC Specification 2.0 (2014)
73. Intel, Intel Xeon Processor W3550 (2009)
74. J. Jeddeloh, B. Keeth, Hybrid memory cube: new DRAM architecture increases density and performance, in *VLSIT* (2012)
75. JEDEC, High bandwidth memory (HBM) DRAM, Standard No. JESD235 (2013)
76. J. Joao, O. Mutlu, Y.N. Patt, Flexible reference-counting-based hardware acceleration for garbage collection, in *ISCA* (2009)
77. R. Jones, R. Lins, *Garbage Collection: Algorithms for Automatic Dynamic Memory Management* (Wiley, New York, 1996)
78. D. Joseph, D. Grunwald, Prefetching using Markov predictors, in *ISCA* (1997)
79. S. Kanev, J.P. Darago, K. Hazelwood, P. Ranganathan, T. Moseley, G.-Y. Wei, D. Brooks, Profiling a warehouse-scale computer, in *ISCA* (2015)
80. Y. Kang, W. Huang, S.-M. Yoo, D. Keen, Z. Ge, V. Lam, P. Pattnaik, J. Torrellas, FlexRAM: toward an advanced intelligent memory system, in *ICCD* (1999)
81. M. Kang, M.-S. Keel, N.R. Shanbhag, S. Eilert, K. Curewitz, An energy-efficient VLSI architecture for pattern recognition via deep embedding of computation in SRAM, in *ICASSP* (2014)
82. U. Kang, H.-S. Yu, C. Park, H. Zheng, J. Halbert, K. Bains, S. Jang, J. Choi, Co-architecting controllers and DRAM to enhance DRAM process scaling, in *The Memory Forum* (2014)
83. M. Karlsson, F. Dahlgren, P. Stenström, A prefetching technique for irregular accesses to linked data structures, in *HPCA* (2000)

84. S. Khan, D. Lee, Y. Kim, A.R. Alameldeen, C. Wilkerson, O. Mutlu, The efficacy of error mitigation techniques for DRAM retention failures: a comparative experimental study, in *SIGMETRICS* (2014)
85. S. Khan, D. Lee, O. Mutlu, PARBOR: an efficient system-level technique to detect data dependent failures in DRAM, in *DSN* (2016)
86. S. Khan, C. Wilkerson, D. Lee, A.R. Alameldeen, O. Mutlu, A case for memory content-based detection and mitigation of data-dependent failures in DRAM, in *CAL* (2016)
87. S. Khan, C. Wilkerson, Z. Wang, A. Alameldeen, D. Lee, O. Mutlu, Detecting and mitigating data-dependent DRAM failures by exploiting current memory content, in *MICRO* (2017)
88. T. Kilburn, D.B.G. Edwards, M.J. Lanigan, F.H. Sumner, One-level storage system. IRE Trans. Electron Comput. **2**, 223–235 (1962)
89. Y. Kim, Architectural techniques to enhance DRAM scaling. Ph.D. dissertation, Carnegie Mellon University, 2015
90. Y. Kim, V. Seshadri, D. Lee, J. Liu, O. Mutlu, A case for exploiting subarray-level parallelism (SALP) in DRAM, in *ISCA* (2012)
91. Y. Kim, R. Daly, J. Kim, C. Fallin, J.H. Lee, D. Lee, C. Wilkerson, K. Lai, O. Mutlu, Flipping bits in memory without accessing them: an experimental study of DRAM disturbance errors, in *ISCA* (2014)
92. Y. Kim, W. Yang, O. Mutlu, Ramulator: a fast and extensible DRAM simulator, in *CAL* (2015)
93. D. Kim, J. Kung, S. Chai, S. Yalamanchili, S. Mukhopadhyay, Neurocube: a programmable digital neuromorphic architecture with high-density 3D memory, in *ISCA* (2016)
94. Y. Kim, R. Daly, J. Kim, C. Fallin, J.H. Lee, D. Lee, C. Wilkerson, K. Lai, O. Mutlu, RowHammer: reliability analysis and security implications (2016). arXiv:1603.00747 [cs:AR]
95. G. Kim, N. Chatterjee, M. O'Connor, K. Hsieh, Toward standardized near-data processing with unrestricted data placement for GPUs, in *SC* (2017)
96. J.S. Kim, D. Senol, H. Xin, D. Lee, S. Ghose, M. Alser, H. Hassan, O. Ergin, C. Alkan, O. Mutlu, GRIM-Filter: fast seed filtering in read mapping using emerging memory technologies. arXiv:1708.04329 [q-bio.GN] (2017)
97. J. Kim, M. Patel, H. Hassan, O. Mutlu, The DRAM latency PUF: quickly evaluating physical unclonable functions by exploiting the latency–reliability tradeoff in modern DRAM devices, in *HPCA* (2018)
98. J.S. Kim, D. Senol, H. Xin, D. Lee, S. Ghose, M. Alser, H. Hassan, O. Ergin, C. Alkan, O. Mutlu, GRIM-Filter: fast seed location filtering in DNA read mapping using processing-in-memory technologies, in *BMC Genomics* (2018)
99. Y.O. Koçberber, B. Grot, J. Picorel, B. Falsafi, K.T. Lim, P. Ranganathan, Meet the walkers: accelerating index traversals for in-memory databases, in *MICRO* (2013)
100. P.M. Kogge, EXECUBE-a new architecture for scaleable MPPs, in *ICPP* (1994)
101. E. Kültürsay, M. Kandemir, A. Sivasubramaniam, O. Mutlu, Evaluating STT-RAM as an energy-efficient main memory alternative, in *ISPASS* (2013)
102. L. Kurian, P.T. Hulina, L.D. Coraor, Memory latency effects in decoupled architectures with a single data memory module, in *ISCA* (1992)
103. S. Kvatinsky, A. Kolodny, U.C. Weiser, E.G. Friedman, Memristor-based IMPLY logic design procedure, in *ICCD* (2011)
104. S. Kvatinsky, D. Belousov, S. Liman, G. Satat, N. Wald, E.G. Friedman, A. Kolodny, U.C. Weiser, MAGIC—memristor-aided logic, in *IEEE TCAS II: Express Briefs* (2014)
105. S. Kvatinsky, G. Satat, N. Wald, E.G. Friedman, A. Kolodny, U.C. Weiser, Memristor-based material implication (IMPLY) logic: design principles and methodologies, in *TVLSI* (2014)
106. L. Lamport, How to make a multiprocessor computer that correctly executes multiprocess programs, in *IEEE TC* (1979)
107. D. Lee, Reducing DRAM latency at low cost by exploiting heterogeneity. Ph.D. dissertation, Carnegie Mellon University, 2016
108. J. Lee, Y. Solihin, J. Torrettas, Automatically mapping code on an intelligent memory architecture, in *HPCA* (2001)

109. C.J. Lee, O. Mutlu, V. Narasiman, Y.N. Patt, Prefetch-aware DRAM controllers, in *MICRO* (2008)
110. B.C. Lee, E. Ipek, O. Mutlu, D. Burger, Architecting phase change memory as a scalable DRAM alternative, in *ISCA* (2009)
111. B.C. Lee, E. Ipek, O. Mutlu, D. Burger, Phase change memory architecture and the quest for scalability, in *CACM* (2010)
112. B.C. Lee, P. Zhou, J. Yang, Y. Zhang, B. Zhao, E. Ipek, O. Mutlu, D. Burger, Phase-change technology and the future of main memory, in *IEEE Micro* (2010)
113. C.J. Lee, O. Mutlu, V. Narasiman, Y.N. Patt, Prefetch-aware memory controllers, in *IEEE TC* (2011)
114. D. Lee, Y. Kim, V. Seshadri, J. Liu, L. Subramanian, O. Mutlu, Tiered-latency DRAM: a low latency and low cost DRAM architecture, in *HPCA* (2013)
115. D. Lee, F. Hormozdiari, H. Xin, F. Hach, O. Mutlu, C. Alkan, Fast and accurate mapping of complete genomics reads, in *Methods* (2014)
116. D. Lee, Y. Kim, G. Pekhimenko, S. Khan, V. Seshadri, K. Chang, O. Mutlu, Adaptive-latency DRAM: optimizing DRAM timing for the common-case, in *HPCA* (2015)
117. D. Lee, L. Subramanian, R. Ausavarungnirun, J. Choi, O. Mutlu, Decoupled direct memory access: isolating CPU and IO traffic by leveraging a dual-data-port DRAM, in *PACT* (2015)
118. J.H. Lee, J. Sim, H. Kim, BSSync: processing near memory for machine learning workloads with bounded staleness consistency models, in *PACT* (2015)
119. D. Lee, S. Ghose, G. Pekhimenko, S. Khan, O. Mutlu, Simultaneous multi-layer access: improving 3D-stacked memory bandwidth at low cost, in *TACO* (2016)
120. D. Lee, S. Khan, L. Subramanian, S. Ghose, R. Ausavarungnirun, G. Pekhimenko, V. Seshadri, O. Mutlu, Design-induced latency variation in modern DRAM chips: characterization, analysis, and latency reduction mechanisms, in *SIGMETRICS* (2017)
121. Y. Levy, J. Bruck, Y. Cassuto, E.G. Friedman, A. Kolodny, E. Yaakobi, S. Kvatinsky, Logic operations in memory using a memristive Akers array. Microelectron. J. **45**, 1429–1437 (2014)
122. S. Li, J.H. Ahn, R.D. Strong, J.B. Brockman, D.M. Tullsen, N.P. Jouppi, The McPAT framework for multicore and manycore architectures: simultaneously modeling power, area, and timing, in *TACO* (2013)
123. S. Li, C. Xu, Q. Zou, J. Zhao, Y. Lu, Y. Xie, Pinatubo: a processing-in-memory architecture for bulk bitwise operations in emerging non-volatile memories, in *DAC* (2016)
124. S. Li, D. Niu, K.T. Malladi, H. Zheng, B. Brennan, Y. Xie, DRISA: a DRAM-based reconfigurable in-situ accelerator, in *MICRO* (2017)
125. K. Lim, J. Chang, T. Mudge, P. Ranganathan, S.K. Reinhardt, T.F. Wenisch, Disaggregated memory for expansion and sharing in blade servers, in *ISCA* (2009)
126. K.T. Lim, D. Meisner, A.G. Saidi, P. Ranganathan, T.F. Wenisch, Thin servers with smart pipes: designing SoC accelerators for memcached, in *ISCA* (2013)
127. Linaro, 64-Bit Linux Kernel for ARM (2014)
128. M.H. Lipasti, W.J. Schmidt, S.R. Kunkel, R.R. Roediger, SPAID: software prefetching in pointer- and call-intensive environments, in *MICRO* (1995)
129. J. Liu, B. Jaiyen, R. Veras, O. Mutlu, RAIDR: retention-aware intelligent DRAM refresh, in *ISCA* (2012)
130. J. Liu, B. Jaiyen, Y. Kim, C. Wilkerson, O. Mutlu, An experimental study of data retention behavior in modern DRAM devices: implications for retention time profiling mechanisms, in *ISCA* (2013)
131. Z. Liu, I. Calciu, M. Harlihy, O. Mutlu, Concurrent data structures for near-memory computing, in *SPAA* (2017)
132. G.H. Loh, 3D-stacked memory architectures for multi-core processors, in *ISCA* (2008)
133. G.H. Loh, N. Jayasena, M. Oskin, M. Nutter, D. Roberts, M. Meswani, D.P. Zhang, M. Ignatowski, A processing in memory taxonomy and a case for studying fixed-function PIM, in *WoNDP* (2013)

134. P. Lotfi-Kamran, B. Grot, M. Ferdman, S. Volos, Y.O. Koçberber, J. Picorel, A. Adileh, D. Jevdjic, S. Idgunji, E. Özer, B. Falsafi, Scale-out processors, in *ISCA* (2012)
135. C. Luk, Tolerating memory latency through software-controlled pre-execution in simultaneous multithreading processors, in *ISCA* (2001)
136. C. Luk, T.C. Mowry, Compiler-based prefetching for recursive data structures, in *ASPLOS* (1996)
137. Y. Luo, S. Govindan, B. Sharma, M. Santaniello, J. Meza, A. Kansal, J. Liu, B. Khessib, K. Vaid, O. Mutlu, Characterizing application memory error vulnerability to optimize datacenter cost via heterogeneous-reliability memory, in *DSN* (2014)
138. Y. Luo, S. Ghose, T. Li, S. Govindan, B. Sharma, B. Kelly, A. Boroumand, O. Mutlu, Using ECC DRAM to adaptively increase memory capacity (2017). arXiv:1706.08870 [cs:AR]
139. D. Lustig, A. Bhattacharjee, M. Martonosi, TLB improvements for chip multiprocessors: inter-core cooperative prefetchers and shared last-level TLBs, in *ACM TACO* (2013)
140. K. Mai, T. Paaske, N. Jayasena, R. Ho, W.J. Dally, M. Horowitz, Smart memories: a modular reconfigurable architecture, in *ISCA* (2000)
141. J.A. Mandelman, R.H. Dennard, G.B. Bronner, J.K. DeBrosse, R. Divakaruni, Y. Li, C.J. Radens, Challenges and future directions for the scaling of dynamic random-access memory (DRAM), in *IBM JRD* (2002)
142. Y. Mao, E. Kohler, R.T. Morris, Cache craftiness for fast multicore key-value storage, in *EuroSys* (2012)
143. S.A. McKee, Reflections on the memory wall, in *CF* (2004)
144. MemSQL, Inc., MemSQL. http://www.memsql.com
145. M.R. Meswani, S. Blagodurov, D. Roberts, J. Slice, M. Ignatowski, G.H. Loh, Heterogeneous memory architectures: a HW/SW approach for mixing die-stacked and off-package memories, in *HPCA* (2015), pp. 126–136
146. J. Meza, Y. Luo, S. Khan, J. Zhao, Y. Xie, O. Mutlu, A case for efficient hardware-software cooperative management of storage and memory, in *WEED* (2013)
147. J. Meza, Q. Wu, S. Kumar, O. Mutlu, Revisiting memory errors in large-scale production data centers: analysis and modeling of new trends from the field, in *DSN* (2015)
148. N. Mirzadeh, O. Kocberber, B. Falsafi, B. Grot, Sort vs. hash join revisited for near-memory execution, in *ASBD* (2007)
149. A. Morad, L. Yavits, R. Ginosar, GP-SIMD processing-in-memory, in *ACM TACO* (2015)
150. J. Mukundan, H. Hunter, K.H. Kim, J. Stuecheli, J.F. Martínez, Understanding and mitigating refresh overheads in high-density DDR4 DRAM systems, in *ISCA* (2013)
151. O. Mutlu, Memory scaling: a systems architecture perspective, in *IMW* (2013)
152. O. Mutlu, The RowHammer problem and other issues we may face as memory becomes denser, in *DATE* (2017)
153. O. Mutlu, J. Stark, C. Wilkerson, Y.N. Patt, Runahead execution: an alternative to very large instruction windows for out-of-order processors, in *HPCA* (2003)
154. O. Mutlu, J. Stark, C. Wilkerson, Y.N. Patt, Runahead execution: an effective alternative to large instruction windows, in *IEEE Micro* (2003)
155. O. Mutlu, H. Kim, Y.N. Patt, Address-value delta (AVD) prediction: increasing the effectiveness of runahead execution by exploiting regular memory allocation patterns, in *MICRO* (2005)
156. O. Mutlu, H. Kim, Y.N. Patt, Techniques for efficient processing in runahead execution engines, in *ISCA* (2005)
157. O. Mutlu, H. Kim, Y.N. Patt, Address-value delta (AVD) prediction: a hardware technique for efficiently parallelizing dependent cache misses, in *TC* (2006)
158. O. Mutlu, H. Kim, Y.N. Patt, Efficient runahead execution: power-efficient memory latency tolerance, in *IEEE Micro* (2006)
159. O. Mutlu, T. Moscibroda, Parallelism-aware batch scheduling: enhancing both performance and fairness of shared DRAM systems, in *ISCA* (2008)
160. O. Mutlu, L. Subramanian, Research problems and opportunities in memory systems, in *SUPERFRI* (2014)

161. A. Muzahid, D. Suárez, S. Qi, J. Torrellas, SigRace: signature-based data race detection, in *ISCA* (2009)
162. H. Naeimi, C. Augustine, A. Raychowdhury, S.-L. Lu, J. Tschanz, STT-RAM scaling and retention failure. Intel Technol. J. **17**, 54–75 (2013)
163. L. Nai, R. Hadidi, J. Sim, H. Kim, P. Kumar, H. Kim, GraphPIM: enabling instruction-level PIM offloading in graph computing frameworks, in *HPCA* (2017)
164. B. Naylor, J. Amanatides, W. Thibault, Merging BSP trees yields polyhedral set operations, in *SIGGRAPH* (1990)
165. OProfile, http://oprofile.sourceforge.net/
166. M. Oskin, F.T. Chong, T. Sherwood, Active pages: a computation model for intelligent memory, in *ISCA* (1998)
167. M.S. Papamarcos, J.H. Patel, A low-overhead coherence solution for multiprocessors with private. Cache memories, in *ISCA* (1984)
168. M. Patel, J. Kim, O. Mutlu, The reach profiler (REAPER): enabling the mitigation of DRAM retention failures via profiling at aggressive conditions, in *ISCA* (2017)
169. Y.N. Patt, W.-M. Hwu, M. Shebanow, HPS, a new microarchitecture: rationale and introduction, in *MICRO* (1985)
170. Y.N. Patt, S.W. Melvin, W.-M. Hwu, M.C. Shebanow, Critical issues regarding HPS, a high performance microarchitecture, in *MICRO*, (1985)
171. D. Patterson, T. Anderson, N. Cardwell, R. Fromm, K. Keeton, C. Kozyrakis, R. Thomas, K. Yelick, A case for intelligent RAM, in *IEEE Micro* (1997)
172. A. Pattnaik, X. Tang, A. Jog, O. Kayiran, A.K. Mishra, M.T. Kandemir, O. Mutlu, C.R. Das, Scheduling techniques for GPU architectures with processing-in-memory capabilities, in *PACT* (2016)
173. B. Pichai, L. Hsu, A. Bhattacharjee, Architectural support for address translation on GPUs: designing memory management units for CPU/GPUs with unified address spaces, in *ASPLOS* (2014)
174. G. Pokam, C. Pereira, K. Danne, R. Kassa, A.-R. Adl-Tabatabai, Architecting a chunk-based memory race recorder in modern CMPs, in *MICRO* (2009)
175. J. Power, M.D. Hill, D.A. Wood, Supporting x86-64 address translation for 100s of GPU lanes, in *HPCA* (2014)
176. S.H. Pugsley, J. Jestes, H. Zhang, R. Balasubramonian, V. Srinivasan, A. Buyuktosunoglu, A. Davis, F. Li, NDC: analyzing the impact of 3D-stacked memory+logic devices on mapreduce workloads, in *ISPASS* (2014)
177. M.K. Qureshi, M.A. Suleman, Y.N. Patt, Line distillation: increasing cache capacity by filtering unused words in cache lines, in *HPCA* (2007)
178. M.K. Qureshi, A. Jaleel, Y.N. Patt, S.C. Steely Jr., J. Emer, Adaptive insertion policies for high-performance caching, in *ISCA* (2007)
179. M.K. Qureshi, V. Srinivasan, J.A. Rivers, Scalable high performance main memory system using phase-change memory technology, in *ISCA* (2009)
180. M.K. Qureshi, D.H. Kim, S. Khan, P.J. Nair, O. Mutlu, AVATAR: a variable-retention-time (VRT) aware refresh for DRAM systems, in *DSN* (2015)
181. J. Ren, J. Zhao, S. Khan, J. Choi, Y. Wu, O. Mutlu, ThyNVM: enabling software-transparent crash consistency in persistent memory systems, in *MICRO* (2015)
182. S. Rixner, W.J. Dally, U.J. Kapasi, P. Mattson, J.D. Owens, Memory access scheduling, in *ISCA* (2000)
183. O. Rodeh, C. Mason, J. Bacik, BTRFS: the Linux B-tree filesystem, in *TOS* (2013)
184. A. Rogers, M. C. Carlisle, J.H. Reppy, L.J. Hendren, Supporting dynamic data structures on distributed-memory machines, in *TOPLAS* (1995)
185. P. Rosenfeld, E. Cooper-Balis, B. Jacob, DRAMSim2: a cycle accurate memory system simulator, in *CAL* (2011)
186. A. Roth, G.S. Sohi, Effective jump-pointer prefetching for linked data structures, in *ISCA* (1999)

187. A. Roth, A. Moshovos, G.S. Sohi, Dependence based prefetching for linked data structures, in *ASPLOS* (1998)
188. SAFARI Research Group, IMPICA (in-memory pointer chasing accelerator) – GitHub repository. https://github.com/CMU-SAFARI/IMPICA/
189. SAFARI Research Group, Ramulator: A DRAM simulator – GitHub repository. https://github.com/CMU-SAFARI/ramulator/
190. SAFARI Research Group, SAFARI software tools – GitHub repository. https://github.com/CMU-SAFARI/
191. SAFARI Research Group, SoftMC v1.0 – GitHub repository. https://github.com/CMU-SAFARI/SoftMC/
192. D. Sanchez, L. Yen, M.D. Hill, K. Sankaralingam, Implementing signatures for transactional memory, in *MICRO* (2007)
193. SAP SE, SAP HANA. http://www.hana.sap.com/
194. B. Schroeder, E. Pinheiro, W.-D. Weber, DRAM errors in the wild: a large-scale field study, in *SIGMETRICS* (2009)
195. V. Seshadri, D. Lee, T. Mullins, H. Hassan, A. Boroumand, J. Kim, M.A. Kozuch, O. Mutlu, P.B. Gibbons, T.C. Mowry, Buddy-RAM: improving the performance and efficiency of bulk bitwise operations using DRAM (2016). arXiv:1611.09988 [cs:AR]
196. V. Seshadri, D. Lee, T. Mullins, H. Hassan, A. Boroumand, J. Kim, M.A. Kozuch, O. Mutlu, P.B. Gibbons, T.C. Mowry, Ambit: in-memory accelerator for bulk bitwise operations using commodity DRAM technology, in *MICRO* (2017)
197. V. Seshadri, Simple DRAM and virtual memory abstractions to enable highly efficient memory systems. Ph.D. dissertation, Carnegie Mellon University, 2016
198. V. Seshadri, O. Mutlu, The processing using memory paradigm: In-DRAM bulk copy, initialization, bitwise AND and OR (2016). arXiv:1610.09603 [cs:AR]
199. V. Seshadri, O. Mutlu, Simple operations in memory to reduce data movement. Adv. Comput. **106**, 107–166 (2017)
200. V. Seshadri, Y. Kim, C. Fallin, D. Lee, R. Ausavarungnirun, G. Pekhimenko, Y. Luo, O. Mutlu, M.A. Kozuch, P.B. Gibbons, T.C. Mowry, RowClone: fast and energy-efficient in-DRAM bulk data copy and initialization, in *MICRO* (2013)
201. V. Seshadri, A. Bhowmick, O. Mutlu, P.B. Gibbons, M.A. Kozuch, T.C. Mowry, The dirty-block index, in *ISCA* (2014)
202. V. Seshadri, K. Hsieh, A. Boroumand, D. Lee, M.A. Kozuch, O. Mutlu, P.B. Gibbons, T.C. Mowry, Fast bulk bitwise AND and OR in DRAM, *CAL* (2015)
203. V. Seshadri, T. Mullins, A. Boroumand, O. Mutli, P.B. Gibbons, M.A. Kozuch, T.C. Mowry, Gather-scatter DRAM: in-DRAM address translation to improve the spatial locality of non-unit strided accesses, in *MICRO* (2015)
204. V. Seshadri, S. Yedkar, H. Xin, O. Mutlu, P.B. Gibbons, M.A. Kozuch, T.C. Mowry, Mitigating prefetcher-caused pollution using informed caching policies for prefetched blocks. ACM TACO **11**(4), 51:1–51:22 (2015)
205. A. Shafiee, A. Nag, N. Muralimanohar, R. Balasubramonian, J.P. Strachan, M. Hu, R.S. Williams, V. Srikumar, ISAAC: a convolutional neural network accelerator with in-situ analog arithmetic in crossbars, in *ISCA* (2016)
206. J.S. Shapiro, J. Adams, Design evolution of the EROS single-level store, in *USENIX ATC* (2002)
207. J.S. Shapiro, J.M. Smith, D.J. Farber, EROS: a fast capability system, in *SOSP* (1999)
208. D.E. Shaw, S.J. Stolfo, H. Ibrahim, B. Hillyer, G. Wiederhold, J. Andrews, The NON-VON database machine: a brief overview. IEEE Database Eng. Bull. **4**, 41–52 (1981)
209. J. Shun, G.E. Blelloch, Ligra: a lightweight graph processing framework for shared memory, in *PPoPP* (2013)
210. J.E. Smith, Decoupled access/execute computer architectures, in *ISCA* (1982)
211. J.E. Smith, Dynamic instruction scheduling and the astronautics ZS-1, in *Computer* (1986)
212. J.E. Smith, S. Weiss, N.Y. Pang, A simulation study of decoupled architecture computers, in *IEEE TC* (1986)

213. Y. Solihin, J. Torrellas, J. Lee, Using a user-level memory thread for correlation prefetching, in *ISCA* (2002)
214. V. Sridharan, N. DeBardeleben, S. Blanchard, K.B. Ferreira, J. Stearley, J. Shalf, S. Gurumurthi, Memory errors in modern systems: the good, the bad, and the ugly, in *ASPLOS* (2015)
215. S. Srikantaiah, M. Kandemir, Synergistic TLBs for high performance address translation in chip multiprocessors, in *MICRO* (2010)
216. S. Srinath, O. Mutlu, H. Kim, Y.N. Patt, Feedback directed prefetching: improving the performance and bandwidth-efficiency of hardware prefetchers, in *HPCA* (2007)
217. Stanford Network Analysis Project, http://snap.stanford.edu/
218. H.S. Stone, A logic-in-memory computer, in *TC* (1970)
219. M. Stonebraker, A. Weisberg, The VoltDB main memory DBMS. IEEE Data Eng. Bull. **36**, 21–27 (2013)
220. D.B. Strukov, G.S. Snider, D.R. Stewart, R.S. Williams, The missing memristor found. Nature **453**, 80 (2008)
221. Z. Sura, A. Jacob, T. Chen, B. Rosenburg, O. Sallenave, C. Bertolli, S. Antao, J. Brunheroto, Y. Park, K. O'Brien, R. Nair, Data access optimization in a processing-in-memory system, in *CF* (2015)
222. R.M. Tomasulo, An efficient algorithm for exploiting multiple arithmetic units, in *IBM JRD* (1967)
223. Transaction Processing Performance Council, TPC benchmarks. http://www.tpc.org
224. M. Waldvogel, G. Varghese, J. Turner, B. Plattner, Scalable high speed IP routing lookups, in *SIGCOMM* (1997)
225. L. Wang, J. Zhan, C. Luo, Y. Zhu, Q. Yang, Y. He, W. Gao, Z. Jia, Y. Shi, S. Zhang, C. Zheng, G. Lu, K. Zhan, X. Li, B. Qiu, BigDataBench: a big data benchmark suite from internet services, in *HPCA* (2014)
226. M.V. Wilkes, The memory gap and the future of high performance memories, in *CAN* (2001)
227. P.R. Wilson, Uniprocessor garbage collection techniques, in *IWMM* (1992)
228. H.-S.P. Wong, S. Raoux, S. Kim, J. Liang, J.P. Reifenberg, B. Rajendran, M. Asheghi, K.E. Goodson, Phase change memory. Proc. IEEE **98**, 2201–2227 (2010)
229. H.-S.P. Wong, H.-Y. Lee, S. Yu, Y.-S. Chen, Y. Wu, P.-S. Chen, B. Lee, F.T. Chen, M.-J. Tsai, Metal-oxide RRAM. Proc. IEEE **100**, 1951–1970 (2012)
230. L. Wu, R.J. Barker, M.A. Kim, K.A. Ross, Navigating big data with high-throughput, energy-efficient data partitioning, in *ISCA* (2013)
231. L. Wu, A. Lottarini, T.K. Paine, M.A. Kim, K.A. Ross, Q100: the architecture and design of a database processing unit, in *ASPLOS* (2014)
232. Y. Wu, Efficient discovery of regular stride patterns in irregular programs, in *PLDI* (2002)
233. W.A. Wulf, S.A. McKee, Hitting the memory wall: implications of the obvious, *CAN* (1995)
234. S.L. Xi, O. Babarinsa, M. Athanassoulis, S. Idreos, Beyond the wall: near-data processing for databases, in *DaMoN* (2015)
235. H. Xin, D. Lee, F. Hormozdiari, S. Yedkar, O. Mutlu, C. Alkan, Accelerating read mapping with FastHASH, in *BMC Genomics* (2013)
236. H. Xin, J. Greth, J. Emmons, G. Pekhimenko, C. Kingsford, C. Alkan, O. Mutlu, Shifted hamming distance: a fast and accurate SIMD-friendly filter to accelerate alignment verification in read mapping. Bioinformatics **31**, 1553–1560 (2015)
237. J. Xue, Z. Yang, Z. Qu, S. Hou, Y. Dai, Seraph: an efficient, low-cost system for concurrent graph processing, in *HPDC* (2014)
238. C. Yang, A.R. Lebeck, Push vs. pull: data movement for linked data structures, in *ICS* (2000)
239. H. Yoon, R.A.J. Meza, R. Harding, O. Mutlu, Row buffer locality aware caching policies for hybrid memories, in *ICCD* (2012)
240. H. Yoon, J. Meza, N. Muralimanohar, N.P. Jouppi, O. Mutlu, Efficient data mapping and buffering techniques for multilevel cell phase-change memories, in *ACM TACO* (2014)
241. X. Yu, G. Bezerra, A. Pavlo, S. Devadas, M. Stonebraker, Staring into the abyss: an evaluation of concurrency control with one thousand cores, in *VLDB* (2014)
242. X. Yu, C.J. Hughes, N. Satish, S. Devadas, IMP: indirect memory prefetcher, in *MICRO* (2015)

243. D.P. Zhang, N. Jayasena, A. Lyashevsky, J.L. Greathouse, L. Xu, M. Ignatowski, TOP-PIM: throughput-oriented programmable processing in memory, in *HPDC* (2014)
244. J. Zhao, O. Mutlu, Y. Xie, FIRM: fair and high-performance memory control for persistent memory systems, in *MICRO* (2014)
245. P. Zhou, B. Zhao, J. Yang, Y. Zhang, A durable and energy efficient main memory using phase change memory technology, in *ISCA* (2009)
246. Q. Zhu, T. Graf, H.E. Sumbul, L. Pileggi, F. Franchetti, Accelerating sparse matrix-matrix multiplication with 3D-stacked logic-in-memory hardware, in *HPEC* (2013)
247. C.B. Zilles, Benchmark health considered harmful, in *CAN* (2001)
248. C.B. Zilles, G.S. Sohi, Execution-based prediction using speculative slices, in *ISCA* (2001)
249. W.K. Zuravleff, T. Robinson, Controller for a synchronous DRAM that maximizes throughput by allowing memory requests and commands to be issued out of order. US Patent No. 5,630,096 (1997)

# Chapter 6
# Emerging Steep-Slope Devices and Circuits: Opportunities and Challenges

**Xueqing Li, Moon Seok Kim, Sumitha George, Ahmedullah Aziz, Matthew Jerry, Nikhil Shukla, John Sampson, Sumeet Gupta, Suman Datta, and Vijaykrishnan Narayanan**

## 6.1 Introduction

Supply voltage reduction for dynamic power reduction has been the key enabler in Dennard scaling of semiconductors to lower the power density in the past years. While the technology is approaching toward the node of a few nanometers, further supply voltage scaling has become extremely challenging. This is essentially because of the scaling switching characteristics of CMOS transistors. With conventional CMOS technologies, either in planar or FinFET structure, the device current as a function of the gate control voltage, as shown in Fig. 6.1, has a fundamental bottleneck in the subthreshold swing (SS), which is defined as the amount of required gate voltage to change the device current by a decade in the subthreshold region. This definition could be numerically expressed as

$$SS = \frac{\mathrm{d}V_{GS}}{\mathrm{d}\log_{10} I_{DS}}. \tag{6.1}$$

X. Li · M. S. Kim · S. George · J. Sampson · V. Narayanan (✉)
Department of Computer Science and Engineering, The Pennsylvania State University, University Park, PA, USA
e-mail: lixueq@cse.psu.edu; mqk5211@cse.psu.edu; sug241@cse.psu.edu; sampson@cse.psu.edu; vijay@cse.psu.edu

A. Aziz · S. Gupta
Department of Electrical Engineering, The Pennsylvania State University, University Park, PA, USA
e-mail: afa5191@psu.edu; skg157@engr.psu.edu

M. Jerry · N. Shukla · S. Datta
Department of Electrical Engineering, The University of Notre Dame, Notre Dame, IN, USA
e-mail: mjerry@nd.edu; nshukla@nd.edu; sdatta@nd.edu

 
R. O. Topaloglu, H.-S. P. Wong (eds.), *Beyond-CMOS Technologies for Next Generation Computer Design*, https://doi.org/10.1007/978-3-319-90385-9_6

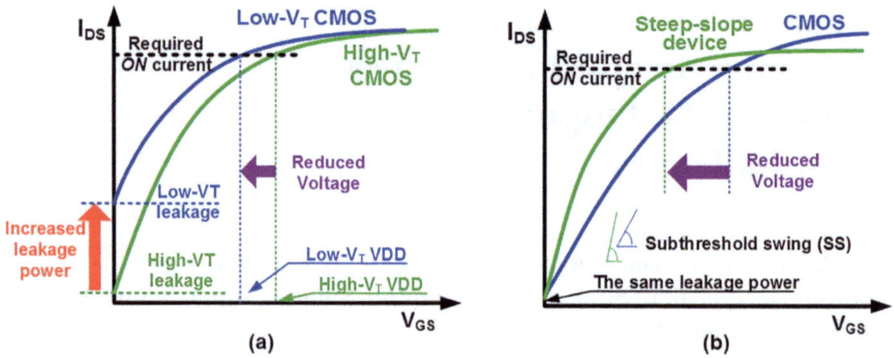


**Fig. 6.1** Shifting the transistor $I_{DS}$–$V_{GS}$ curves with threshold voltage ($V_T$) tuning in (**a**) and with a steep switching slope in (**b**)

In conventional MOSFETs, the thermionic emission of carriers, in which only the high energy carriers with energy exceeding the source-channel energy barrier contribute to the overall current, limits SS to be no less than $kT/q*\ln 10$, or 60 mV/dec at the room temperature, regardless of process optimizations applied such as High-K, 3D FinFET, etc.

Through the transistor threshold $V_{TH}$ tuning, it is yet still practical to shift the curves in Fig. 6.1a horizontally, so that the transistor is able to work with a different supply voltage and the same performance of speed or delay. When shifted to the left with a lower $V_{TH}$, the transistor's ON current, $I_{ON}$, could be obtained at a lower voltage, but the problem is the exponentially increased OFF-current, or leakage power by ten times per reduction of supply voltage equal to SS. While various optimizations could be applied in the process, circuit, and architecture tuning levels, the SS is now fundamentally discontinuing the voltage and power scaling in CMOS systems.

Fortunately, the advent of steep-slope devices has brought new opportunities for continuing the voltage scaling beyond CMOS. A steep-slope device has the intrinsic SS lower than 60 mV/dec. As shown in Fig. 6.1b, with a steep slope, the supply voltage could be lowered while keeping the ON current and the OFF current the same. At this point of time, reported steep-slope devices include tunneling FETs (TFETs) [1–10], ferroelectric negative capacitance FETs (FerroFETs or NCFETs) [11–21], mechanical gates [22–24], impact ionization [25–28], and phase transition material and FETs [29, 30], for example, Hyper-FETs based on $VO_2$, etc. While exhibiting a steep slope for lower digital power consumption, these devices also present other different features that not only change the story of existing conventional analog/RF and memory design, but also enable a set of emerging circuits and applications. Meanwhile, there are still challenges on the way toward replacing CMOS using these emerging devices, from device fabrication and integration, characterization and modeling, to circuit design and architecture optimizations. Thus, it is still of significance to continue the device–circuit co-design to mitigate side effects and make use of new features.

In this chapter, we focus on TFETs, NCFETs, and $VO_2$ devices, which are the three most promising emerging steep-slope devices that have the potential to replace CMOS. We will briefly summarize the state-of-the-art device research results in Sect. 6.2, review their operation mechanisms and circuits designs in Sects. 6.3, 6.4, and 6.5, discuss the modeling and benchmarking methodology in Sect. 6.6, and discuss the application opportunities and challenges in Sect. 6.7.

## 6.2 Steep-Slope Devices: State-of-the-Art

### 6.2.1 *Mechanism Toward Steep Slope*

There are a few different ways to achieve a steep slope. For a silicon-based MOSFET, the ideal subthreshold swing SS defined in Eq. 6.1 could be further investigated as

$$\mathrm{SS} = \frac{\mathrm{d}V_{\mathrm{GS}}}{\mathrm{dlog}_{10} I_{\mathrm{DS}}} = \frac{\mathrm{d}V_{\mathrm{GS}}}{\mathrm{d}\psi_S}\frac{\partial\psi_S}{\mathrm{dlog}_{10} I_{\mathrm{DS}}} = \left(1+\frac{C_s}{C_{\mathrm{inv}}}\right)\frac{\mathrm{kT}}{q}\ln 10, \qquad (6.2)$$

where $\psi_S$ is the silicon surface potential, $C_s$ is the silicon capacitance, and $C_{\mathrm{inv}}$ is the gate insulator capacitance [31, 32]. Tunneling diodes operate with a band-to-band-tunneling (BTBT)-based different switching mechanism and is thus not limited by the kT/$q$ Boltzmann thermodynamics. Negative capacitance materials provide a negative gate insulator capacitance $C_{\mathrm{inv}}$ which can lead to the internal gate voltage amplification. Correlation or phase-transition diodes introduce additional current enhancement when the applied voltage crosses the phase-transition threshold. Those three approaches are now being explored in TFETs, NCFETs, and phase transition devices, respectively. Sections 6.3, 6.4, and 6.5 will describe the device operating mechanism in more detail so as to understand how steep slope is achieved. While fabricated devices with the concurrent use of two or more of these effects are still being developed [33], the understanding of these effects alone has improved significantly recently. In [34, 35], the non-idealities that lead to degradation of ON current and SS are analyzed, such as the impact of tunnel junction abruptness and source dopant fluctuations.

### 6.2.2 *State-of-the-Art Devices*

After the first discussions of TFETs around 2004 [2–4], <60 mV/dec TFET devices have been reported based on carbon nanotube [4], Si [5], Ge [6], III-V materials [7, 8], etc. In [36], reported <60 mV/dec TFET results till the year 2014 were

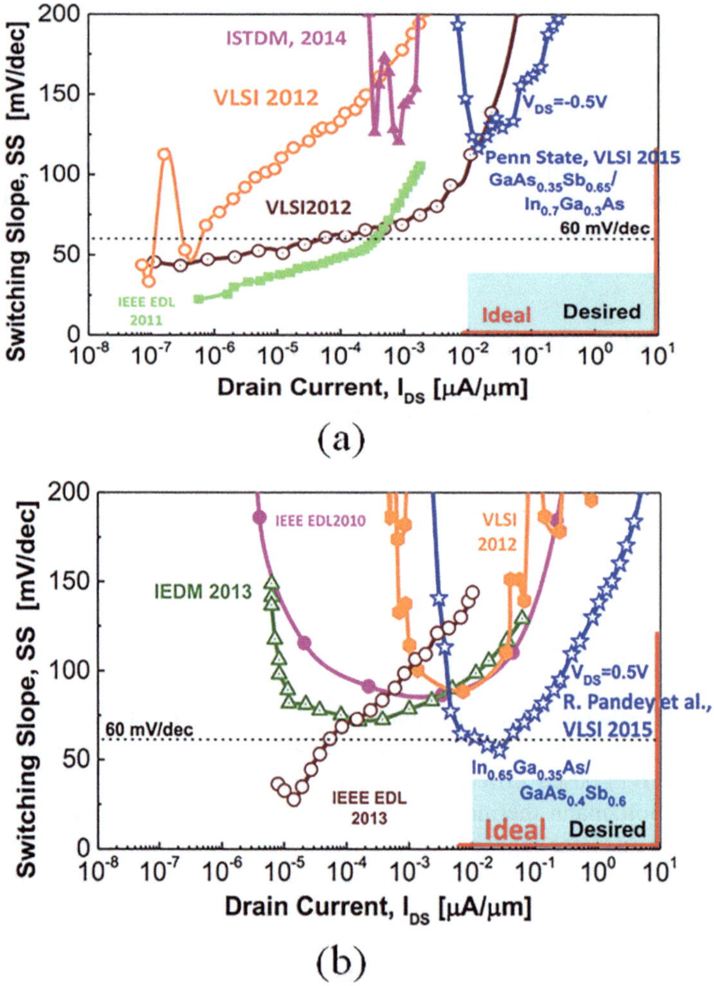


**Fig. 6.2** Comparisons of recent published steep-slope TFETs [8], with P-type TFET in (**a**) and N-type TFET in (**b**)

summarized, showing ON current up to $10^{-2}$ μA/μm for n-type TFET [9], and a few $10^{-4}$ μA/μm for p-type TFET [10]. In 2015, the first demonstration of complementary TFET was shown with 275 μA/μm $I_{ON}$ and sub-60 mV/dec up to a few $10^{-2}$ μA/μm current for n-type TFET, and 10 μA/μm $I_{ON}$ and 115 mV/dec minimum SS for p-type TFET [8]. The comparisons are shown in Fig. 6.2.

The NCFET research emerged more recently. In 2008, when Salahuddin et al. proposed the concept of using negative differential capacitance as a gate capacitor in a MOSFET [12], it was shown that it could potentially reduce SS to below 60 mV/dec. Since then, a few demonstrations of negative capacitance-based FETs are shown. Theoretical NCFET projections were reported, such as those in [11] and [13]. In [14], the anti-ferroelectricity in HfZrO2 (HZO) annealed at 600 °C with an abrupt turn ON of FET characteristics with $SS_{min}$ = 23 mV/dec and $SS_{avg}$ = 50 mV/dec over four decades of $I_{DS}$ was experimentally demonstrated. In [15], NCFET with organic/ferroelectric materials in a MOS gate stack was experimentally demonstrated with SS ~ 18 mV/dec at 300 K. In [16], SS as low as 8.5 mV/decade over as eight orders of magnitude of drain current in 100 nm gate length FinFET structure was experimentally measured with wide hysteresis. In [17], a nearly hysteresis-free 48 mV/dec NCFET in a p-type bulk MOSFET externally connected to a ferroelectric capacitor was reported. In [18], the first negative-capacitance FinFET with ALD $Hf_{042}Zr_{058}O_2$ added on top of the FinFET's gate stack was reported, showing 55–87 mV/dec SS and 25% $I_{ON}$ improvement. In [19], Ferroelectric HfZrOx (FE-HZO) FETs were experimentally demonstrated with small hysteresis window shift <0.1 V, forward SS = 42 mV/dec, and reverse SS = 28 mV/dec.

The Hyper-FET was initially proposed [29], after the introduction of vanadium dioxide (VO2) in [30]. In a Hyper-FET, the metallic state VO2 has little impact on the ON current, while the insulator state VO2 significantly reduces the OFF current. By proper threshold voltage tuning of the MOSFET, the Hyper-FET could have the same OFF current, while presenting higher ON current. More details would be discussed in Sect. 6.5. In [29], the ON current has improved by 20% and 60% for n-type and p-type Hyper-FETs over conventional MOSFETs. In [30], vanadium dioxide was monolithically integrated with silicon MOSFET to demonstrate a steep-slope transition of 8 mV/dec, leading to 36% increase in ON current over the baseline MOSFET.

## 6.3 TFET

### *6.3.1 Device Introduction*

A III-V HTFET is essentially a gated tunnel diode with asymmetric source/drain doping [1]. A double-gate conceptual implementation is illustrated in Fig. 6.3a, b [37], in which the field-effect gate control, band-to-band tunneling interface, and the asymmetrical doping are shown. The energy band diagram is shown in Fig. 6.3c, d for OFF state and ON state, respectively [38]. At a low gate bias voltage, the energy barrier is wide enough to suppress the band-to-band tunneling probability, and the device current is small. As the gate voltage increases, the bands are lowered in energy, narrowing the tunneling barrier and allowing more tunneling current to flow. Such a quantum-mechanical phenomenon of band-to-band tunneling (BTBT) provides an abrupt transition between the ON and OFF states, which has been theoretically and experimentally demonstrated to be less than the 60 mV/decade MOSFET limit at the room temperature.

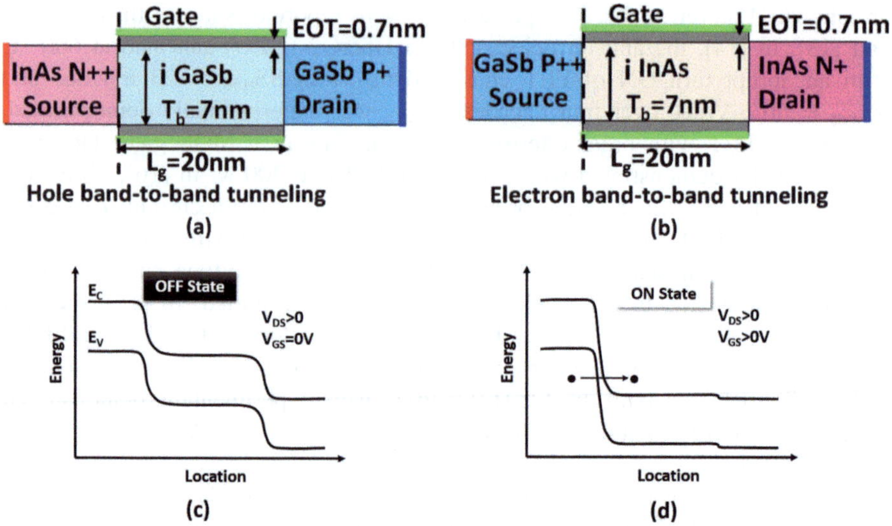


**Fig. 6.3** HTFET schematic in (**a**) for P-type and (**b**) for N-type with the energy band diagram in (**c**) and (**d**) for OFF state and ON state, respectively [37, 38]

**Table 6.1** HTFET model parameters [37]

| | Si FinFET | N-HTFET | P-HTFET |
|---|---|---|---|
| Gate length (Lg) | 20 nm | 20 nm | 20 nm |
| EOT (HfO2) | 0.7 nm | 0.7 nm | 0.5 nm |
| Body thickness (Tb) | 10 nm | 7 nm | 7 nm |
| Source doping concentration | $1 \times 10^{20}$ cm$^{-3}$ | $4 \times 10^{19}$ cm$^{-3}$ | $5 \times 10^{18}$ cm$^{-3}$ |
| Drain doping concentration | $1 \times 10^{20}$ cm$^{-3}$ | $2 \times 10^{17}$ cm$^{-3}$ | $5 \times 10^{19}$ cm$^{-3}$ |
| Gate work-function | 4.55 eV | 4.85 eV | 4.285 eV |
| Hetero-junction band alignment | – | Eg, GaSb = 0.845 eV, Eg, InAs = 0.49 eV, ΔEc = 0.439 eV | |

The conceptual TFET device structure in Fig. 6.3a does not necessarily restrict a TFET to be built in a lateral fashion like conventional CMOS transistors. In fact, the state-of-the-art III-V HTFETs are fabricated with source, channel, and drain deposited vertically [7, 8, 35]. This is mainly to help build high-quality tunnel interface with fewer traps so as to enable a steep slope [35].

In order to do circuit and architecture evaluations, a model that is supported by SPICE is necessary. A Verilog-A model based on these parameters was built with Synopsys TCAD Sentaurus tools for SPICE circuit simulations [37]. Sentaurus is a tool that simulates the fabrication, operation, and reliability of semiconductor devices, using physical models to represent the wafer fabrication steps and device operation [39]. Table 6.1 gives the device parameters for a projected HTFET process in comparison with conventional Si FinFET [37]. The parameter values have been calibrated according to experimental results, and verified through atomistic simulations.

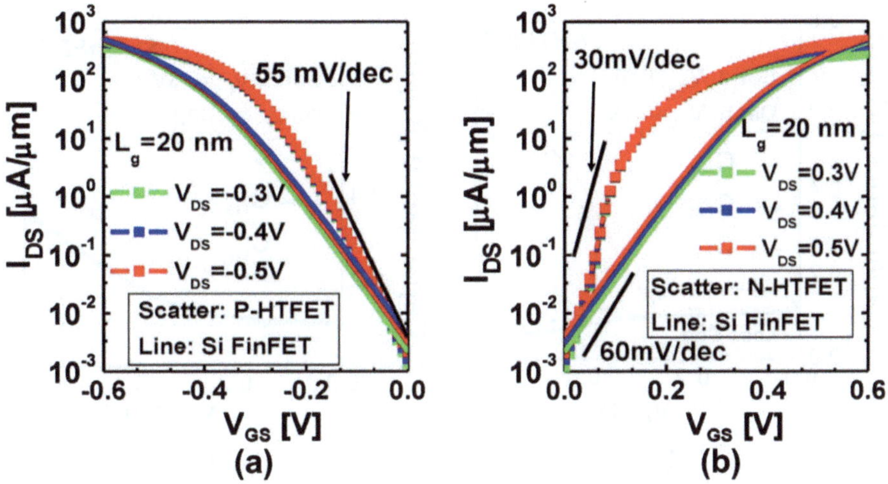


**Fig. 6.4** HTFET current–voltage (IV) curves for P-type HTFET in (**a**) and N-type HTFET in (**b**) [37]

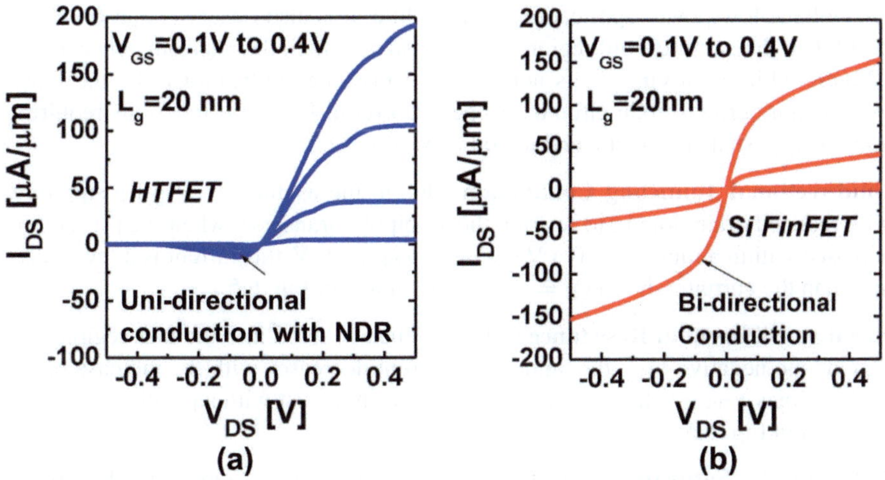


**Fig. 6.5** HTFET IV curve in (**a**) in comparison with Si FinFET in (**b**) [37]

The modeled HTFET device characteristics are shown in Fig. 6.4 and Fig. 6.5 for DC current, and Fig. 6.6 for intrinsic capacitance. Comparing with the conventional 20 nm Si FinFET performance, HTFETs show different features as discussed below.

**Steep-Slope Switching** As shown in Fig. 6.4, the HTFET shows an average SS of 30 mV/decade over two decades of current, and 7× improvement of on-state current over FinFET at a 0.30 V power supply [37, 40]. Accordingly, transconductance, that is, $g_m$, is also larger. Due to such characteristics, HTFET has the potential for

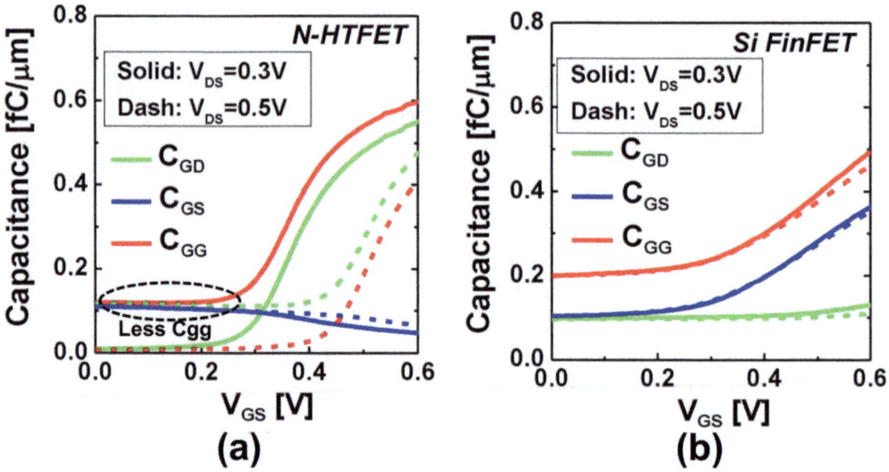


**Fig. 6.6** HTFET C-V curves in (**a**) in comparison with Si FinFET in (**b**) [37]

low-voltage low-power applications. Meanwhile, it is also observed that the SS of HTFET in the subthreshold region is not constant, which differs from conventional CMOS. While a varying SS is not preferred in some applications like the linear log-domain current-mode circuits, it opens up room for some functions requiring nonlinearity, such as mixers, and neural networks.

**Unidirectional Tunneling Conduction** Due to the asymmetrical p-i-n structure, with reduced drain doping to restrain the ambipolar transport, when the p-i-n diode is biased within a range of $-0.4\ \text{V} < V_{DS} = V_{NEG} < 0\ \text{V}$, the current is a few orders less than the current when $V_{DS} = |V_{NEG}|$, as shown in Fig. 6.5.

**Negative Differential Resistance (NDR)** When zoomed into the bias region with a moderate negative $V_{DS}$, the amount of current decreases with the increase of the absolute bias $V_{DS}$. Such negative resistance does not exist with a positive $V_{DS}$. This is shown in Fig. 6.5.

**Late and Flat Saturation** It is also observed that in order to make the HTFET to saturate, the required $V_{DS}$ is larger than what is required in conventional MOSFET.[1] This is shown in Fig. 6.5. Meanwhile, after the HTFET enters saturation, the output resistance is considerably larger than MOSFET, reducing the cost of a high-gain amplifier and high-accuracy current mirror.

[1]For conventional MOSFETs, the required minimum $V_{DS}$ to meet the saturation requirement is around $V_{GS} - V_{TH}$. For HTFETs, a widely-accepted $V_{TH}$ is not yet defined. This statement of comparison, however, is still meaningful, because for HTFETs, the required $V_{DS}$ needs to be larger than $V_{GS}$ to enable saturation, which is intuitively much larger than that of conventional MOSFETs [37, 41].

**Capacitance** On the one hand, HTFETs show higher $C_{GD}$ than conventional MOSFETs, at a moderate to high voltage bias. One result of this in the digital logic is the increase of glitches during switching because of enhanced miller effect. In comparison, the $C_{GS}$ in this range is much smaller than conventional MOSFET. On the other hand, during a low voltage bias < 0.25 V, the $C_{GS}$ and thus the total gate capacitance $C_{GG}$ is much less (close to 50%) than conventional MOSFET, as shown in Fig. 6.6. Such a feature could provide superior high-frequency performance regarding output impedance and cutoff frequency $f_T$.

### *6.3.2 Circuits Design*

**Digital Logics and Current-Mode Logic** When used to build complementary logic gates or current-mode logic gates, HTFETs present significant energy-delay improvement for low-voltage applications. This has been confirmed at different designs [40, 42–45], such as gate, memory, flip-flop, and processors. Such a lower-power feature also extends the design space considering power, yield, and thermal limits [44], with the consideration of variation [46] and parasitics [47]. It is noted that for pass-transistor logic that requires bidirectional current conduction, the unidirectional TFET feature should be backed up with another transistor for current conduction in the other direction. For low-voltage digital logic, variation can have a much more significant impact. For example, a work function shifting of TFETs operating at a low voltage can cause significant increase in the off-state leakage current or decrease in the on-state current. Some reports have shown that, with this impact considered, TFET-based digital logics and memories may still enable better trade-offs for lower power [48].

**Rectifiers and DC-DC Charge Pumps** A rectifier is an AC-to-DC converter and a DC-DC charge pump converts a DC voltage to another. For energy harvesting applications, they are both designed so as to get high power conversion efficiency (PCE) [37, 49]. Some conventional structures for rectifiers and DC-DC charge pumps are shown in Fig. 6.7a, b. While there are other structures making use of off-chip conductors, these structures get rid of them and are more suitable for system integration. However, in some scenarios where the input power is low, their PCE is seriously degraded because of low input voltage that makes the transistors hard to fully switch on. Making use of the steep-slope switching characteristics of HTFETs, these switches could be turned on at a lower voltage, leading to less resistive power loss. Meanwhile, the unidirectional tunneling conduction helps prevent reverse current, in a way that the harvested charge at the output node and internal nodes becomes less likely to be drained to the input. Such a unidirectional tunneling conduction feature could be further explored, leading to a new HTFET DC-DC charge pump as shown in Fig. 6.7c, where the gate control of the output PMOS pair is connected to the bottom plate of the coupling capacitors Cc. Such a new topology is able to provide twice the gate driving voltage, which could further

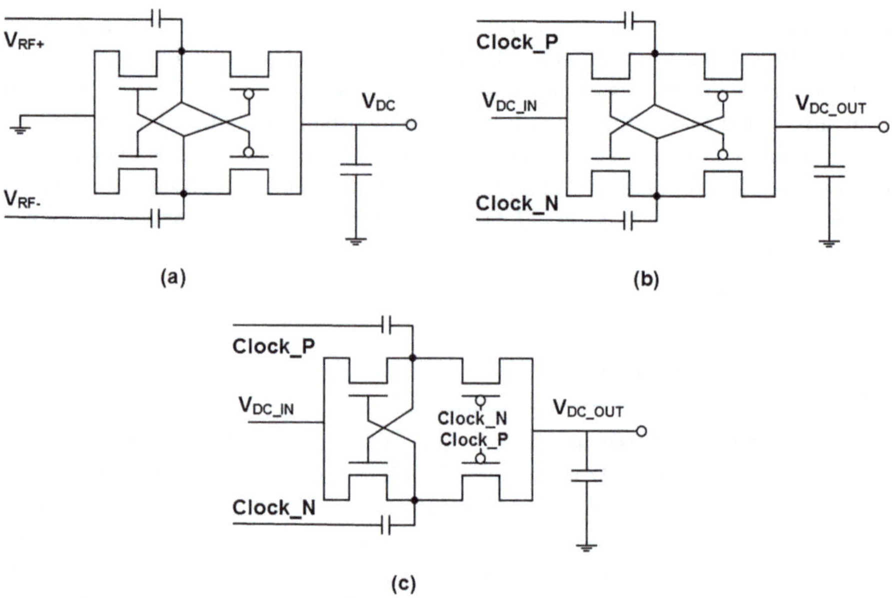


**Fig. 6.7** Rectifier and DC-DC charge pumps circuit topologies [37, 49, 50]. (**a**) Conventional rectifier; (**b**) Conventional DC-DC charge pump; (**c**) New topology for DC-DC charge pump in HTFET

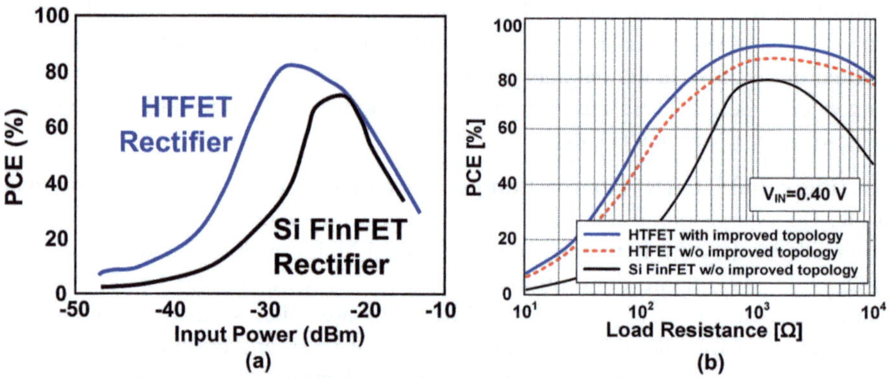


**Fig. 6.8** Rectifier and DC-DC charge pump performance comparison between HTFET and Si FinFET designs [37, 49]. (**a**) Rectifier; (**b**) DC-DC charge pump

reduce the resistive loss. As a result, as reported in the circuit simulations in Fig. 6.8, HTFET is very useful for energy harvester designs for higher efficiency [51].

**D/A Converters** Current-steering D/A converters are capable of high-speed, medium-to-high-resolution conversion. For the existing current-steering D/A converter implemented in conventional MOSFET, their key spectral performance, such as spurious-free dynamic range (SFDR), is limited by a few factors. While

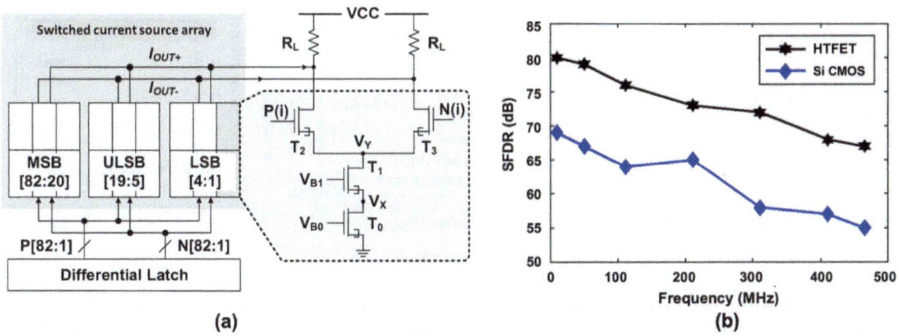


**Fig. 6.9** Current-steering D/A converter design using HTFET [52]. (**a**) Circuit topology; (**b**) SFDR comparison

some D/A converters have a bottleneck given by the current source mismatches, calibration or trimming can help to mitigate this impact. Other factors include the output impedance of the current source and switches (circuit structure shown in Fig. 6.9a), as well as the switching distortions because of nonideal switching of the switches. Conventional CMOS-based designs need complex techniques to mitigate these effects [52–55]. When switched to HTFET, the capability of low-voltage operation enables a smaller device size that leads to less capacitance. In addition, the low-capacitance density, as shown in Fig. 6.6, makes the total capacitance even smaller when operating at a low voltage. As a result, with less capacitance, HTFET helps to not only improve the output impedance, but also reduce the coupling glitches during switching [52]. Reported circuit simulation results in [52] confirm the SFDR performance advantage with HTFET, as shown in Fig. 6.9b.

**A/D Converters and Low-Noise Amplifiers** The benefit of using HTFETs in A/D converters and amplifiers originates from their high gain, especially with a low-voltage supply. The gain of transconductance operational amplifiers (OTA) relies on the transconductance of the input transistor pair, and the output resistance of the output current branch. When implemented in HTFET, the output resistance is higher after saturation, and a higher transconductance is obtained directly from the steep-slope switching. It is noted that, with HTFET, the transistors biased in saturation should be provided with sufficient drain–source voltage to guarantee saturation. While the HTFET single-device noise performance is reported to be comparable with conventional MOSFET [56], the noise with HTFET at the amplifier level is superior [57]. This is because of the higher gain by HTFET that turns out to suppress the equivalent noise at the input port (usually called the input referred noise) [57]. Figure 6.10 shows the evaluation results of an example design of an A/D converter [58], and a low-noise bio-signal amplifier in which the HTFET device noise was considered in the simulation [57].

**Cellular Neural Network (CNN)** CNN is useful when dealing with computation-intensive information processing application, such as pattern recognition and motion

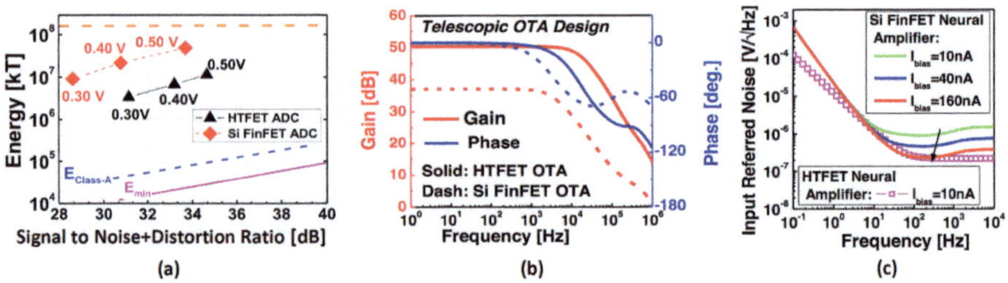


**Fig. 6.10** Performance evaluation of a 6-bit 10 M-S/s SAR A/D converter in (**a**) for energy versus signal to noise+distortion ratio (SNDR), a device-noise considered low-noise bio-signal amplifier in (**b**) for gain and (**c**) for input referred noise [57, 58]

detection [59, 60]. Turning to HTFET enables low-voltage operation and also obviates the necessity of nonlinear transfer function required in conventional CNN systems. This reduces both the hardware footprint and the power of CNN systems [59–61].

**Radio Frequency (RF) Applications** Using HTFET in microwave and mm-wave circuits in low-power applications has been evaluated in [62]. In this work, HTFET device noise was modeled and considered in the circuit evaluation. The high transconductance and microwave gain at low current bias and low voltage reduces the power dissipation. The smaller capacitance at low voltage also enables higher cutoff frequency ($f_T$) which is critical for high-frequency operation. The nonlinearity embedded in HTFET (see varying SS in Fig. 6.4) also enable circuits such as detectors with unprecedented sensitivity. While the work of [62] has provided some preliminary results, further work on HTFET RF circuit design and optimization considering device non-idealities, such as noise, variation, and reliability, is critical.

## 6.4 NCFET

### 6.4.1 Device Introduction

Steep slope in negative capacitance FETs (NCFETs) is achieved by a unique mechanism known as negative capacitance [12] provided by its unique device structure. NCFETs are also known as ferroelectric FETs (FEFETs). They also exhibit hysteresis in their $I_{DS}$–$V_{GS}$ transfer characteristics [2] with certain device design. The position and width of the hysteresis is tunable which can lead to non-volatility and noise immunity in the circuits.

The structure of the NCFET contains a ferroelectric (FE) layer in its gate stack as shown in Fig. 6.11a [11]. The steep slope behavior corresponds to lower SS in the $I_{DS}$–$V_{GS}$ characteristics. SS for NCFET in Eq. (6.1) could be further rewritten to be

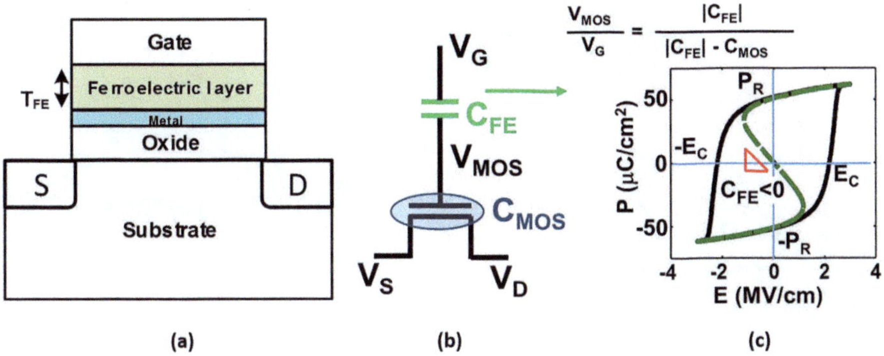


**Fig. 6.11** NCFET. (**a**) Device structure [11]; (**b**) Transistor capacitance model [63]; (**c**) PE loop [64]

$$SS \approx \frac{dV_{GS}}{d\psi_S}\frac{d\psi_S}{d\log 10\,(I_{DS})}, \tag{6.3}$$

where $V_G$ is the gate voltage and $\psi_S$ is the channel surface potential. $dV_{GS}/d\psi_S$ and $d\psi_S/d\log 10(I_{DS})$ are usually called body factor and transport factor, respectively. In TFET-like technologies, the lower SS is obtained by modifying the transport factor. In NCFET, the lower subthreshold is achieved by changing the body factor, using the concept of negative capacitance to get voltage step-up action of the gate voltage [12].

NCFET can be modeled as a combination of the ferroelectric (FE) layer capacitance in series with a MOSFET (Fig. 6.11b. Ferroelectrics are highly polarizable materials. The polarization (P) vs electric field (E) relation of the ferroelectrics can be represented using time-dependent Landau–Khalatnikov (LK) equation [12] given in

$$E = \alpha P + \beta P3 + \gamma P5 + \rho dP/dt. \tag{6.4}$$

The PE loop of the ferroelectric capacitor is shown in Fig. 6.11c. The dotted line shows the part of PE loop where $C_{FE}$ is negative. The entire system is made stable by adding the positive capacitance from the MOSFET ($C_{MOS}$). The voltage in the intermediate node is amplified by the factor $|C_{FE}|/(|C_{FE}| - C_{MOS})$ [64].

FinFETs are the preferable devices for sub-20 nm technologies due their higher potential for scalability. Building NCFETs using FinFET as the underlying transistor is expected to perform as ultralow voltage devices [13]. The suggested schematic of the NC-FinFET is given in Fig. 6.12 [64]. The experimental validation of the NC-FinFET concept is given in [18].

The change in device parameters can bring a significant impact in the characteristics of NCFET devices. For example, tuning the new FE layer thickness parameter which is absent in traditional transistors makes the device operate as a low power

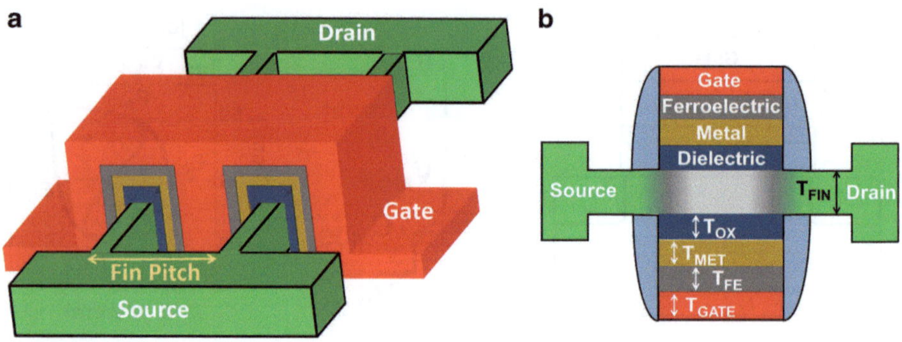


**Fig. 6.12** (**a**) NCFinFET structure [64]; (**b**) NCFinFET layers [64]

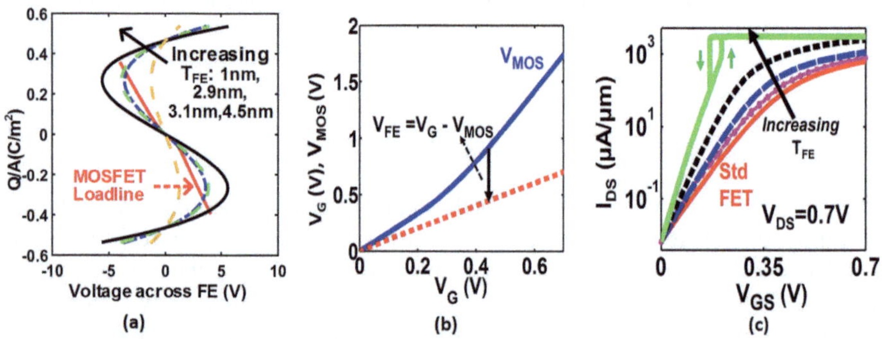


**Fig. 6.13** (**a**) Load line Analysis [12, 65]; (**b**) Voltage step-up action of NCFET [63]; (**c**) $I_{DS}$–$V_{GS}$ characteristics of NCFET with different FE layer thickness [63]

logic device or a nonvolatile memory device. So device analysis is crucial for the design of efficient, functionally valid NCFET circuits.

Device design plays a crucial role in determining the characteristics of the NCFET device. For example, increasing the FE layer thickness beyond a point introduces hysteresis in the $I_{DS}$–$V_{GS}$ characteristics of the NCFET device. The hysteretic jump which occurs when the magnitude of the FE layer capacitance equals the MOSFET capacitance. The occurrence of hysteresis can be predicted by doing a load line analysis of the device. An example of a load line plot is given in Fig. 6.13a. The 'S' shaped curve of the FE capacitor is plotted by solving the LK equation. It can be predicted that hysteresis will not be occurred when the FE layer thickness is small ($T_{FE} = 1$ nm in the Fig. 6.13a) as there is only one intersection point with the MOSFET load line [12, 65]. With increase in FE layer thickness, the FE capacitance is lowered due to which two intersection points are occurred resulting in hysteresis. Non-hysteretic operation is preferred for low power logic devices while hysteretic mode of operation is preferred for memory and noise immune logic.

### 6.4.2 Low Power Logic

Reduction of supply voltage is one crucial factor to minimize the power density (dynamic power proportional to $CV^2$). NCFET's negative capacitance mechanism enables steep slope operation and thereby reduces the power supply voltage. Figure 6.13b shows the steep slope operation in NCFETs. With the application of $V_{GS}$, a negative voltage gets developed across the FE layer capacitor. This leads to the development of a higher potential larger than the applied voltage $V$ in the intermediate node ($V_{MOS}$) [63, 66] (Fig. 6.13b). Increase in the ferroelectric layer thickness leads to decrease in $|C_{FE}|$ which amplifies the potential in the intermediate gate leading to high gain in $I_{DS}$. Increasing the thickness of the FE layer further leads to hysteresis. [13] has predicted around six orders magnitude of difference in the $I_{ON}$ to $I_{OFF}$ ratio at 0.2 V operation. For logic devices, it is preferable to have hysteresis free mode of operation. However, it has to be noted that the gate capacitance of the NCFET is higher compared to that of conventional CMOS. The speed of operation of a device is roughly related to $CV/I$ metric where $C$ is the device gate capacitance, $V$ supply voltage operation, and $I$ is the drive current. So it is essential to optimize the $C$, $V$, $I$ components for the optimum performance [47]. Figure 6.14 shows the energy delay of a simulated 8-bit NCFinFET-based Kogge–Stone adder. NCFinFET shows better performance at low voltages compared to the regular FinFET adders [66].

For applications where noise is a key concern, using the hysteresis feature of NCFET to build logic helps the noise immunity. This is because with the hysteresis in the positive voltage region, the gate output does not update until the input voltage exceeds beyond the hysteresis window, indicating more noise margin [67].

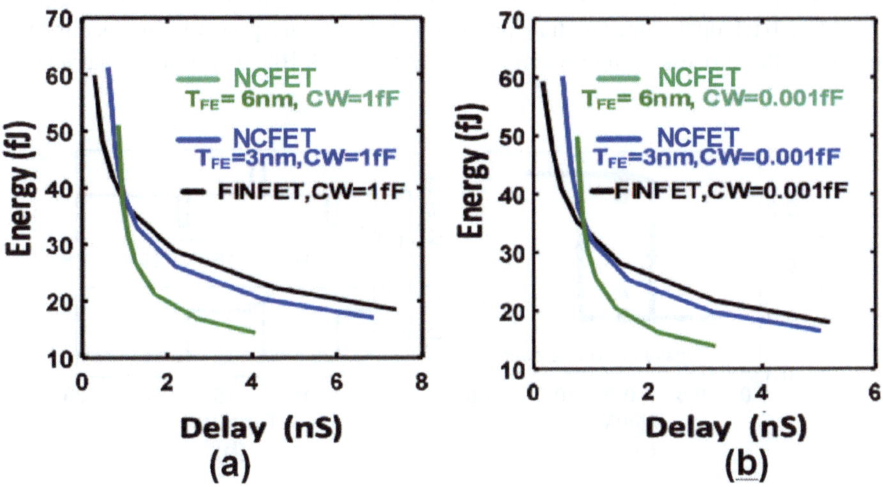


**Fig. 6.14** Energy vs carry delay diagram of the 8-bit KSA: (**a**) 1fF wireload; (**b**) 0.001fF wireload [66]

### 6.4.3 NCFET Memory, DFF, and Processor

NCFET devices can be used as nonvolatile devices as the polarization can be retained in the FE layer of the gate stack even in the absence of electric field, with the proper device design. Hysteresis is an essential condition for non-volatility. For the NCFET to be stable, the overall capacitance ($1/C_{TOT} = 1/C_{FE} + 1/C_{MOSFET}$) of the device should be positive and hysteretic jump occurs to keep the $|C_{FE}|$ to be higher than $C_{MOS}$ such that overall capacitance is positive [11, 65]. Therefore, it is crucial to maintain the proper ratio of $C_{FE}/C_{MOS}$ for non-volatility.

In a nonvolatile design, increasing the $V_{GS}$ beyond coercive voltage of the device switches the polarization stored in the gate stack and subsequent withdrawal of $V_{GS}$ to zero volts retains the polarity. Similar application of $V_{GS}$ in the negative direction beyond the coercive voltage reverses the polarization direction and retains the polarity even when $V_{GS}$ is removed to zero volts. Figure 6.15a shows an example of hysteresis in the transfer characteristics centered origin, which is an essential condition for non-volatility, and Fig. 6.15b shows polarization is retained in the FE layer when applied $V_{GS}$ is brought to zero volts leading to nonvolatile design [65]. Such characteristics lead to two solutions for the $I_{DS}$ when $V_{GS}$ is zero volts. The two distinctive resistance points (A and B in Fig. 6.15a) can be interpreted as logic high and logic low states of the memory [11, 65]. Figure 6.16 shows that position of the hysteresis is also important for nonvolatile designs as hysteresis, which is not centered zero gate voltage, does not lead to non-volatility. [16] has reported more than 6 orders of magnitude difference in the $I_{ON}$ to $I_{OFF}$ ratio of drain current in the hysteresis region. The huge difference in the drain current can be translated to high distinguishability of logic states by using current sensing in the memory design scheme [65, 68].

Nonvolatile processors (NVP) are processors with built-in nonvolatile memory (NVM) to back up the intermediate processor states when a power failure occurs and to restore the processor state when power comes back [69, 70]. Energy harvesting

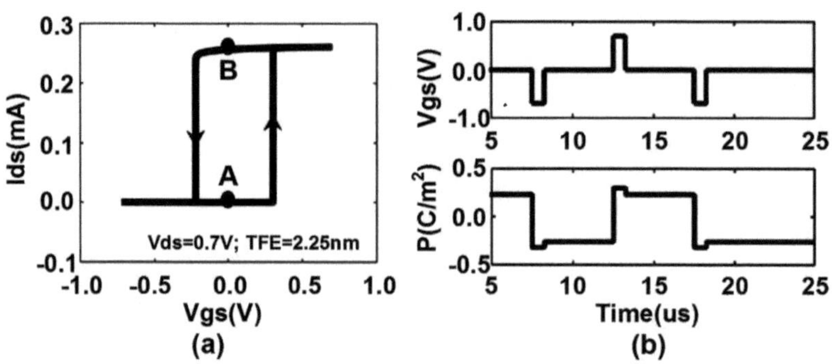


**Fig. 6.15** 65 nm N-type FEFET with 2.25 nm ferroelectric layer thickness: (**a**) hysteresis; (**b**) non-volatility [65]

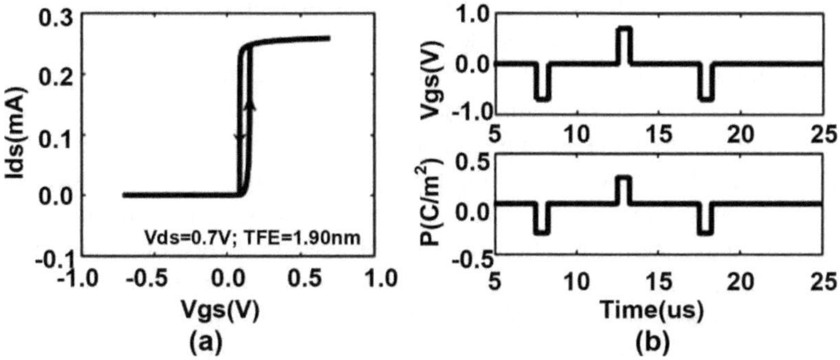


**Fig. 6.16** 65 nm N-type FEFET with 1.90 nm ferroelectric layer thickness: (**a**) hysteresis; (**b**) no non-volatility [65]

systems typically use NVPs as they have to work with frequent power interruption [71]. Both the memory and logic should be optimized to accomplish more forward progress in computation in NVPs. For example, energy savings from a low energy memory can be utilized to make more forward progress [65] in computation. As discussed in the previous section, NCFET has the potential to be utilized as memory with its polarization retention in the gate. A simulation study shows that replacing FERAM memory with NCFET memory in an NVP can achieve better forward progress due to the energy savings from NCFET memory [65].

An alternate method to the backup and restoration of the processor states to the memory in NVPs is using a nonvolatile flip-flop. Integrating non-volatility inside or nearby the flip-flop reduces the overheads of data transfer to the memory. Nonvolatile flip-flops can also be used for power gating in general-purpose processors in order to eliminate leakage power consumption. Spin-based memories [72] and ferroelectric capacitors [70] have been proposed to use in the design nonvolatile flip-flops. One of the advantages of NCFET-based NVFF design over other technologies is the high distinguishability between two logic states that the NCFET inherently offers. Also using the three-terminal NCFET device with built-in storage mechanism significantly reduces design complexity of the backup/restore circuits [73–75].

With more architectural optimizations, it has shown that NVP exhibits more advantage over conventional volatile processors [76–79], which highlights the benefits of using NCFET for future processor design.

### *6.4.4 Logic in Memory and Security Applications*

Tunable hysteresis property of the NCFET opens up the possibility of potential circuit features like memory in logic, where logic operations could be done close to or at the memory cells to reduce the number of memory accesses by fetching only the results of logic operations [80].

The potential to build simpler and power-efficient memory in logic/logic in memory circuits can find immediate applications in the field of hardware security [68]. The potential dynamic tuning from a nonvolatile cell to a volatile cell can help to achieve logic obfuscation. Also employing logic in memory cell to store the keys reduces the communication overhead for key accessing and verification between memory and logic [68]. Retention time of the data is a very important aspect in security applications. The retention time is dependent on multiple device factors like coercive voltage, remnant polarization, area of the design etc.; the cell can be optimized for energy and retention by tuning the cell parameters [65, 81].

## 6.5 Phase Transition Devices and Hyper-FET

### *6.5.1 Device Introductions*

A recent addition to the family of steep switching devices is the Hyper-FET [29] or the Phase-transition-FET [82]. Hyper-FET can be visualized as a transistor with a phase transitioning correlated material (CM) attached to the *source* terminal (Fig. 6.17). The steep switching behavior of the Hyper-FET arises from the collective carrier dynamics of the assisting CM which undergoes selective phase switching during transistor operation. A discussion on Hyper-FET requires a prior discussion on correlated materials which is as follows.

Correlated materials are unique entities which exhibit a strong correlation between their inherent electrons. Therefore, they possess exotic electronic and magnetic properties such as metal–insulator transitions, spin–charge separation, and half-metallicity. A group of such materials (e.g., $VO_2$, $V_2O_3$, $NbO_2$, TiO, $Ti_2O_3$, doped chalcogenide, Cu-doped $HfO_2$, etc. [83–86]) manifest insulator to metal phase transition driven by external perturbations like temperature, pressure, or electrical stimulus. These materials have highly nonlinear electrical behavior as shown in Fig. 6.18. In response to a sufficiently high applied electric field (or

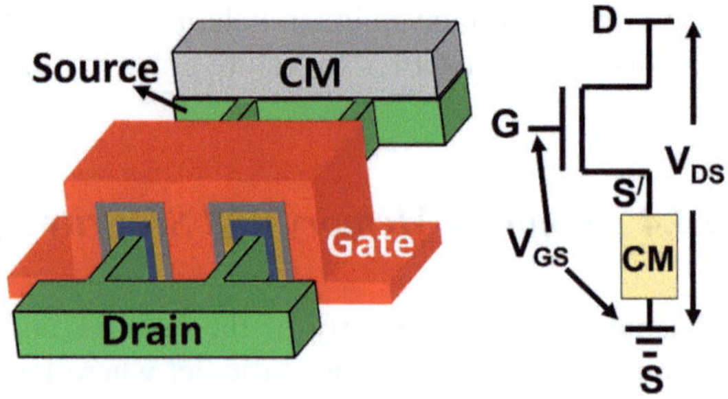


**Fig. 6.17** Hyper-FET structure and schematics

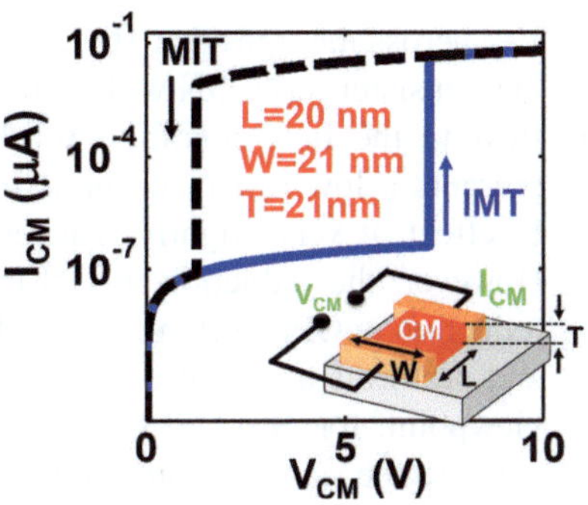


**Fig. 6.18** Typical I–V characteristics of a correlated material

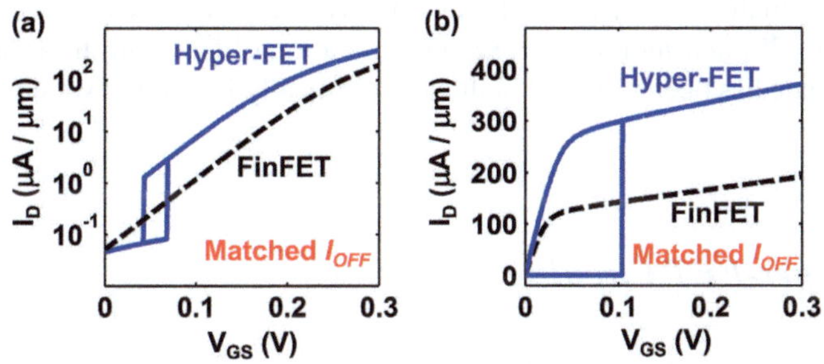


**Fig. 6.19** (**a**) Transfer and (**b**) output characteristics of Hyper-FET, in comparison with FinFET

current), the CM undergoes abrupt (not instantaneous) insulator–metal transition (IMT) and illustrates an immense rise in current attributed to the significant change in resistivity. On the other hand, below a critical level of electric field (or current), the CM switches back to the insulating state (naturally default). Since, the two transitions occur at different threshold levels, the *I-V* response becomes hysteretic. A wide variety of such phase transitioning CMs (having diverse resistivity and switching thresholds) have already been reported, and further exploration is still going on. There has been demonstration of effective tuning of the properties of existing CM using physical agitation like strain [87].

Hyper-FET ingeniously utilizes the above mentioned characteristics of CM to operate well below the Boltzmann limit for SS (60 mV/decade). Its principle of operation can be explained using the transfer and output characteristics presented in Fig. 6.19.

- During the OFF state of the transistor, the CM stays in insulating state (due to absence of stimuli). The high insulating state resistivity of the CM substantially suppresses the OFF current of the transistor ($I_{OFF}$).
- With rising voltage at the *gate* terminal, *drain* current of the transistor increases and as the current exceeds a critical level for triggering IMT, the CM turns metallic. While manifesting the transition, the CM generates a negative differential resistance (NDR) effect at the *source* terminal. By dint of this NDR, effective

$V_{GS}$ across the transistor structure increases beyond the applied value and hence the transistor turns ON with a sharper subthreshold slope.

- Beyond the bias condition for IMT, CM operates in its metallic phase and presents a low resistance in series with the innate resistance of the transistor. In effect, it very slightly reduces the ON state current of the transistor ($I_{ON}$). However, the reduction in ON current is negligibly small compared to strong reduction in OFF current. Thus, Hyper-FET eventually achieves higher $I_{ON}/I_{OFF}$ and performs as a better switch than the conventional transistor. From another viewpoint, for matched $I_{OFF}$ condition, Hyper-FET will have higher $I_{ON}$ compared to the standard transistor.

The functionality of Hyper-FET has already been shown experimentally both with discrete [29] and monolithic [82] incorporation of CM with transistor. An SS of 8 mV/decade has been reported in the monolithically integrated version of Hyper-FET.

## 6.5.2 *Hyper-FET Circuits*

Being an emerging device with a completely new phenomenon, Hyper-FET demands proper device–circuit co-design for optimum performance. Due to the hysteretic behavior of the CM itself, Hyper-FET exhibits hysteresis in its transfer characteristics. This hysteresis has a strong influence on the performance of Hyper-FET-based circuits. The size and position of the hysteresis window are important aspects to consider while designing Hyper-FET-based logic devices. But it is assuring to note that the hysteresis can be tuned and modified by changing the geometry of the CM. In addition, as mentioned earlier, a wide variety of CMs are available and there is ample scope to choose the best-suited material or artificially tune the material properties to some extent.

Implementation of logic devices with Hyper-FET is one of the immediate steps to verify their feasibility in circuit domain. Figure 6.20a shows transient waveforms for a ring oscillator (RO) made of a Hyper-FET-based inverter along with that with regular CMOS. The energy-delay benchmark for Hyper-FET-based RO (Fig. 6.20b) shows superior performance over regular CMOS at low $V_{DD}$ ($<0.35$ V) [82]. At high $V_{DD}$, cutting-edge CMOS transistors exhibit faster operation but the Hyper-FET suffers from additional delay infusing from the phase switching dynamics of the CM. However, at low $V_{DD}$, as transistors tend to incur higher intrinsic delay, such transition time no longer remains as a bottleneck. [82] reported 1.86× reduction in energy (compared to 14 nm ULP FinFET) in a Hyper-FET-based ring oscillator at *iso*-delay (2.18 ns).

The unique properties of Hyper-FET or more precisely the inherent CM make them a favorable candidate for brain-inspired neuromorphic computing. The current driven phase transition in between two states of CM can be utilized to design circuits that behave like biological *neuron* [88, 89]. The *neuron* cells generate spikes

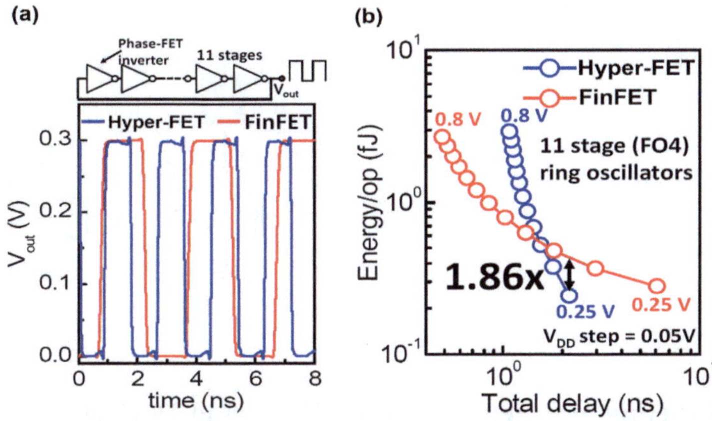


**Fig. 6.20** (**a**) Transient waveforms for RO based on Hyper-FET and FinFET (**b**) Energy delay characteristics of these two versions of RO [82].

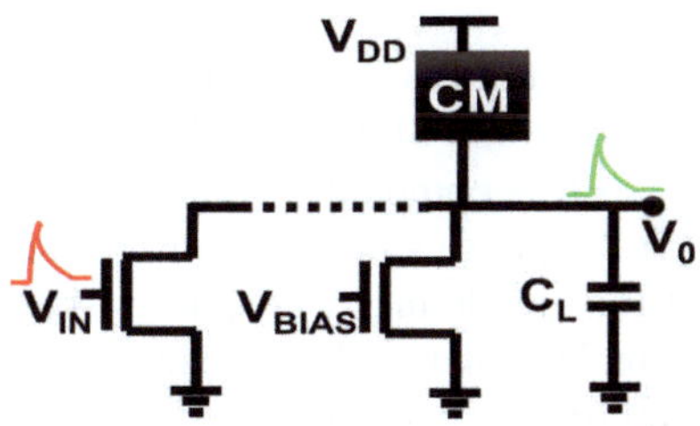


**Fig. 6.21** Circuit diagram of a Hyper-FET-based neuron cell [88]

in response to incoming excitements exceeding a certain threshold. Likewise, as discussed earlier, CMs exhibit phase transition driven by electrical agitation beyond a certain level. This selective phase transition can give rise to voltage spikes in an electrical neuron circuit. The concept has been illustrated in Fig. 6.21. The CM shown in this circuit diagram undergoes phase transition depending on the intensity of the input(s) and the transition gives rise to a spike in the output node voltage ($V_0$). While a single neuron has very limited computational capability, a collective network of them (known as neural network) is of more practical interest.

In addition to the possibility of neuromorphic computing, Hyper-FET can also be used to design nano-oscillator circuits [30, 88], as shown in Fig. 6.22. With appropriate design, it is possible to invoke the CM of a Hyper-FET to keep switching its phase for a certain *gate* bias of the host transistor. The relentless phase transitions result in continuous charging and discharging of the corresponding node. Thus, an oscillatory voltage waveform is generated. For sustained oscillation, the metal–insulator transition (MIT) and insulator–metal transition (IMT) should occur before the output voltage stabilizes. The gate bias of the Hyper-FET can be tuned to provide configurable impedance to control the frequency of oscillation. Moreover, two or more relaxation oscillators can be used to make coupled oscillator systems which are more capable for computation.

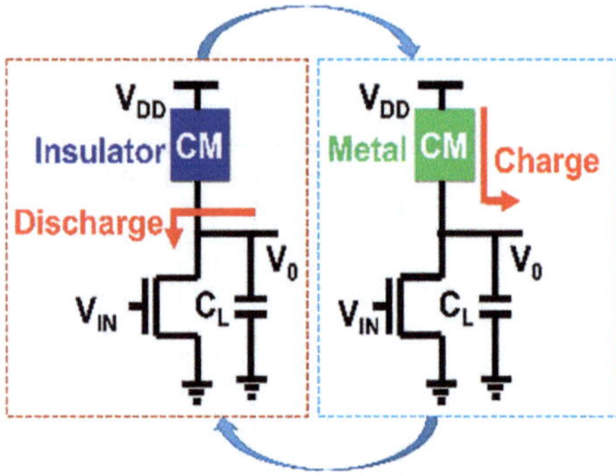


**Fig. 6.22** Circuit diagram of a Hyper-FET-based relaxation oscillator [88]

## 6.6 Modeling and Benchmarking Emerging Devices

### *6.6.1 Modeling Emerging Devices*

Compact device models have been widely used to capture their electrical behaviors within circuit SPICE simulators [47, 63, 88, 90–93]. To describe electrical behaviors in *SPICE*-compatible compact-device models, there are two broad approaches: (a) the physics-based compact modeling and (b) numerical-based look-up tables (LUT) modeling. The physics-based compact modeling approach includes two complementary roles in nanoscale electronic devices [94, 95]. First, an analytical description in a physics-based compact model helps interpret its electrical or behavioral characteristics; second, this approach forms the basis of a simple *SPICE* model for simulators. This physics-based modeling is predictive and has a small number of parameters, incorporating with the process-defined device structure and well-known material properties while the Berkeley Spice (BSIM) model requires such a large number of model parameters. Another modeling approach is the numerical-based look-up tables (LUT) modeling. It provides an alternative to analytical models and is particularly useful for emerging technologies if the electrical characteristics are based on simulations and measurements and the basis of a simple SPICE model like physics-based modeling. This modeling approach employs the predefined numerical values with respect to operating conditions. For the SPICE device modeling, two types of modeling approaches can be combined. For example, ideal electrical device behavior can be modeled with LUT-based modeling and geometry and material-dependent *RC* parasitics can be modeled with analytical descriptions.

Among two types of approaches, as briefly depicted above, if physics-based mathematical expressions go along with the precise electrical behaviors for an emerging device technology in a broad range of operational conditions, the physics-based compact modeling methodology is commonly employed in not only examinations of a device itself but also device–circuit *SPICE* simulations because it generally provides an efficient way to predict behaviors and a shorter simulation time for device–circuit interactions compared to numerical-based LUT modeling

[47, 93, 96]. However, unfortunately, some of emerging technologies are strenuous to model with the physics-based compact modeling since some of the devices such as TFETs, NCFETs, Hyper-FET, etc. are still under development and have only valid mathematical expressions in a short range of operating conditions. In this case, computer-aided simulations of devices and circuits are usually carried out with a LUT-based approach, which is computationally demanding for simulation and testing of large device count circuits because this approach can provide more accurate results than physics-based compact modeling [63, 88, 96].

BSIM models and Verilog-A models are widely used for device modeling to support SPICE circuit simulations [90, 91, 97–99]. Over the past few decades, amongst numerous types of compact modeling methodologies, the analog subset of Verilog-AMS [91] has become the preferred compact modeling language for both academic and industrial research groups [29, 30, 63, 88, 90, 99]. Many of emerging devices have been continuously modeled by using Verilog-A [29, 30, 88, 90, 92, 93, 100]. The Verilog-A is a naturally explicit language for an electromechanical device in not only physics-based but also numerical LUT-based compact models and it has been demonstrated as a promising language for analog device model developments [90, 91, 101]. An important key feature of emerging device modeling with Verilog-A is that this language can be expanded with BSIM and other SPICE models. Verilog-A models can also support device features of noise, variation, and parasitics. Some prior modeling such as gate-all-around nanowire FETs, NCFETs, and Hyper-FETs uses BSIM as a base of a transistor and uses Verilog-A language to add unique features of emerging devices, geometry- and material-dependent *RC* parasitics, and tweak parameters of BSIM models [63, 88, 92, 102, 103].

While building the LUT-based Verilog-A model for emerging devices, providing a dense LUT helps to improve the accuracy of circuit simulation results at the cost of slower simulation speed. It also helps to reduce the chances of simulation convergence failures. Sometime the LUT is based on measured results which can be more or less distorted by the measuring accuracy; some preprocessing of the LUT data can also be helpful to prevent simulation convergence failures. Such preprocessing can include data smoothing, and also correction. For example, the drain–source current $I_{\mathrm{DS}}$ direction should match that of the voltage applied. Also, $I_{\mathrm{DS}}$ be zero with a zero $V_{\mathrm{DS}}$.

Unlike the simple ideal modeling for emerging devices to capture intrinsic electrical behaviors [90–95, 99, 101–107], the practical device modeling needs to include such subthreshold characteristics, ballisticity dependency, noise, parasitics, and process variations terms into the emerging device model since the increased device sensitivity to noise, parasitics, and process variations at reduced VDD and physical device dimension pose further challenge on robust circuit operation using emerging devices [40, 102, 107]. In particular, electrical noise and paratisics poses a growing reliability concern for optimal system design at emerging technologies [40, 56, 102, 107]. Hence, the detailed analysis of electrical noise and parasitics in emerging devices is required and the accurate device modeling taking into account electrical noise and process variations needs to be applied to evaluate emerging technology because the noise figure and parasitics is one of the key design factors in determining the performance for not only analog/mixed-signal and RF circuits but also memories and processors [40, 56, 102, 107–110].

### 6.6.2 Benchmarking: From Device to Architecture

The post-CMOS device research has continuously brought the immense benefits of higher integration density, higher energy efficiency, and superior performance [40, 46, 47, 63, 88, 93, 102, 104, 107]. Such benefits at the device-level and small-scale circuits have been widely investigated with *SPICE* device modeling [29, 30, 40, 63, 88, 91–93, 102]. Aside from benefits at the device level, the examinations of architecture designs accompany challenges with practical layout issues such as complexity and restrictions, and interconnect routing, noise, and increased susceptibility of circuits to process variations [47]. In order to understand applicability of emerging devices across a broad design space of architecture-level designs, the abstraction through a device-logic-architecture benchmarking framework is crucial in such large-scale designs. The benchmark processing is broadly classified into three different paradigms [46, 104]: (1) general power estimation in use of architecture simulators such as GEMS [46, 105] and McPAT [46, 106], (2) examination of performance metrics in use of hardware synthesis process, and (3) variation studies in architectures with different datapath configurations by varying parameters (e.g., the issue width, pipeline depth, etc.) within EDA tools (e.g., Fabscalar [46, 111] and plus hardware synthesis process). The appropriate paradigm, depending on application and benchmark characteristics, has to be chosen to capture the impacts with emerging devices at processor architecture level.

Figure 6.23 presents the succinct flow for cell characterization process. Logic gate designs and cell characterization are the critical portion in architecture-level designs because the logic gates are a key factor to determine the performance metrics (e.g., energy, critical path delay, and chip area) in architecture designs through a synthesis process with synthesizable hardware description language (HDL). Logic gate designs and cell characterization are strongly dependent on cell layouts because of considerations of floorplanning and routing for full-chip designs and benchmarks. At sub-10 nm technologies and emerging devices, generating layouts for cells becomes immensely difficult because of challenges with regard to smaller physical dimension, different device structures, routing complexity, and complex design rules [47]. Hence, the cell designs with geometry parameters should be carried out carefully while maintaining the full-functionality of each cell.

Several works have already explored the potential design spaces in medium-to-large-scale systems such as adders, ring-oscillators, application-specific accelerators, processors nonvolatile memory, and logics with emerging devices [29, 30, 40, 46, 63, 88, 105, 107]. This device-to-architecture framework reveals that the *SPICE* modeling-to-logic and logic-architecture benchmarking provide profound insights into impacts and help predict unique behavior and potentiality of future computing systems with emerging devices.

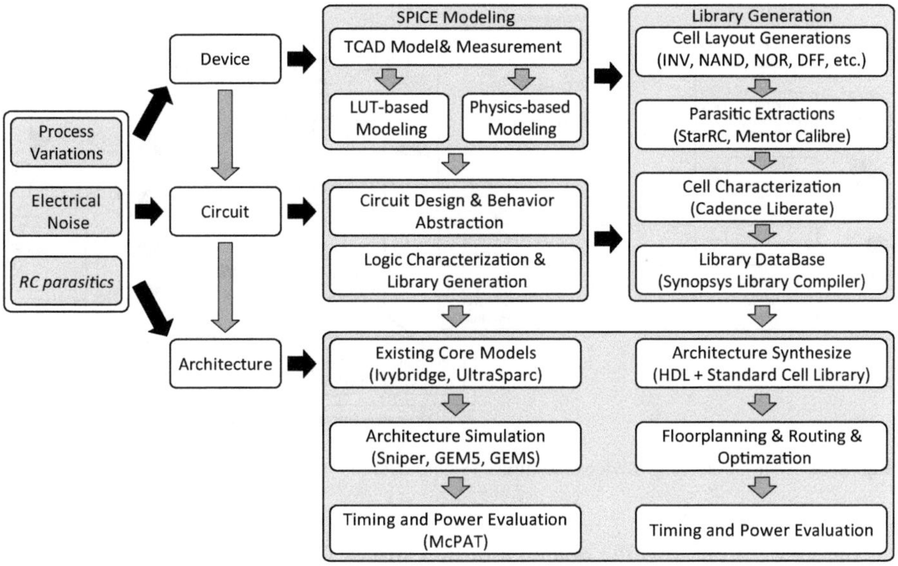


**Fig. 6.23** Brief flow of the device-to-architecture benchmarking

## 6.7 Opportunities and Challenges of Emerging Devices in Circuits and Systems

### *6.7.1 Opportunities*

Steep-slope emerging technologies have brought the unique opportunities to beat limitations in the performance and energy efficiency of traditional CMOS transistors at subthreshold voltages. Exploring the potential design space with such steep-slope technologies has been demonstrated through the detailed device modeling and implementation to circuits in many prior works [29, 30, 40, 42, 46, 47, 58, 63, 88, 104, 107] as outlined in Sects. 6.3, 6.4, 6.5, and 6.6. The most direct application scenario for these steep-slope devices is similar to that of the traditional CMOS, in that they could be used as three-terminal switching transistor with On-Off drain–source mechanism by the gate controls [29, 88]. These block level examinations in various fields such as analog/RF/mixed-signal domains (e.g., Rectifiers, DC-DC charge pumps, DAC, ADC, LNA, etc.), Boolean and non-Boolean logics, memory, brain-inspired neural networks, and specific-purpose processor (e.g., nonvolatile processor) bring broad insights to explore the optimum design space in the scenario of energy-efficient and low-voltage systems for not only portable battery-powered systems and highly energy-efficient applications (i.e., implantable medical devices and wearable bio-activity monitoring platforms, etc.) but also general-purpose processors.

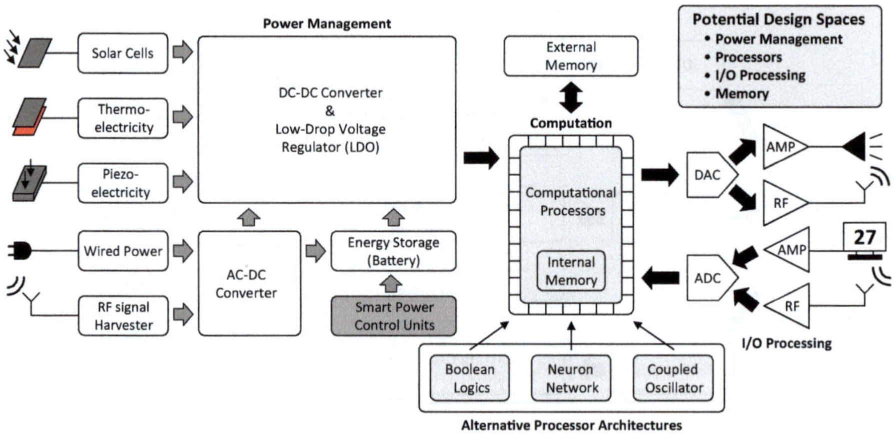


**Fig. 6.24** Using emerging devices for future system designs

Figure 6.24 presents the potential design spaces for emerging technologies in a system. An integrated circuit (IC) generally comprises multiple small components such as the controlling logics, memory, digital signal processor (DSP), mixed-signal circuits, etc. The block-level advantages through a detailed examination of these components with such steep-slope emerging technologies in terms of energy and delay at a range of low voltages directly deliver magnificent benefits and opportunities with superior energy-efficiency and low-voltage operations. Energy harvesting systems, for example, require the high power conversion efficiency in the limited ambient RF power and low energy computations to extend functionality and to operational time for use at a given energy and consist of multiple circuits for enabling the utilization of the ambient RF signal power and signal processing units such as Rectifier, LNA, ADC, DAC, DSP, etc. [37, 51, 57].

As depicted in Fig. 6.24, the compatibility of emerging technologies is not limited to small signal processing components. A broad range of large-scale digital computation units such as general-purpose processors, hardware-specific accelerator, and brain-inspired neural network and nonvolatile memory can employ such steep-slope emerging technologies (TFET, NCFET, Hyper-FET, VO2, etc.) for energy-efficient computing at low voltages. Compared to CMOS-based computation hardware, scaling supply voltages to reduce the power consumption leads to a significant increase in delay performance. Also, to retain performance, another attempt such as scaling $V_T$ affects not only noise performance but also overall standby leakage power dissipations on ICs. In contrast, steep switching mechanism at low voltages can provide the remarkable delay reductions in computation hardware while maintaining low power consumption of both dynamic switching and standby leakage, leading to high energy-efficiency in such large-scale integrated circuits. In addition, integration with nonvolatile memory by taking hysteresis in ferroelectric emerging technologies also provides additional opportunities for significant reduction of power and replacement of conventional memory systems in large-scale ICs. Some benchmarks using these emerging devices in system-level designs [40, 44, 46, 88, 104, 107] reveal that steep-slope emerging technologies can provide the significant reduction in energy and become the promising candidates for numerous types of applications and systems while keeping the same functionality and better noise immunity compared to conventional CMOS [30, 42, 43, 58, 88, 107, 112].

### *6.7.2 Challenges*

Although numerous emerging devices (i.e., TFETs, NCFETs, Hyper-FETs, etc.) hint at a bright future [40, 42, 46, 47, 104, 107], given the practical device challenges such as the unidirectional conduction behavior, biasing- and material-dependent hysteresis and fabrication complexity and cost in circuits and system-level designs still remain. This chapter covers how these challenges affect in circuits and systems not only at device level.

**Unidirectional Conduction** Foremost challenges in logic gates stemming from the unidirectional conduction can be broadly categorized by three aspects: (a) the routing complexity, (b) chip area overhead, and (c) more power consumption by increased capacitances. Although the routing complexity and chip area overhead are unavoidable, the power consumption and energy can be reduced by VDD scaling because TFETs are inherently capable of operating at very low voltages (e.g., VDD $<$ 0.3 V) compared to non-steep switching devices. Unidirectional conduction induced from the asymmetrical device architecture of TFETs become a key limitation factor for the full-functionality in TFET gates (i.e., SRAMs, latch, DFF, pass-transistor, and transmission gate logics) and connectivity at over two stacked transistors because of incomplete charging/discharging paths and limitations of sharing contacts/active regions originating from this unique characteristic [47]. In order to make logic gates operate with full-functionality, modifications of logic designs are required for TFET circuit implementation. The simple way to modify logics for charging/discharging paths is adding an additional transistor on a path, resulting in additional power dissipation and area overhead of logics. Since sequential logics like DFFs, for example, comprise a large fraction of the total gates in a design, the additional transistors for each charging/discharging path affect significant chip area and power dissipation overheads. Fortunately, adding an additional transistor is not an only option for DFFs. The static $C^2$MOS DFF, for instance, can be employed instead of the use of transmission gates-based master/slave DFF. However, to retain the read/write functions in TFET SRAM, the implementation of additional transistors is inevitable. For the TFET SRAM, numerous works have already explored its alternative options to acquire read/write operations fully [40, 42, 107]. At an iso-area perspective in processors, the capacity of TFET SRAM cells is smaller than that of other devices, which have bidirectional conduction behaviors.

**SRAM Noise Margin** Another challenge in TFET SRAM is lower static noise margin (SNM) from a simultaneous turned-on transmission gate in operation of read/write [40, 107]. To alleviate this SNM issue, alternative schemes have been explored in [40, 107] while sacrificing SRAM area by using additional transistors to apply additional controlling signals. Another challenge induced from unidirectional conduction is the penalty with isolations between two transistors. As mentioned above, TFET cannot share contact and diffusion area at >2 stacked transistors due to unidirectional conduction, resulting in additional isolation by placing shallow trench isolation (STI) between two adjunct transistors and area overhead in logic gates. This is becoming more severe when the driving ability of logic gates needs to be high enough to meet timing requirements in processors at a harsh operation environment (e.g., high frequency range, low voltage ranges, etc.) because the cell should be bigger to minimize the computation delay, resulting in an area penalty with a number of additional isolations.

**Undesirable Hysteretic Characteristic and Speed in Boolean NCFET and Hyper-FET Logics** As described in Sects. 6.4 and 6.5, NCFET and Hyper-FET exhibit the unique hysteresis switching and hysteretic polarization in presenting two states in certain biasing voltages. The hysteresis switching and hysteretic polarization characteristics provide superb nonvolatile characteristics in memory applications [113]. Hysteresis can be ingeniously used to harness additional benefits in holding the stored data. The possibility of hysteresis, for instance, in the transfer characteristics ($I_D$–$V_{GS}$) of NCFET is actually governed by the relative capacitive matching between the gate of the transistor and the ferroelectric layer. Load line analysis considering conservation of charge between these two mutually connected capacitances yields two stable physical solutions (origin of hysteresis) beyond a critical thickness of the ferroelectric. Also, these hysteresis characteristics can be employed in some specific-purpose non-Boolean logics. However, except in specific-purpose non-Boolean logics and memory applications, these hysteresis characteristics in Hyper-FETs and NCFETs associated with the negative differential capacitance generally pose adverse effects in Boolean functioned logics because of the metastable states, although they provide superior device performance such as steep-subthreshold switching, high $I_{ON}$ while maintaining low $I_{OFF}$ current. Hence, some literatures have aimed to minimize the hysteresis by using various techniques such as by threshold $V_T$ shift adapting different ferroelectric materials [19], adjustment of thickness of the thin ferroelectric layer, annealing [18, 19], etc. while retaining device performance. For example, the adjustment of critical thickness works as upper bound for hysteresis free operation. Hence, geometry (especially thickness) of the ferroelectric layer is a vital aspect to traverse between hysteretic and non-hysteretic modes. The critical thickness differs among materials and is directly linked to the corresponding remnant polarization and coercive field. These numerous techniques result in the increase of additional fabrication process control, complexity, and cost. Hence, the minimizing hysteresis and development of efficient fabrication process control to augment the current state-of-the-art technology will be key factors in realizing a NCFET and Hyper-FET-based hardware platform [29, 88], albeit sacrificing some degree of current fabrication complexity and cost.

**Additional Phase/State Change Delay** Another concern is the delay stemming from the introduction of the additional layer stacks and extra delay for state/phase changes within transistors. For example, this delay challenge of NCFET arises from the considerable increase in gate capacitance, which is a direct consequence of introducing the additional layer of ferroelectric material. Although NCFETs provide steeper switching and enhanced $I_{ON}$ (at lower $V_{DD}$), they also incur additional delay due to large gate capacitance. In addition, polarization switching transients demand extra time during dynamic switching operation of logic circuits. Hence, materials with lower kinetic coefficient are desired for optimum operation in NCFET-based logic designs. In spite of these specific concerns, at low $V_{DD}$, as the inherent circuit delay immensely increases in conventional CMOS, these additional aspects of delay no longer remain a bottleneck to performance. In such a scenario, NCFETs garner benefits due to their steep switching operation.

**Variation** For mass production, variation is an important factor that limits the yield and thus cost, especially for circuits with minimized device dimensions. Variation between different devices, or between different operating points of time for the same device, can also cause reliability issues. There have been some recent research reports on these topics [48, 114–116]. However, further modeling, experimental verification, fabrication process optimization, and also circuit design techniques to mitigate the impact can be of great significance.

**Fabrication Complexity and Cost** With enormous efforts (i.e., device concepts, SPICE modeling, single transistor measurement, etc.) in emerging device researches, examinations of the device impacts on circuit and systems have been explored by using computer-aided simulations in many literatures. However, there are still many challenges facing the fabrication complexity of emerging transistors and increased difficulty and cost in the large-scale integration of the technology [117–119]. For example, the introduction of new materials such as III–V compounds, ferroelectric, insulator-to-metal transition (IMT) materials leads to other fabrication steps and techniques to guarantee device functionality and performance, resulting in increased fabrication complexity and immense cost.

NCFETs and Hyper-FETs are compatible to CMOS or FinFET technologies [29, 30, 63, 88]. They stack certain materials for exploring unique electrical characteristics onto the transistor. NCFET and FEFETs, for example, add the piezoelectric and ferroelectric materials (i.e., ALD $HfZrO_2$, $Pb(Zr,Ti)O_3$, BaTiO3, $SrBi_2Ta\ _2O_9$, etc.) [18, 80] to ferroelectric layers between gate electrode and dielectric, and Hyper-FET uses the insulator-to-metal transition in vanadium dioxide ($VO_2$) onto the source of CMOS to design a hybrid-phase-transition FET. However, for measurements and evaluations in [29, 30, 63, 88], these newly formed materials and structure are connected to CMOS using metals rather than fabricated into a single transistor due to the fabrication complexity and cost, leading to challenges in large-scale integration for circuits and systems.

The fabrication of III-V materials in TFETs has been reported in literatures with various techniques [40, 107]. The doping and electrical behaviors are well controlled by the fabrication process for a single transistor [40, 107]. However, they do not follow the conventional CMOS fabrication process since they are naturally vertical-oriented devices and use III–V and other materials. Another issue for fabrication in TFETs is that the integration of such n-type and p-type transistors on the same wafer is still questionable because there are some fabrication ranging from growing III–V materials, different material stacks, and other fabrication process issues such as extreme ultraviolet lithography (EUV), X-ray, imprint, projection electron-beam, or projection ion-beam lithography. These issues, nevertheless, tackle the advent of emerging technology in realistic applications like other emerging devices mentioned above.

As emerging devices offer the unique opportunities over conventional CMOS, the restrictions and features of new fabrication technologies are becoming the key challenges to bring these emerging technologies to apply in real circuits and systems. To advance the appearance of circuits and systems with emerging devices, such limitations and challenges of new fabrication technologies should be resolved in the future.

## 6.8 Conclusion

Three types of emerging devices, that is, TFET, NCFET, and Hyper-FET, have been investigated for next-generation low-power applications. The device mechanism, device–circuit co-design, as well as related architecture implications are discussed in detail. Evaluation results show that the steep-slope characteristics, together with new features, have brought great low-power opportunities for both conventional Boolean logic and nonconventional circuits and architectures. The challenges for future implementation with these emerging devices are also discussed in detail.

**Acknowledgment** This work is supported in part by the Center for Low Energy Systems Technology (LEAST), one of the six SRC STARnet centers, sponsored by MARCO and DARPA, by NSF awards 1160483 (ASSIST), and NSF Expeditions in Computing Award-1317560.

## References

1. A.C. Seabaugh, Q. Zhang, Low-voltage tunnel transistors for beyond CMOS logic. Proc. IEEE **98**(12), 2095–2110 (2010)
2. P.-F. Wang, K. Hilsenbeck, T. Nirschl, M. Oswald, C. Stepper, M. Weiss, D. Schmitt-Landsiedel, W. Hansch, Complementary tunneling transistor for low power application. Solid State Electron. **48**(12), 2281–2286 (2004)
3. K.K. Bhuwalka, J. Schulze, I. Eisele, Performance enhancement of vertical tunnel field-effect transistor with SiGe in the pþ layer. Jpn. J. Appl. Phys. **43**, 4073–4078 (2004)

4. J. Appenzeller, Y.-M. Lin, J. Knoch, P. Avouris, Band-to-band tunneling in carbon nanotube field-effect transistors. Phys. Rev. Lett. **93**(19), 196805-1–196805-4 (2004)
5. W.Y. Choi, B.-G. Park, J.D. Lee, T.-J.K. Liu, Tunneling field-effect transistors (TFETs) with subthreshold swing (SS) less than 60 mV/dec. IEEE Electron. Device Lett. **28**, 743–745 (2007)
6. T. Krishnamohan, D. Kim, S. Raghunathan, K. Saraswat, Double-gate strained-Ge heterostructure tunneling FET (TFET) with record high drive currents and G 60 mV/dec subthreshold slope, in *Proc. Int. Electron Devices Meeting (IEDM)* (2008), pp. 947–949
7. R. Pandey et al., Demonstration of p-type $In_{0.7}Ga_{0.3}As/GaAs_{0.35}Sb_{0.65}$ and n-type $GaAs_{0.4}Sb_{0.6}/In_{0.65}Ga_{0.35}As$ complimentary heterojunction vertical tunnel FETs for ultra-low power logic, in *2015 Symposium on VLSI Technology (VLSI Technology)*, Kyoto (2015), pp. T206–T207
8. R. Bijesh et al., Demonstration of In0.9Ga0.1As/GaAs0.18Sb0.82 near broken-gap tunnel FET with ION=740μA/μm, GM=70μS/μm and gigahertz switching performance at VDS=0.5V, in *2013 IEEE International Electron Devices Meeting (IEDM)*, Washington, DC (2013), pp. 28.2.1–28.2.4
9. G. Dewey et al., Fabrication, characterization, and physics of III-V heterojunction tunneling field effect transistors (H-TFET) for steep sub-threshold swing, in *Proc. IEDM Tech. Dig.*, Washington, DC (2011), pp. 33.6.1–33.6.4
10. R. Gandhi, Z. Chen, N. Singh, K. Banerjee, S. Lee, CMOS-compatible vertical-silicon-nanowire gate-all-around p-type tunneling FETs with ≤50-mV/decade subthreshold swing. IEEE Electron Device Lett. **32**(11), 1504–1506 (2011)
11. A.I. Khan, C.W. Yeung, C. Hu, S. Salahuddin, Ferroelectric negative capacitance MOSFET: Capacitance tuning & antiferroelec- tric operation, in *Proc. IEEE IEDM, Dec.* (2011), pp. 11.3.1–11.3.4
12. S. Salahuddin, S. Datta, Use of negative capacitance to provide voltage amplification for low power nanoscale devices. Nano Lett. **8**(2), 405–410 (2008)
13. H. Chenming, S. Salahuddin, C.-I. Lin, A. Khan, 0.2 V adiabatic NC-FinFET with 0.6 mA/μm $I_{ON}$ and 0.1 nA/μm $I_{OFF}$, in *Device Research Conference (DRC)* (2015), pp. 39–40
14. M.H. Lee et al., Steep slope and near non-hysteresis of FETs with antiferroelectric-like HfZrO for low-power electronics. IEEE Electron Device Lett. **36**(4), 294–296 (2015)
15. J. Jo, W.Y. Choi, J.-D. Park, J.W. Shim, H.-Y. Yu, C. Shin, Negative capacitance in organic/ferroelectric capacitor to implement steep switching MOS devices. Nano Lett. **15**(7), 4553–4556 (2015)
16. A.I. Khan et al., Negative capacitance in short-channel FinFETs externally connected to an epitaxial ferroelectric capacitor. IEEE Electron Device Lett. **37**(1), 111–114 (2016)
17. J. Jo, C. Shin, Negative capacitance field effect transistor with hysteresis-free Sub-60-mV/decade switching. IEEE Electron Device Lett. **37**(3), 245–248 (2016)
18. K.S. Li et al., Sub-60mV-swing negative-capacitance FinFET without hysteresis, in *2015 IEEE International Electron Devices Meeting (IEDM)*, Washington, DC (2015), pp. 22.6.1–22.6.4
19. M.H. Lee et al., Prospects for ferroelectric HfZrOx FETs with experimentally CET=0.98nm, $SS_{for}$=42mV/dec, $SS_{rev}$=28mV/dec, switch-off <0.2V, and hysteresis-free strategies, in *2015 IEEE International Electron Devices Meeting (IEDM)*, Washington, DC (2015), pp. 22.5.1–22.5.4
20. S. Dasgupta, A. Rajashekhar, K. Majumdar, N. Agrawal, A. Razavieh, S. Trolier-Mckinstry, S. Datta, Sub-kT/q switching in strong inversion in PbZr0.52Ti0.48O3 gated negative capacitance FETs. IEEE J. Explor. Solid-State Computat. Devices Circuits **1**, 43–48 (2015)
21. P.-G. Chen, Y.-T. Wei, M. Tang, M.H. Lee, Experimental demonstration of ferroelectric gate-stack AlGaN/GaN-on-Si MOS-HEMTs with voltage amplification for power applications. IEEE Trans. Electron Devices **61**(8), 3014–3017 (2014)
22. H. Nathanson, W. Newell, R. Wickstro, J. Davis, Resonant gate transistor. IEEE Trans. Electron Devices **ED-14**(3), 117–133 (1967)

23. A.M. Ionescu, V. Pott, R. Fritschi, K. Banerjee, M.J. Declercq, P. Renaud, C. Hilbert, P. Fluckiger, G.A. Racine, Modeling and design of a low-voltage SOI suspended-gate MOSFET (SG-MOSFET) with a metal-over-gate architecture, in *Proc. Int. Symp. Quality Elecron. Design* (2002), pp. 496–501
24. H. Kam, D.T. Lee, R.T. Howe, T.-J. King, A new nano-electro-mechanical field effect transistor (NEMFET) design for low-power electronics, in *Proc. Int. Electron Devices Meeting (IEDM)* (2005), pp. 463–466
25. K. Gopalakrishnan, P.B. Griffin, J.D. Plummer, I-MOS: A novel semiconductor device with a subthreshold slope lower than kT/q, in *Proc. Int. Electron Devices Meeting (IEDM)* (2002) pp. 289–292
26. W.Y. Choi, J.D. Lee, B.-G. Park, Integration process of impact ionization metal-oxide-semiconductor devices with tunneling field-effect transistors and metal-oxide-semiconductor field-effect transistors. Jpn. J. Appl. Phys. **46**(1), 122–124 (2007)
27. C.-W. Lee, A.N. Nazarov, I. Ferain, N.D. Akhavan, R. Yan, P. Razavi, R. Yu, R.T. Doria, J.-P. Colinge, Low subthreshold slope in junctionless multigate transistors. Appl. Phys. Lett. **96**, 102106 (2010)
28. E.-H. Toh, G.H. Wang, L. Chan, D. Weeks, M. Bauer, J. Spear, S.G. Thomas, G. Samudra, Y.-C. Yeo, Cointegration of in situ doped silicon-carbide source and silicon-carbon i-region in p-channel silicon nanowire impact-ionization transistor. IEEE Electron Device Lett. **29**(7), 731–733 (2008)
29. N. Shukla, A. Thathachary, A. Agrawal, H. Paik, A. Aziz, D. Schlom, S. Gupta, R. Engel-Herbert, S. Datta, A steep-slope transistor based on abrupt electronic phase transition. Nat. Commun. **6**, 7812-1–7812-6 (2015)
30. N. Shukla et al., Pairwise coupled hybrid vanadium dioxide-MOSFET (HVFET) oscillators for non-boolean associative computing, in *2014 IEEE International Electron Devices Meeting (IEDM)*, San Francisco, CA (2014), pp. 28.7.1–28.7.4
31. J.H. Park, G.S. Jang, H.Y. Seok, et al., Sub-kT/q subthreshold-slope using negative capacitance in low-temperature polycrystalline-silicon thin-film transistor. Sci. Rep. **6**, 24734 (2016)
32. J.P. Colinge, Subthreshold slope of thin-film SOI MOSFET's. IEEE Electron Device Lett. **7**(4), 244–246 (1986)
33. M.H. Lee et al., Ferroelectric negative capacitance hetero-tunnel field-effect-transistors with internal voltage amplification, in *2013 IEEE International Electron Devices Meeting (IEDM)*, Washington, DC (2013), pp. 4.5.1–4.5.4
34. U.E. Avci et al., Study of TFET non-ideality effects for determination of geometry and defect density requirements for sub-60mV/dec Ge TFET, in *2015 IEEE International Electron Devices Meeting (IEDM)*, Washington, DC (2015), pp. 34.5.1–34.5.4
35. R. Pandey et al., Tunnel junction abruptness, source random dopant fluctuation and PBTI induced variability analysis of GaAs0.4Sb0.6/In0.65Ga0.35As heterojunction tunnel FETs, in *2015 IEEE International Electron Devices Meeting (IEDM)*, Washington, DC (2015), pp. 14.2.1–14.2.4
36. H. Lu, A. Seabaugh, Tunnel field-effect transistors: State-of-the-art. IEEE J. Electron Devices Soc. **2**(4), 44–49 (2014)
37. H. Liu, X. Li, R. Vaddi, K. Ma, S. Datta, V. Narayanan, Tunnel FET RF rectifier design for energy harvesting applications. IEEE Trans. Emerg. Sel. Topics Circuits Syst. **4**(4), 400–411 (2014)
38. A.M. Ionescu, Tunnel FETs and emerging device concepts for subthermal switching, in *2013 IEEE International Electron Devices Meeting (IEDM)* (2013)
39. Synopsys TCAD, http://www.synopsys.com/tools/tcad
40. S. Datta et al., Tunnel transistors for energy efficient computing, in *IEEE Int. Reliability Physics Symp. (IRPS)* (2013), pp. 6A.3.1–6A.3.7
41. B. Sedighi, X.S. Hu, H. Liu, J.J. Nahas, M. Niemier, Analog circuit design using tunnel-FETs. IEEE Trans. Circuits Syst. I **62**(1), 39–48 (2015)

42. V. Saripalli, S. Datta, V. Narayanan, J.P. Kulkarni, Variation-tolerant ultra low-power heterojunction tunnel FET SRAM design, in *2011 IEEE/ACM International Symposium on Nanoscale Architectures*, San Diego, CA (2011), pp. 45–52
43. M.S. Kim, H. Liu, K. Swaminathan, X. Li, S. Datta, V. Narayanan, Enabling power-efficient designs with III-V tunnel FETs, in *2014 IEEE Compound Semiconductor Integrated Circuit Symposium (CSICS)*, La Jolla, CA (2014), pp. 1–4
44. K. Swaminathan, H. Liu, J. Sampson, V. Narayanan, An examination of the architecture and system-level tradeoffs of employing steep slope devices in 3D CMPs, in *2014 ACM/IEEE 41st International Symposium on Computer Architecture (ISCA)*, Minneapolis, MN (2014), pp. 241–252
45. W.Y. Tsai, H. Liu, X. Li, V. Narayanan, Low-power high-speed current mode logic using tunnel-FETs, in *2014 22nd International Conference on Very Large Scale Integration (VLSI-SoC)*, Playa del Carmen (2014), pp. 1–6
46. K. Swaminathan, H. Liu, X. Li, M.S. Kim, J. Sampson, V. Narayanan, Steep slope devices: Enabling new architectural paradigms, in *2014 51st ACM/EDAC/IEEE Design Automation Conference (DAC)*, San Francisco, CA (2014), pp. 1–6
47. M.S. Kim, W. Cane-Wissing, X. Li, J. Sampson, S. Datta, S.K. Gupta, V. Narayanan, Comparative area and parasitics analysis in FinFET and Heterojunction vertical TFET standard cells. ACM J. Emerg. Technol. Comput. Syst. **12**(4), 38 (2016)
48. S. Datta, H. Liu, V. Narayanan, Tunnel FET technology: A reliability perspective. Microelectron. Reliab. **54**(5), 861–874 (2014)
49. U. Heo, X. Li, H. Liu, S. Gupta, S. Datta, V. Narayanan, A high-efficiency switched-capacitance HTFET charge pump for low-input-voltage applications, in *2015 28th International Conference on VLSI Design*, Bangalore (2015), pp. 304–309
50. X. Li, K. Ma, S. George, J. Sampson, V. Narayanan, Enabling internet-of-things: Opportunities brought by emerging devices, circuits, and architectures, in *2016 IFIP/IEEE International Conference on Very Large Scale Integration (VLSI-SoC)* (Springer, Cham, 2016), pp. 1–23
51. X. Li, U. Dennis Heo, K. Ma, V. Narayanan, H. Liu, S. Datta, Rf-powered systems using steep-slope devices, in *New Circuits and Systems Conference (NEWCAS), 2014 IEEE 12th International*, Trois-Rivieres, QC (2014), pp. 73–76
52. M.S. Kim, X. Li, H. Liu, J. Sampson, S. Datta, V. Narayanan, Exploration of low-power high-SFDR current-steering D/A converter design using steep-slope heterojunction tunnel FETs. IEEE Trans. VLSI Syst. **24**(6), 2299–2309 (2016)
53. X. Li, Q. Wei, Z. Xu, J. Liu, H. Wang, H. Yang, A 14 bit 500 MS/s CMOS DAC using complementary switched current sources and time-relaxed interleaving DRRZ. IEEE Trans. Circuits Syst. I **61**(8), 2337–2347 (2014)
54. X. Li, Q. Wei, H. Yang, Code-independent output impedance: A new approach to increasing the linearity of current-steering DACs, in *2011 18th IEEE International Conference on Electronics, Circuits and Systems (ICECS)*, Beirut (2011), pp. 216–219
55. J. Liu, X. Li, Q. Wei, H. Yang, A 14-bit 1.0-GS/s dynamic element matching DAC with >80 dB SFDR up to the Nyquist, in *2015 IEEE International Symposium on Circuits and Systems (ISCAS)*, Lisbon (2015), pp. 1026–1029
56. R. Pandey, B. Rajamohanan, H. Liu, V. Narayanan, S. Datta, Electrical noise in heterojunction interband tunnel FETs. IEEE Trans. Electron Devices **61**(2), 552–560 (2014)
57. H. Liu, S. Datta, M. Shoaran, A. Schmid, X. Li, V. Narayanan, Tunnel FET-based ultra-low power, low-noise amplifier design for bio-signal acquisition, in *2014 IEEE/ACM International Symposium on Low Power Electronics and Design (ISLPED)*, La Jolla, CA (2014), pp. 57–62
58. M.S. Kim, H. Liu, X. Li, S. Datta, V. Narayanan, A steep-slope tunnel FET based SAR analog-to-digital converter. IEEE Trans. Electron Devices **61**(11), 3661–3667 (2014)
59. P. Mazumder et al., Tunneling-based cellular nonlinear network architectures for image processing. IEEE Trans. VLSI Syst. **17**(4), 487–495 (2009)
60. A.R. Trivedi, S. Datta, S. Mukhopadhyay, Application of silicon-germanium source tunnel-FET to enable ultralow power cellular neural network-based associative memory. IEEE Trans. Electron Devices **61**(11), 3707–3715 (2014)

61. A.R. Trivedi, S. Mukhopadhyay, Potential of ultralow-power cellular neural image processing with Si/Ge tunnel FET. IEEE Trans. Nanotechnol. **13**(4), 627–629 (2014)
62. P.M. Asbeck, K. Lee, J. Min, Projected performance of heterostructure tunneling FETs in low power microwave and mm-wave applications. IEEE J. Electron Devices Soc. **3**(3), 122–134 (2015)
63. A. Aziz, S. Ghosh, S. Datta, S.K. Gupta, Physics-based circuit-compatible SPICE model for ferroelectric transistors. IEEE Electron Device Lett. **37**(6), 805–808 (2016)
64. Aziz, S. Ghosh, S. Datta, S.K. Gupta, Polarization charge and coercive field dependent performance of negative capacitance FETs, in *2016 74th Annual Device Research Conference (DRC)* (2016), pp. 1–2
65. S. George, K. Ma, A. Aziz, X. Li, A. Khan, S. Salahuddin, M.-F. Chang, et al., Nonvolatile memory design based on ferroelectric FETs, in *Proceedings of the 53rd Annual Design Automation Conference (DAC)* (2016), p. 118
66. S. George, A. Aziz, X. Li, et al., Device circuit co design of FEFET based logic for low voltage processors, in *ISVLSI* (2016), pp. 649–654
67. S. George, X. Li, et al., NCFET based logic for energy harvesting systems, in *SRC TECHCON 2015* (SRC, Durham, 2015)
68. X. Li et al., Design of Nonvolatile SRAM with ferroelectric FETs for energy-efficient backup and restore. IEEE Trans. Electron Devices **64**(7), 3037–3040 (2017)
69. K. Ma, Y. Zheng, S. Li, K. Swaminathan, X. Li, Y. Liu, J. Sampson, Y. Xie, V. Narayanan, Architecture exploration for ambient energy harvesting nonvolatile processors, in *2015 IEEE 21st International Symposium on High Performance Computer Architecture (HPCA)* (2015), pp. 526–537
70. Y. Wang, Y. Liu, S. Li, D. Zhang, B. Zhao, M.-F. Chiang, Y. Yan, B. Sai, H. Yang, A 3us wake-up time nonvolatile processor based on ferroelectric flip-flops, in *2012 Proceedings of the ESSCIRC (ESSCIRC)* (2012), pp. 149–152
71. Y. Liu et al., Ambient energy harvesting nonvolatile processors: From circuit to system, in *Proceedings of the 52nd Annual Design Automation Conference* (2015), p. 150
72. K.-W. Kwon, S.H. Choday, Y. Kim, X. Fong, S.P. Park, K. Roy, SHE-NVFF: Spin hall effect-based nonvolatile flip-flop for power gating architecture. IEEE Electron Device Lett. **35**(4), 488–490 (2014)
73. D. Wang et al., Ferroelectric transistor based non-volatile flip-flop, in *Proceedings of the 2016 International Symposium on Low Power Electronics and Design (ISLPED'16)* (2016), pp. 10–15
74. X. Li, J. Sampson, A. Khan, K. Ma, S. George, A. Aziz, S. Gupta, S. Salahuddin, M.-F. Chang, S. Datta, V. Narayanan, Enabling energy-efficient nonvolatile computing with negative capacitance FET. IEEE Trans. Electron Devices **64**(8), 3452–3458 (2017)
75. X. Li, S. George, K. Ma, W.-Y. Tsai, A. Aziz, J. Sampson, S. Gupta, M.-F. Chang, Y. Liu, S. Datta, V. Narayanan, Advancing nonvolatile computing with nonvolatile NCFET latches and flip-flops. IEEE Trans. Circuits Syst. I **64**(11), 2907–2919 (2017)
76. K. Ma et al., Nonvolatile processor architecture exploration for energy-harvesting applications. IEEE Micro **35**(5), 32–40 (2015)
77. K. Ma et al., Nonvolatile processor architectures: Efficient, reliable progress with unstable power. IEEE Micro **36**(3), 72–83 (2016)
78. K. Ma, X. Li, et al., Nonvolatile processor optimization for ambient energy harvesting scenarios, in *The 15th Non-volatile Memory Technology Symposium (NVMTS)* (2015)
79. K. Ma et al., Dynamic machine learning based matching of nonvolatile processor microarchitecture to harvested energy profile, in *Proceedings of the IEEE/ACM International Conference on Computer-Aided Design* (2015), pp. 670–675
80. M.H. Park, Y.H. Lee, H.J. Kim, Y.J. Kim, T. Moon, K.D. Kim, J. Müller, A. Kersch, U. Schroeder, T. Mikolajick, C.S. Hwang, Ferroelectricity and antiferroelectricity of doped thin $HfO_2$-based films. Adv. Mat. **27**(11), 1811–1831 (2015)
81. A. Chen, X. Sharon Hu, Y. Jin, M. Niemier, X. Yin, Using emerging technologies for hardware security beyond PUFs, in *2016 Design, Automation & Test in Europe Conference & Exhibition (DATE)* (2016), pp. 1544–1549

82. J. Frougier, N. Shukla, D. Deng, M.J. Jerry, A. Aziz, L. Liu, G. Lavallee, T.S. Mayer, S.K. Gupta, S. Datta, Phase-transition-FET exhibiting steep switching slope of 8mV/decade and 36% enhanced ON current, in *2016 Symposia on VLSI Technology and Circuits* (2016)
83. F.J. Morin, Oxides which show a metal-to-insulator transition at the Neel temperature. Phys. Rev. Lett. **3**, 34–36 (1959)
84. Y. Tomioka et al., Phase diagrams of perovskite-type manganese oxides. J. Phys. Chem. Solids **67**(9–10), 2214–2221 (2006)
85. Q. Luo et al., Cu BEOL compatible selector with high selectivity (>107), extremely low off-current (pA) and high endurance (>1010), in *2015 IEEE International Electron Devices Meeting (IEDM)*, Washington, DC (2015), pp. 10.4.1–10.4.4
86. Yang et al., Novel selector for high density non-volatile memory with ultra-low holding voltage and 107 on/off ratio, in *2015 Swymposium on VLSI Technology (VLSI Technology)*, Kyoto (2015), pp. T130–T131
87. J. Cao et al., Strain engineering and one-dimensional organization of metal-insulator domains in single-crystal vanadium dioxide beams. Nature Nanotechnol. **4**(11), 732–737 (2009)
88. W.Y. Tsai, X. Li, M. Jerry, B. Xie, N. Shukla, H. Liu, N. Chandramoorthy, M. Cotter, A. Raychowdhury, D.M. Chiarulli, S.P. Levitan, S. Datta, J. Sampson, N. Ranganathan, V. Narayanan, Enabling new computation paradigms with HyperFET - an emerging device. IEEE Trans. Multi-Scale Comput. Syst. **2**(1), 30–48 (2016)
89. M. Jerry et al., Phase transition oxide neuron for spiking neural networks, in *DRC 2016 74th Annual* (2016)
90. C.C. McAndrew et al., Best practices for compact modeling in Verilog-A. IEEE J. Electron Devices Soc. **3**(5), 383–396 (2015)
91. Verilog-AMS language reference manual [Online]. Available: http://www.accellera.org/downloads/standards/v-ams. Accessed Jun 2015
92. M.A. Wahab, M.A. Alam, A Verilog-A Compact Model for Negative Capacitance FET [Online]. (nanoHUB, 2016), https://doi.org/10.4231/D3PV6B79V. Available: https://nanohub.org/publications/95/2
93. H. Liu, V. Saripalli, V. Narayanan, S. Datta, III-V Tunnel FET Model [Online]. (nanoHUB, 2015), https://doi.org/10.4231/D30Z70X8D. Available: https://nanohub.org/publications/12/2
94. S. Rakheja, D. Antoniadis, Physics-based compact modeling of charge transport in nanoscale electronic devices, in *2015 IEEE International Electron Devices Meeting (IEDM)*, Washington, DC (2015), pp. 28.6.1–28.6.4
95. V.P. Trivedi, G. Fossum, L. Mathew, M.M. Chowdhury, W. Zhang, G.O. Workman, B.-Y. Nguyen, Physics-based compact modeling for nonclassical CMOS, in *ICCAD-2005. IEEE/ACM International Conference on Computer-Aided Design* (2005), pp. 211–216
96. J.U. Mehta, W.A. Borders, H. Liu, R. Pandey, S. Datta, L. Lunardi, III–V tunnel FET model with closed-form analytical solution. IEEE Trans. Electron Devices **63**(5), 2163–2168 (2016)
97. J.D. Harms, F. Ebrahimi, X. Yao, J.P. Wang, SPICE macromodel of spin-torque-transfer-operated magnetic tunnel junctions. IEEE Trans. Electron Devices **57**(6), 1425–1430 (2010)
98. G.D. Panagopoulos, C. Augustine, K. Roy, Physics-based SPICE-compatible compact model for simulating hybrid MTJ/CMOS circuits. IEEE Trans. Electron Devices **60**(9), 2808–2814 (2013)
99. M. Mierzwinsk, P.O. Halloran, B. Troyanovsky, R. Dutton, Changing the paradigm for compact model integration in circuit simulators using Verilog-A, in *Tech. Proceeding of 2003 Nanotechnology Conference and Trade Show*, Vol. 2, Chap. 7 (Compact Modeling, San Francisco, CA, 2003), pp. 376–379
100. .V.K. Chavali, J. Joseph, V.K. Chaubey, A.K. Saini, Compact drain current modeling in long channel SOI double gate FET for sub 40nm gate width, in *15th International Workshop on Physics of Semiconductor Devices* (2009)
101. L. Lemaitre, C. McAndrew, S. Hamm, ADMS-automatic device model synthesizer, in *Custom Integrated Circuits Conference, 2002. Proceedings of the IEEE 2002* (2002), pp. 27–30

102. D. Yakimets, G. Eneman, P. Schuddinck, T.H. Bao, M.G. Bardon, P. Raghavan, A. Veloso, N. Collaert, A. Mercha, D. Verkest, A.V.-Y. Thean, K.D. Meyer, Vertical GAAFETs for the ultimate CMOS scaling. IEEE Trans. Electron Devices **62**(5), 1433–1439 (2015)
103. A. Akturka, M. Peckerara, K. Engb, J. Hamletb, S. Potbharea, E. Longoriab, R. Youngb, T. Gurrierib, M.S. Carrollb, N. Goldsmana, Compact modeling of 0.35μm SOI CMOS technology node for 4K DC operation using Verilog-a. J. Microelectron. Eng. **87**(12), 2518–2524 (2010)
104. K. Swaminathan, M.S. Kim, N. Chandramoorthy, B. Sedighi, R. Perricone, J. Sampson, V. Narayanan, Modeling steep slope devices: From circuits to architectures, in *2014 Design, Automation & Test in Europe Conference & Exhibition (DATE), Dresden* (2014), pp. 1–6
105. M. Martin, D.J. Sorin, B.M. Beckmann, M.R. Marty, M. Xu, A.R. Alameldeen, K.E. Moore, M.D. Hill, D.A. Wood, Multifacet's general execution-driven multiprocessor simulator (GEMS) toolset. ACM SIGARCH Comput. Archit. News **33**, 92–99 (2005)
106. S. Li, J.H. Ahn, R.D. Strong, J.B. Brockman, D.M. Tullsen, N.P. Jouppi, McPAT: An integrated power, area, and timing modeling framework for multicore and manycore architectures, in *2009 42nd Annual IEEE/ACM International Symposium on Microarchitecture (MICRO)*, New York, NY (2009), pp. 469–480
107. S. Datta, H. Liu, V. Narayanan, Tunnel FET technology: A reliability perspective. Microelectron. Reliab. **54**, 861–874 (2014)
108. K. Hung, P.-K. Ko, C. Hu, Y. Cheng, A physics-based MOSFET noise model for circuit simulators. IEEE Trans. Electron Devices **37**(5), 1323–1333 (1990)
109. M. Agostinelli, J. Hicks, J. Xu, B. Woolery, K. Mistry, K. Zhang, et al., Erratic fluctuations of SRAM cache VMIN at the 90 nm process technology node, in *Electron Devices Meeting, 2005. IEDM Technical Digest* (IEEE International, 2005), pp. 655–658
110. M.S. Kim et al., Comparative area and parasitics analysis in FinFET and heterojunction vertical TFET standard cells. ACM J. Emerg. Technol. Comput. Syst. **4**, 38 (2016)
111. N. Choudhary et al., Fabscalar: Composing synthesizable RTL designs of arbitrary cores within a canonical superscalar template, in *International Symposium on Computer Architecture (ISCA)* (2011), pp. 11–22
112. M.S. Kim, W. Cane-Wissing, J. Sampson, S. Datta, V. Narayanan, S.K. Gupta, Comparing energy, area, delay tradeoffs in going vertical with CMOS and asymmetric HTFETs, in *2015 IEEE Computer Society Annual Symposium on VLSI (ISVLSI)*, Montpellier (2015), pp. 303–308
113. D. Damjanovic, Hysteresis in piezoelectric and ferroelectric materials, in *The Science of Hysteresis*, ed. by I. Mayergoyz, G. Bertotti (Eds), vol. 3 (Elsevier, New York, 2005)
114. K.J. Kuhn, Reducing variation in advanced logic technologies: Approaches to process and design for manufacturability of nanoscale CMOS, in *2007 IEEE International Electron Devices Meeting, Washington, DC* (2007), pp. 471–474
115. U.E. Avci, R. Rios, K. Kuhn, I.A. Young, Comparison of performance, switching energy and process variations for the TFET and MOSFET in logic, in *2011 Symposium on VLSI Technology - Digest of Technical Papers*, Honolulu, HI (2011), pp. 124–125
116. N. Damrongplasit, S.H. Kim, T.J.K. Liu, Study of random dopant fluctuation induced variability in the raised-Ge-source TFET. IEEE Electron Device Lett. **34**(2), 184–186 (2013)
117. S. Bampi, R. Reis, Challenges and emerging technologies for system integration beyond the end of the roadmap of Nano-CMOS, in *VLSI-SoC: Technologies for Systems Integration Vol. 360 of the Series IFIP Advances in Information and Communication Technology* (Springer, Hiedelberg, 2011), pp. 21–33
118. M.H. Ben Jamaa, *Regular Nanofabrics in Emerging Technologies: Design and Fabrication Methods for Nanoscale Digital Circuits*, vol 82, 1st edn. (Springer, Amsterdam, 2011)
119. R. Compaño, L. Molenkamp, D.J. Paul, Roadmap for Nanoelectronics, in *Future and Emerging Technologies. Microelectronics Advanced Research Initiative Melari Nano* (2001)

# Chapter 7
# Spin-Based Majority Computation

**Odysseas Zografos, Adrien Vaysset, Bart Sorée, and Praveen Raghavan**

## 7.1 Introduction

The exploration and study of novel non-charge-based logic devices has been a main research focus in the past decade [1]. The purpose is to identify concepts that can extend the semiconductor industry roadmap beyond the complementary metal oxide semiconductor (CMOS) technology [1]. Since CMOS scaling, dictated by Moore's Law [2], will reach its limits in the following decade [3], there is a need for logic components that can operate at high frequencies, be extremely compact, and also consume ultralow power [4]. A variety of magnetic devices have been benchmarked as promising candidates for low-power applications [4–8].

The goal of this chapter is to introduce, analyze, and discuss two spin-based logic concepts that utilize majority-based computation. Namely, the Spin Wave Device (SWD) and Spin Torque Majority Gate (STMG) concepts, which were first proposed by Khitun et al. in [9] and Nikonov et al. in [6] respectively. This section introduces some terms and concepts which are useful for establishing the framework of this chapter. More specifically, Sect. 7.1.1 introduces basic physics terminology. Important spin-based logic concepts, different from SWD and STMG, are described in Sect. 7.1.2. Finally, Sect. 7.1.3 introduces majority-based logic, which is used extensively by SWD and STMG logic as elaborated in Sects. 7.2 and 7.3.

O. Zografos (✉) · A. Vaysset · P. Raghavan
imec, Leuven, Belgium
e-mail: odysseas.zografos@imec.be

B. Sorée
KU Leuven, ESAT, Leuven, Belgium

imec, Leuven, Belgium

Universiteit Antwerpen, Physics Department, Condensed Matter Theory, Antwerpen, Belgium

 
R. O. Topaloglu, H.-S. P. Wong (eds.), *Beyond-CMOS Technologies for Next Generation Computer Design*, https://doi.org/10.1007/978-3-319-90385-9_7

### *7.1.1 Spin and Magnetism Basics*

In this part, we will introduce the most important physics terms required to understand the rest of the chapter. This introduction is by no means exhaustive and we invite the interested reader to look for further information and insight at the referenced manuscripts.

#### 7.1.1.1 From Angular Momentum to Ferromagnetism

The angular momentum (**L**) of a particle is generally defined as the cross-product of the particle's position (**r**) and momentum (**p**) [10]:

$$\mathbf{L} = \mathbf{r} \times \mathbf{p}, \tag{7.1}$$

where $\mathbf{p} = m\mathbf{v}$ is the linear momentum of the particle. When the particle carries a charge $q$ and is flowing in a circular loop with radius $R$ and area $A = \pi R^2$, carrying a current $I = qv/2\pi R$ where $v$ is the velocity of the charged particle, the magnitude of the orbital angular momentum $|\mathbf{L}| = mvR$, while the associated magnetic moment is

$$\boldsymbol{\mu}_\mathrm{L} = IA\mathbf{e}_n = \frac{qv}{2\pi R}\pi R^2\mathbf{e}_n = \frac{qvR}{2}\mathbf{e}_n = \frac{q}{2m}\mathbf{L} = \gamma_\mathrm{L}\mathbf{L}, \tag{7.2}$$

where $\gamma_\mathrm{L} = q/2m$ is the gyromagnetic ratio.

In classical mechanics, a rigid object admits two kinds of angular momentum: *orbital* (see Eq. (7.1)), associated with the motion of the center of mass, and *spin*, associated with motion about the center of mass [10]. Similarly, in quantum mechanics there is also another kind (other than the orbital) of angular momentum occurring called spin angular momentum [11]. However, the spin angular momentum of a particle cannot be decomposed into orbital angular momenta of constituent parts [10] and should not be pictured as due to some internal motion of the particle. In that sense the description of a particle's spin angular momentum is completely different from the one of a rigid object (in classical mechanics) and hence it is considered a purely quantum mechanical phenomenon.

As a quantum mechanical entity, the spin of a particle is described by a set of complete Hermitian operators corresponding to the magnitude of the spin $\hat{S}$ and one component of the spin vector conventionally along the $z$-direction, that is, $\hat{S}_z$. The spin state is then completely described by a state vector $|sm_s\rangle$ being the eigenstate of the aforementioned Hermitian operators, that is

$$\hat{S}^2|sm_s\rangle = \hbar^2 s(s+1)|sm_s\rangle \tag{7.3}$$

$$\hat{S}_z|sm_s\rangle = \hbar m_s|sm_s\rangle, \tag{7.4}$$

where the eigenvalue $s$ assumes an integer value for Bosons and half integer number for Fermions, while the eigenvalue can assume the following values $m_s = -s, -s+1, \ldots, s-1, s$ (e.g., for electrons $s = 1/2$, $m_s = +1/2, -1/2$). The eigenvalues $s$ and $m_s$ are the possible quantized outcomes of a measurement with a probability given by the modulus squared of the corresponding amplitudes. For a spin $1/2$ system, if the spin state is given by

$$|\Psi\rangle = \alpha|1/2, 1/2\rangle + \beta|1/2, -1/2\rangle, \quad (7.5)$$

where $|\alpha|^2 + |\beta|^2 = 1$ due to normalization, the probability of measuring the spin $s = 1/2$ with spin along the $+z$-direction (spin-up), that is, $S_z = \hbar/2$, is given by $|\alpha|^2$, while for spin down, that is, $S_z = -\hbar/2$, the corresponding probability is $|\beta|^2$.

According to Pauli's exclusion principle [12], no two fermions in an atom can have all their quantum numbers equal [13]. This means that in order for two electrons to fill the same orbital in an atom they have to have opposite spins (one will occupy the quantum state of spin-up and the other of spin down).

A charged particle with spin angular momentum constitutes a *magnetic dipole* [10], meaning that an electron acts also as a tiny magnet. The magnetic dipole moment associated with spin, $\boldsymbol{\mu}_S$ is proportional to its spin angular momentum $\mathbf{S}$. The proportionality is defined by a constant called **gyromagnetic ratio** $\gamma_S$ [14], that is

$$\boldsymbol{\mu}_S = \gamma_S \mathbf{S}. \quad (7.6)$$

Both the intrinsic spin as well as the orbital angular momentum will contribute to the total magnetic moment of an electron. The total angular momentum $\mathbf{J} = \mathbf{L} + \mathbf{S}$ is related to the total magnetic moment $\boldsymbol{\mu}$ as follows:

$$\boldsymbol{\mu} = \gamma \mathbf{J}. \quad (7.7)$$

where $\gamma$ is the gyromagnetic ratio relating the total magnetic moment and total angular momentum. Two magnetic dipoles can interact in two different ways: by exchange interaction, when the dipoles are close together or by dipole–dipole interaction (i.e., dipolar coupling), when the dipoles are far from each other. The exchange interaction arises from the overlap of the two particles' wave functions combined with the Coulomb interaction and the Pauli exclusion principle. If these two particles are at a fixed close distance then the interaction tends to change their spin eigenstates so that they either have the same or parallel orientation (both spin-up or spin down) or an opposite or antiparallel orientation. On the other hand, dipolar coupling tends to align spins of far-away magnetic moments in an antiparallel fashion. The competition between the long-range dipolar interaction and the short-range exchange interaction gives rise to magnetic configurations that strongly depend on the size of the magnet. For submicron sizes, exchange interaction dominates, which leads to uniform magnetization distribution at equilibrium. For larger magnets, the dipolar interaction favors multidomain states.

**Table 7.1** Ferromagnetic crystals [13]

| Substance | Curie temperature in K |
|---|---|
| Fe | 1043 |
| Co | 1388 |
| Ni | 627 |
| Gd | 292 |
| Dy | 88 |
| MnAs | 318 |
| $CrO_2$ | 386 |
| EuO | 69 |
| $Y_3Fe_5O_{12}$ | 560 |

Ferromagnetism is the basic mechanism by which certain materials (see Table 7.1) form permanent magnets due to the aforementioned exchange interaction, which means that they can be magnetized in the absence of an external magnetic field. Ferromagnetism involves the magnetic dipoles associated with the spins of unpaired electrons [10] (electrons that partially fill the outer shell of the atoms). The magnetic dipoles of neighboring atoms interact via exchange interaction and are strongly aligned to the same orientation. So ferromagnetism is one of the macroscopic expressions of the quantum mechanical spin and exchange phenomena. Other possible long-range magnetic ordering driven by the exchange interaction are antiferromagnetism and ferrimagnetism which are not considered here as they are out of scope and for the purpose of this chapter we will refrain for elaborating furthermore. The reader is encouraged to look into [14] and [15] for more insights.

Ferromagnetism is found in the binary and ternary alloys of Fe, Co, and Ni with one another, in alloys of Fe, Co, and Ni with other elements, and in a relatively few alloys which do not contain any ferromagnetic elements [15]. Table 7.1 enumerates a few examples of known ferromagnetic materials, along with their Curie temperature.[1]

A phenomenon occurring to ferromagnetic materials, which is important for the introduction of this chapter, is magnetostriction. Magnetostriction is the phenomenon whereby the shape of a ferromagnetic specimen changes during the process of magnetization [16]. In other words, the dimensions and magnetic properties of magnetostrictive materials are intertwined. This class of materials is significant to any spin-based technology and their integrated application because in combination with piezoelectric materials they offer a way of voltage-induced magnetic control. Table 7.2 enumerates a few examples of magnetostrictive materials along with their respective magnetostriction constants.[2]

[1]Curie temperature of a ferromagnetic material is the temperature over which the material loses its magnetic ordering [14].

[2]Magnetostriction constants define how much the material deforms [16].

**Table 7.2** Magnetostrictive ferrites [16]

| Substance | $\lambda_{100}$ | $\lambda_{111}$ |
|---|---|---|
| $MnFe_2O_4$ | −31 | 6.5 |
| $Fe_3O_4$ | −20 | 78 |
| $Co_{0.8}Fe_{2.2}O_4$ | −590 | 120 |
| $NiFe_2O_4$ | −42 | −14 |
| $CuFe_2O_4$ | −57.5 | 4.7 |
| $MgFe_2O_4$ | −10.5 | 1.7 |

#### 7.1.1.2 Magnetization Dynamics and the Landau–Lifshitz–Gilbert Equation

A magnetic moment $\boldsymbol{\mu}$ subjected to an effective magnetic induction field $\mathbf{B}_{\text{eff}}$ experiences a torque $\boldsymbol{\tau} = \boldsymbol{\mu} \times \mathbf{B}_{\text{eff}}$. Due to $\boldsymbol{\mu} = \gamma \mathbf{J}$, where $\gamma$ is the gyromagnetic ratio, the equation of motion for the magnetic moment is

$$\frac{d\mathbf{J}}{dt} = \boldsymbol{\tau} = \boldsymbol{\mu} \times \mathbf{B}_{\text{eff}} \Rightarrow \frac{d\boldsymbol{\mu}}{dt} = \gamma \boldsymbol{\mu} \times \mathbf{B}_{\text{eff}}. \tag{7.8}$$

When the magnetic moment $\boldsymbol{\mu}$ is at an angle $\varphi$ with the effective magnetic induction field $\mathbf{B}_{\text{eff}}$, it will make a precessional motion around the $\mathbf{B}_{\text{eff}}$ field vector. The effective magnetic induction field accounts for all external applied fields as well as internal fields due to exchange and anisotropy and is related to the magnetic field by $\mathbf{B}_{\text{eff}} = \mu_0 \mathbf{H}_{\text{eff}}$. The magnetization or magnetic moment per unit volume of a magnetic material is given by $\mathbf{M} = N\boldsymbol{\mu}$ where $N$ is the number of magnetic moments per unit volume. As a result, the equation of motion for the magnetization is

$$\frac{d\mathbf{M}}{dt} = -\gamma \mu_0 \mathbf{M} \times \mathbf{H}_{\text{eff}}. \tag{7.9}$$

This equation is known as the dampless Landau equation for the magnetization and does not take into account the presence of damping processes that occur in real magnetic materials. In order to account for such damping, an additional phenomenological term is added to the equation of motion which tends to pull the magnetization in the direction of the effective magnetic field:

$$\frac{d\mathbf{M}}{dt} = -\gamma \mu_0 \mathbf{M} \times \mathbf{H}_{\text{eff}} + \frac{\alpha}{M_s} \times \left(\mathbf{M} \times \frac{d\mathbf{M}}{dt}\right). \tag{7.10}$$

This equation is known as the Landau–Lifshitz–Gilbert equation. The strength of the damping is quantified by the damping parameter $\alpha$ and $M_s$ is the saturation magnetization.

#### 7.1.1.3 Anisotropy

Crystal symmetries can induce preferred magnetization directions. This phenomenon is called *magnetocrystalline anisotropy*, and the preferred directions are referred to as *easy axes*. In the common case of uniaxial anisotropy, the magnetization tends to align along one particular axis. The anisotropy energy is then expressed as

$$E_{\text{anis}} = K_u V \sin^2 \theta, \tag{7.11}$$

where $K_u$ is the anisotropy constant in $\text{J}/\text{m}^3$ and $\theta$ is angle between the magnetization and the easy axis.

Magnetocrystalline anisotropy is a bulk effect. However, in thin films, some interface effects can dominate over bulk. For example, the interfaces MgO–CoFeB and Co–Ni induce a perpendicular anisotropy. The energy of this interface anisotropy is expressed as

$$E_{\text{anis}} = \frac{K_s}{t} V \left(1 - \cos^2 \theta\right), \tag{7.12}$$

where $K_s$ is the surface anisotropy coefficient in $\text{J}/\text{m}^2$, $t$ is the thickness, and $\theta$ is the out-of-plane angle.

In thin films, the aforementioned dipolar interaction favors in-plane magnetization. It is therefore opposite to the interface anisotropy. As a consequence, an effective anisotropy coefficient can be expressed as

$$K_{\text{eff}} = \frac{K_s}{t} - \frac{1}{2}\mu_0 M_s^2. \tag{7.13}$$

At equilibrium, the magnetization is perpendicular to the plane if $K_{\text{eff}} > 0$ (out-of-plane easy axis) and in-plane if $K_{\text{eff}} < 0$ (in-plane easy axis).

#### 7.1.1.4 Spin Transfer Torque

The Spin Transfer Torque (STT) [17] is the effect induced by a spin polarized current on the magnetization. The spin polarization is created by a thick ferromagnetic layer called *polarizer*, or *reference layer*. The torque is exerted on the magnetization of the *free layer*, bringing it along the direction of the spin polarization. In simulations, the spin torque is modeled as an additional term in the Landau–Lifshitz–Gilbert equation (7.10).

In a Magnetic Tunnel Junction (MTJ), a thin oxide layer is sandwiched between the free layer and the polarizer. The tunneling of the current through the oxide barrier is spin-dependent [18], leading to enhanced spin torque efficiency, in particular for coherent tunneling through crystallized MgO barrier [19–21].

#### 7.1.1.5 Tunnel Magnetoresistance

The magnetic state of the free layer can be detected via Tunnel Magnetoresistance (TMR). If the free layer magnetization is parallel to the magnetization of the reference layer, a low-resistance state is detected. In contrast, antiparallel alignment leads to a high-resistance state. The TMR ratio is defined as

$$\mathrm{TMR} = \frac{R_{\mathrm{AP}} - R_{\mathrm{P}}}{R_{\mathrm{P}}}, \tag{7.14}$$

where $R_{\mathrm{P}}$ and $R_{\mathrm{AP}}$ are the resistances of the parallel and antiparallel states.

### *7.1.2 Spin-Based Logic Concepts*

As shown in [5] and [4], there exists a variety of novel spin-based devices and components. Three of the most important concepts, as potential IC applications, are presented here below.

#### 7.1.2.1 SpinFET

The SpinFET was first proposed by Datta and Das in [22]. It consists of a quasi-one-dimensional semiconductor channel with ferromagnetic source and drain contacts Fig. 7.1a. The concept makes use of the Rashba spin–orbit interaction [23], where spin polarized electrons are injected from the source to the channel and then detected at the drain. The electron transmission probability depends on the relative alignment of its spin with the fixed magnetization of the drain. This alignment is controlled by the gate voltage and the induced Rashba interaction, meaning that also the source–drain current is controlled. This first proposal had several impediments toward experimental demonstration, such as low spin-injection efficiency due to resistance mismatch [24], spin relaxation and the spread of spin precession angles, which resulted in alternative proposals such as [25] (see Fig. 7.1b, c). Recently, Chuang et al. in [26] have shown experimentally an all-electric and all-semiconductor spin field-effect transistor in which aforementioned obstacles are overcome by using two quantum point contacts as spin injectors and detectors.

#### 7.1.2.2 Nanomagnetic Logic

Among the most prominent concepts investigated for beyond-CMOS applications is the NanoMagnetic Logic (NML) (also known as Magnetic Quantum Cellular Automata) that was first introduced by Cowburn et al. [27] and Csaba et al. [28]. In NML, the information is encoded in the perpendicular magnetization (along

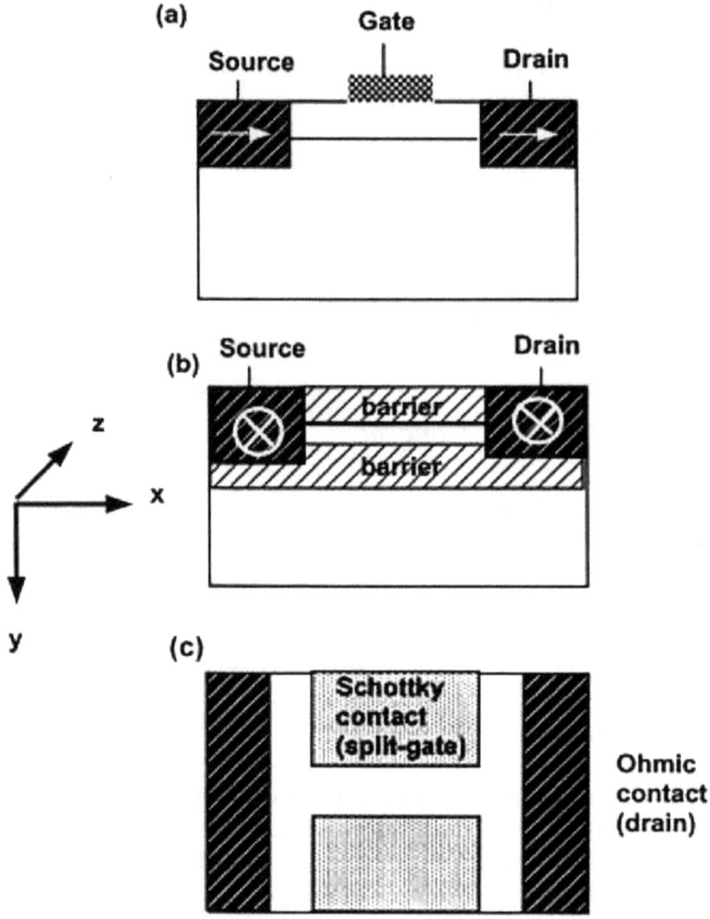


**Fig. 7.1** Schematic of the spintronic modulator of [22]. (**b**) Side view of the spintronic modulator proposed in [25]. (**c**) Top view showing the split gates [25]

+$\hat{\mathbf{z}}$ or $-\hat{\mathbf{z}}$) of ferromagnetic dots. The computation is mediated through dipolar coupling between nanomagnets. Although NML devices can be beneficial in terms of power consumption and non-volatility [4], they have an operating frequency limited to about 3 MHz and an area around 200 nm × 200 nm [29], limitations which are imposed by the nanomagnet material properties. However, a functional 1-bit full adder based on NML majority gates has been shown experimentally in [29] and a schematic is depicted in Fig. 7.2.

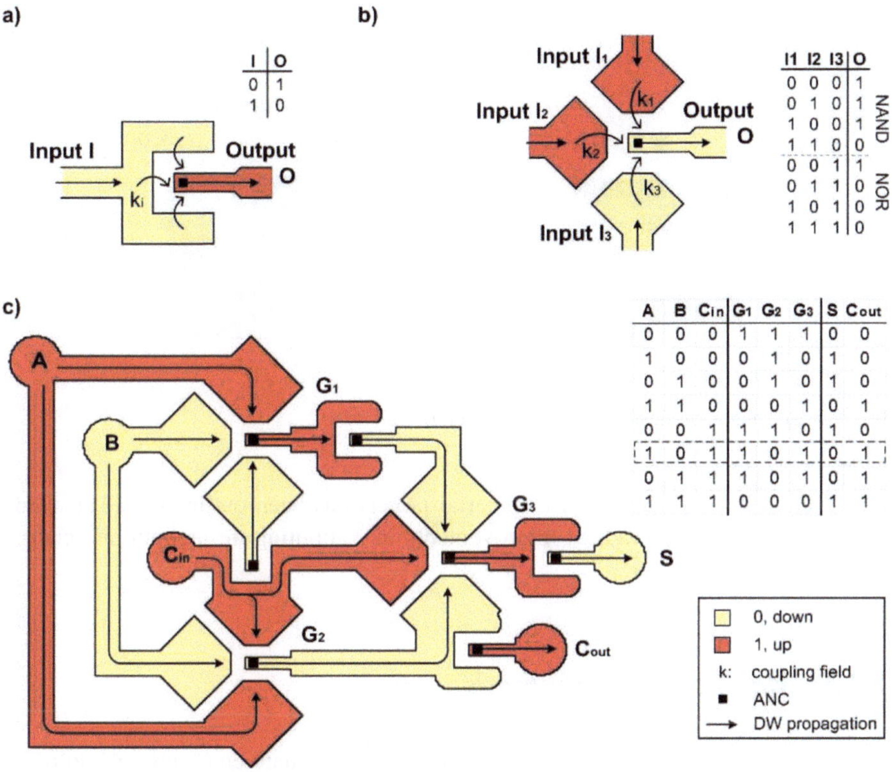


**Fig. 7.2** Inverter (**a**) and majority gate (**b**) as basic building blocks for perpendicular NML. A 1-bit full adder (**c**) with inputs A, B and carry-in $C_{in}$ and outputs sum S and carry-out $C_{out}$ is realized by three majority gates and four inverter structures connected by wires [29]

### 7.1.2.3 All-Spin Logic

Proposed by Behin-Aein et al. in [30] as a logic device with built-in memory, All-Spin Logic (ASL) is a concept that combines magnetization states of nanomagnets and spin injection through spin-coherent channels. A schematic of the device is shown in Fig. 7.3. The input logic bit controls the state of the corresponding output logic bit with the energy coming from an independent source. Information is stored in the bistable states of magnets. Corresponding inputs and outputs communicate with each other via spin currents through a spin-coherent channel, and the state of the magnets is determined by the spin-torque phenomenon. The aforementioned challenges of SpinFET and NML are also present in the ASL concept. Existing nanomagnet material properties and spin-transfer channel properties fall short of the energy and delay targets [31] dictated by modern advanced CMOS devices [32].

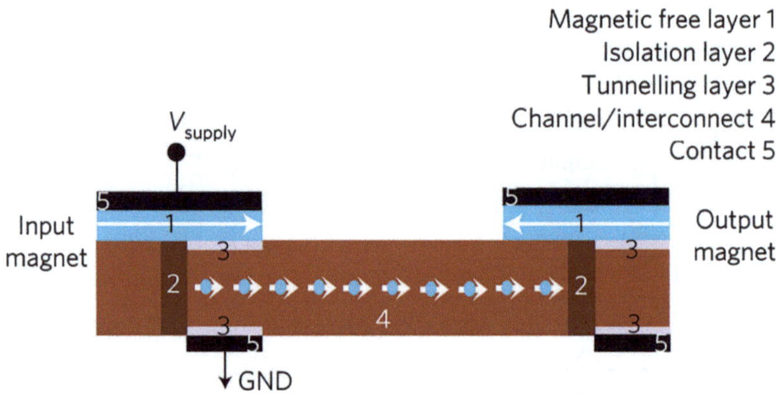


**Fig. 7.3** Schematic of the all-spin logic device [30]

However, a scaling path of ASL material targets has been outlined in [31] which if achieved can enable radical improvements in computing throughput and energy efficiency.

### *7.1.3 Majority Logic Synthesis*

New logic synthesis methods are required to both evaluate emerging technologies and to achieve the best results in terms of area, power, and performance [33]. Majority gates enhance logic power of a design since they can emulate both AND and OR operation and are one of the basis for basic operation of binary arithmetic [34]. In order to build complete circuits composed from MAJ gates, we need to employ specific synthesis methodologies. In the results shown in this chapter, the principle of synthesis is based on Majority-Inverter Graph (MIG) [35]. A novel logic representation structure for efficient optimization of Boolean functions, consisting of three-input majority nodes and regular/complemented edges. This means that only two logic components are required for this representation, a MAJ gate and inverter (INV). In this way, it's possible to reduce the total chip area by utilizing functional scaling [36]. Meaning that instead of scaling down single gates and devices, these single blocks gain functionality.

Also, MIG has proven to be an efficient synthesis methodology for CMOS design optimization [35] and can be further exploited for SWD technology, as shown in [37]. Other novel synthesis tools for majority logic exist, such as [38] but it's specific to a certain technology (QCA), while MIG representation and optimization that is technology-agnostic can be straightforwardly used to evaluate circuit perspectives for any majority-based technology.

## 7.2 Spin Wave Device Circuits

Introduced in 2011 [9], a Spin Wave Device (SWD) is a concept logic device that is based on the propagation and interference of spin waves in a ferromagnetic medium. As a concept that employs wave computing, a SWD circuit consists of (a) wave generators; (b) propagation buses; and (c) wave detectors. One of the most compelling properties of SWDs is the potential of using the same voltage-controlled element for spin wave generation and detection, called Magnetoelectric (ME) Cell. Both spin waves and ME cells are described in Sect. 7.2.1. An overview of experimental results, that can lead to the complete implementation of the SWD circuit concept, is given in Sect. 7.2.2. The potential benefits of such SWD circuits are presented and discussed in Sects. 7.2.3 and 7.2.4.

### *7.2.1 Concept Definition*

#### 7.2.1.1 Spin Waves

Spin waves are usually known as the low-energy dynamic eigen-excitations of a magnetic system [39]. The spin-wave quasi-particle, the magnon, is a boson which carries a quantum of energy $\hbar\omega$ and possesses a spin $\hbar$. Incoherent thermal magnons exist in any magnetically ordered system with a temperature above absolute zero. Here, in the context of spin wave devices the spin waves are rather classical wave excitations of the macroscopic magnetization in a magnetized ferromagnet. In the context of spin-based applications (like SWD), thus, the main interest is not in thermal excitations, but externally excited spin-wave signals: coherent magnetization waves which propagate in ferromagnets over distances which are large in comparison with their characteristic wavelength [40].

Spin wave propagation depends on the nonlinear dispersion relation of the excitation $\omega(k)$, which is strongly affected by the dimensions and geometry of the magnetic medium [40]. This dispersion can be characterized into three distinct regimes, depending on which spin interaction mechanism dominates (dipolar or exchange). These regimes are the magnetostatic (dipolar-dominated) regime [41], exchange regime [42], and an intermediate regime of dipole-exchange waves, where excitations are affected by both contributions [43].

As wave entities, spin waves (or magnons) have a specific wavelength and amplitude. The SWD concept exploits the interference of spin waves, where the logic information is encoded in one of the spin wave properties and two or more waves are combined into an interfered result. Consider two waves $\Psi_A$ and $\Psi_B$ with the same frequency and amplitude and a certain phase shift $\phi$ relative to each other. Interference of these two waves can be elaborated as follows:

$$\Psi_{\text{tot}}(\mathbf{r}, t) = \Psi_A + \Psi_B = A \cdot e^{i(\mathbf{kr}-\omega t)} + B \cdot e^{i(\mathbf{kr}-\omega t+\phi)}. \tag{7.15}$$

If we assume that the two waves have equal amplitude ($A = B$):

$$\text{for } \phi = 0: \qquad \Psi_{tot}(\mathbf{r}, t) = 2A \cdot e^{i(\mathbf{kr}-\omega t)} \tag{7.16a}$$

$$\text{for } \phi = \pi: \qquad \Psi_{tot}(\mathbf{r}, t) = 0 \tag{7.16b}$$

Equation (7.16) show that in a spin wave concept we can define a logic '0' as a spin wave with phase $\phi = 0$ and a logic '1' as a spin wave with phase $\phi = \pi$. This choice is arbitrary but serves as an example of how spin wave (or wave, in general) interference can be used in a logic application, with the information being encoded into the phase of the waves.

#### 7.2.1.2 Magnetoelectric Cell

Aside from the propagation of spin waves, in order for the SWD concept to be integrated as an IC technology, there has to be a way to generate and detect spin waves that is amenable to scaling and is preferably voltage-driven [4]. One of the most prominent concepts that seem to satisfy the above criteria is the Magnetoelectric (ME) cell [44].

The magnetoelectric effect has been studied [45] and applied in several concepts as an interface between the electric and the spin domains [8, 9, 44]. An example of an ME cell is shown in Fig. 7.4. It usually consists of a stack with a magnetostrictive layer at the bottom, a piezoelectric layer above it, and a metal contact on top. When voltage is applied across the stack, the piezoelectric layer is strained and the strain is transferred to the magnetostrictive layer, which modifies its magnetic anisotropy. By modifying the anisotropy, the easy axis goes from out-of-plane to in-plane, which rotates the magnetization and a spin wave is generated and can be propagated through adjacent spin waveguide (stripe of ferromagnetic material). The spin wave detection exploits the inverse phenomenon.

Bistable Magnetization

Basic spin wave generation and detection can be achieved by the aforementioned magnetostrictive/piezoelectric interaction. However, in order to enable SWDs as a complete logic concept, the generators and detectors used need to offer information-

**Fig. 7.4** Schematic view of ME cell stack connected to spin wave ferromagnetic bus [46]

encoding controllability. This means that it should be possible to controllably generate/detect spin waves with phase $\phi = 0$ or $\phi = \pi$. To realize this feature, the ME stacks proposed [8, 44] always include a magnetostrictive material with two stable magnetization states. Each magnetization state is associated with one of the spin wave phases (and also with a logic '0' or '1'). With a bistable magnetization, when a specific (e.g., positive) voltage is applied on the ME cell the magnetostrictive layer's magnetization switches on the associated state (e.g., '0') which generates a spin wave with the equivalent phase (e.g., $\phi = 0$). When an opposite voltage is applied (e.g., negative), then magnetostrictive layer's state will become '1' and a spin wave with phase $\phi = \pi$ will be generated. Hence, bistable magnetization is required to enable the controllable operation of information-encoded spin waves and ME cells. Two options have been proposed for the implementation of bistable magnetization of the magnetostrictive layer each coming with their inherited advantages and disadvantages.

In [9, 37, 46, 47] the bistability of the ME cell magnetization was assumed to be in canted magnetization states, as shown in Fig. 7.5a. Since the two stable states are separated by a relatively small angle (from $\theta_{\mathrm{me}} \simeq 1°$ [9] to $\theta_{\mathrm{me}} \simeq 5°$ [48]), energy required to switch between these states is also small leading to an ultralow-power device [48]. On the other hand, the small state separation indicates that this configuration will be very sensitive to thermal noise.

In [8], the bistability of the ME cell was implemented in in-plane magnetization ($\pm\hat{x}$—Fig. 7.5b). The magnetostrictive layer of the ME cell has two low-energy stable in-plane magnetization states along the $\pm\hat{x}$ direction, favored by the shape anisotropy of the structure. In order for the magnetization to switch, it first has to be put in a meta-stable state (i.e., along $+\hat{z}$). Since this proposal employs two in-plane magnetization states which are well separated, the result is a thermally stable and nonvolatile ME cell. However, the ME cell operation becomes slightly more complicated (compared to the canted state ME cell) since putting the magnetization to the meta-stable state (along $+\hat{z}$) requires an extra "step" before spin wave generation or detection.

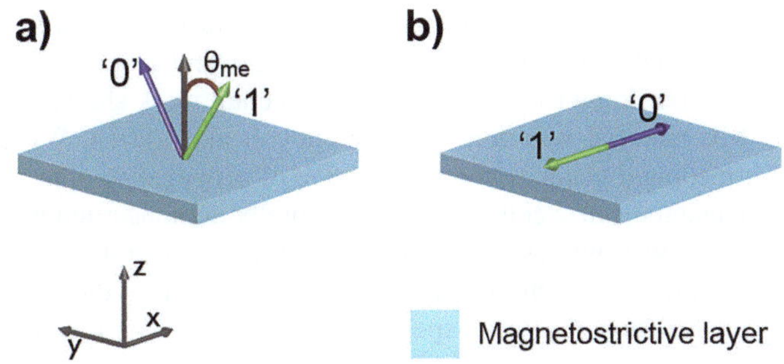


**Fig. 7.5** Proposed bistability of the magnetostrictive layer. (**a**) Canted magnetization states as shown in [9] where $\theta_{\mathrm{me}}$ is the canting angle between a stable magnetization state and $\hat{z}$. (**b**) In-plane bistable magnetization as proposed in [8]

**Table 7.3** Overview of propagation characteristics of different spin wave regimes

| Regime | Propagation length | Waveguide | Reference |
|---|---|---|---|
| Magnetostatic | 6 mm | YIG | [49] |
| Magnetostatic | 7 mm | YIG thin film | [50] |
| Dipole-exchange | 5 μm | Py—2.5 μm wide | [51] |
| Dipole-exchange | 10 μm | CoFeB—0.5 μm wide | [52] |
| Dipole-exchange | Up to 4 μm | Py—500 nm wide | [53] |
| Dipole-exchange | 12 μm | Py—2.5 μm wide | [54] |

### *7.2.2 Experimental Demonstrations*

There has been no experimental proof for the complete SWD concept, containing all necessary parts of excitation, propagation, logic computation, and detection. However, these parts have been separately studied and experimentally shown. Here we give a brief overview of the most relevant experimental work done in spin waves, that is closely related to the realization of SWD circuits. As aforementioned in Sect. 7.2.1, spin waves can be observed in three different regimes, each having different propagation characteristics. Table 7.3 presents a comprehensive overview of these propagation characteristics shown in literature.

Magnetostatic spin waves can propagate for long distances but cannot be confined in nanometer scale structures due to their long wavelengths. Dipole-exchange spin waves have shorter wavelengths and thus can be more confined but also have much shorter propagation lengths. Spin waves in the exchange regime have the shortest wavelengths from the three regimes and that is why it is not possible yet to experimentally observe them.[3] However, the propagation lengths of either dipole-exchange or exchange spin waves do not guarantee signal integrity over more than several circuit stages [46]. This means that in a realistic SWD circuit concept spin wave amplification or regeneration has to be included to enable cascading of SWD gates.

As described in Sect. 7.2.1, ME cells can serve as a generator and a detector, which means they can be used for regeneration of spin waves to ensure propagation. Despite the importance of ME cells to the SWD concept and to spin-based technologies in general, to our knowledge the only experimental work showing spin waves generated by ME material was done by Cherepov et al. in 2014. Where voltage-induced strain-mediated generation and detection of propagating spin waves using multi-ferroic magnetoelectric cells was experimentally demonstrated by fabricating 5 μm wide Ni/NiFe waveguides on top of a piezoelectric substrate, Fig. 7.6.

Although the spin waves amplitudes measured in [55] are rather small, the fact that the ME cell functionality was experimentally proven is significant. However,

[3]Either with an optical measurement setup or with an electrical one, the exchange spin waves would be lower than the resolution of a state-of-the-art measurement setup.

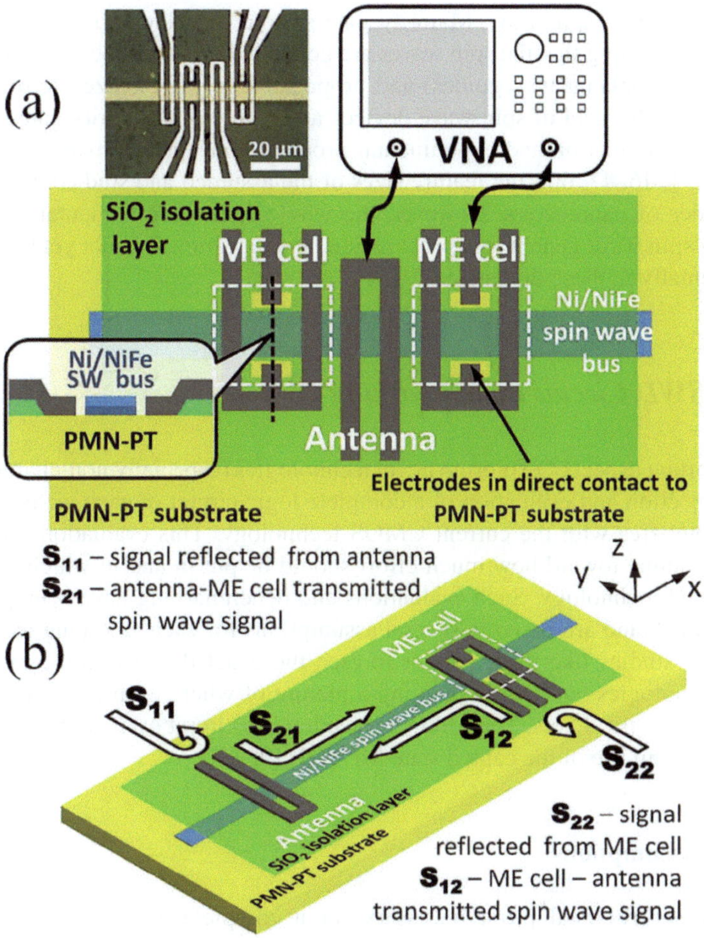


**Fig. 7.6** (**a**) Schematic of the studied device: Spin wave generation and propagation measurements using a vector network analyzer were performed on the 5 µm wide Ni/NiFe bus lithographically defined on a PMN-PT piezoelectric substrate. Inset shows cross-sectional view of the ME cell. (**b**) The schematic of two-port measurements of transmission (S21 and S12) and reflection (S11 and S22) measurements between conventional loop antennas and voltage-driven magnetoelectric cells [55]

the cross section shown in Fig. 7.6a is quite different from the ME cell concept depicted in Fig. 7.4 which means that the ME cell field has to take major strides in order to reach a functional but also IC integrable stack.

The dynamic behavior and propagation is strongly dependent on the geometry of the spin wave structure. In the same way, spin wave interference behavior has a high geometry and material dependence. Several experimental and simulation studies have explored the behavior of spin wave interference [56–59] but are all in the order of microns. More specifically, the work presented in [57, 58] shows simulation of two spin wave Majority gate structures, which can be realistically fabricated. Meaning that the spin waves are generated and detected by micron-sized antennas (or coplanar waveguides) and propagated in micron-sized ferromagnetic waveguides. The field of spin wave devices and spin wave majority gates includes a variety of simulation and experimental proof of concepts. In many publications [8, 9, 37, 44, 46, 47, 60], the feature sizes of the assumed and studied concepts are in the order of nanometers. However, the whole spin wave computation concept (meaning spin wave generation, propagation, and detection) has not yet been shown experimentally in these dimensions.

## 7.2.3 *SWD Circuit Benchmarking*

One important aspect of exploring novel technologies (especially non-charge-based) is the projection and evaluation of a complete logic circuit of each technology and how it compares with the current CMOS technology. This evaluation serves as a useful guideline toward how much effort should be put in and in which aspects of an emerging technology. Such evaluations and benchmarking have been presented in [4] and [5], and are based in several assumptions for each emerging technology. Obviously, studies like these cannot foresee the exact designs and layouts of all novel technologies but help in painting a picture of where each technology stands with respect to the others. The following section is a circuit evaluation of spin wave devices making use of the canted state ME cells [9, 47].

### 7.2.3.1 Assumptions

Since all experimental proof necessary for a complete nanometer-scaled SWD circuit do not exist yet, we need to consider several assumptions in order to evaluate the circuit benefits of SWD. These assumptions include the interface between the spin and electric domains, the geometry of SWD gates, and their cascadability. The block diagram, depicted in Fig. 7.7, provides a frame in which SWD can be integrated with CMOS devices in a realistic IC environment.

We assume that the spin wave domain of the block diagram shown in Fig. 7.7 consists of ME cell gates, presented in Fig. 7.8, and spin wave amplifiers [61]. However, since spin wave amplification is a complex issue, we will ignore the impact of amplifiers for the rest of this evaluation.

In Fig. 7.8a, we present the INV component which is a simple wave bus, with a magnetically pinned layer on top, that inverts the phase of the propagating signal. The MAJ gate (Fig. 7.8b) is the merging of three wave buses. For the gates presented in Fig. 7.8, we assume minimum propagation length equal to one wavelength of the spin wave which in our study is assumed at 48 nm, since the wavelength is defined/confined by the width of the spin wave bus. As aforementioned in

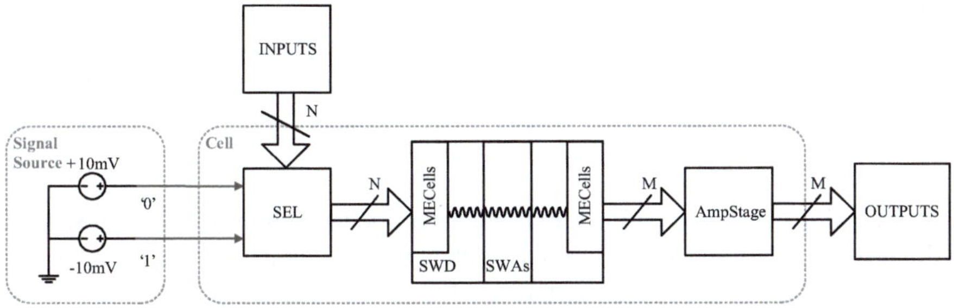


**Fig. 7.7** Block Diagram that integrates SWD with CMOS and digital interfaces [47]

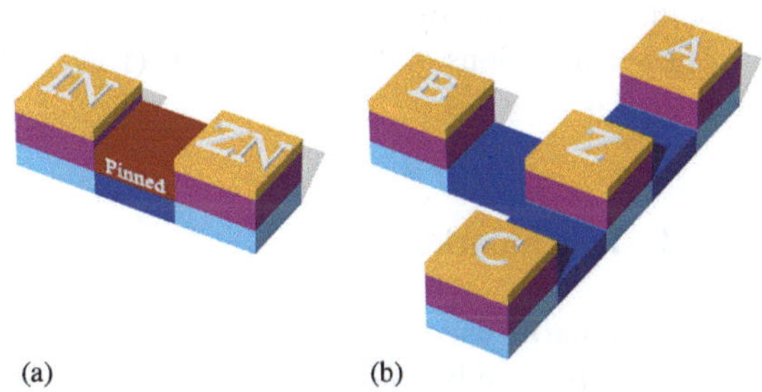


**Fig. 7.8** Gate primitives used for SWD circuits. (**a**) INV. (**b**) MAJ

Sect. 7.1.3, with an inverter and a majority gate as the only primitive components, one can re-create any possible logic circuit that is traditionally (with CMOS technology) composed with NAND/NOR or other gates.

In [9, 47], the operational voltage levels of an ME cell were considered to be $\pm 10\,\text{mV}$. This was because the angle of the canted magnetization was assumed to be $\theta_{\text{me}} \simeq 1^\circ$. However, in [48] a larger and more feasible canted magnetization was calculated and according to that study we assume that the operational voltage level of an ME cell is 119 mV. This means that the *minimum* energy needed to actuate an inverter or a majority operation (Fig. 7.8) is given by

$$E_{\text{INV}} = C_{\text{ME}} \cdot V_{\text{ME}}^2 = 14.4\,\text{aJ} \tag{7.17a}$$

$$E_{\text{MAJ}} = 3 \cdot C_{\text{ME}} \cdot V_{\text{ME}}^2 = 43.3\,\text{aJ}, \tag{7.17b}$$

where $C_{\text{ME}}$ is assumed at 1 fF [48].

For this assumption of ME cell output voltage, the final output stage of the spin wave domain (Fig. 7.7) to the electric domain, a sense amplifier (SA) was designed and used in [48] to accommodate a peak-to-peak input signal of 119 mV with a yield above 1–$10^{-5}$, assuming a Pelgrom constant $A_{\Delta\text{VT}} = 1.25\,\text{mV}\,\mu\text{m}$. The amplifier consists of two stages. The first stage consists of a PMOS differential pair, with one PMOS gate connected to the input signal and the other PMOS gate connected to a 0 mV reference voltage. This first stage operates in a pulsed mode: The current source is activated during only 3 ps. During this time, an amplified version of the

**Table 7.4** Specifications of SWD circuit components, mostly from [48]

| Component | Area (μm$^2$) | Delay (ns) | Energy (fJ) |
|---|---|---|---|
| INV | 0.006912 | 0.42 | $1.44\times10^{-2}$ |
| MAJ | 0.03456 | 0.42 | $4.33\times10^{-2}$ |
| SA | 0.050688 | 0.03 | 2.7 |

input signal is developed on the output nodes of the first stage. The second stage is a drain-input latch-type SA that acts as a latch, amplifying the signals from the first stage to full logic levels. This signal is buffered by two minimal-size inverters to drive amplifier's outputs. Better options might be possible with calibration or offset compensation. The sensing circuitry and the core SWDs of the circuit are considered to be integrated side by side.

The specifications of the components (INV, MAJ, SA) described above are given in Table 7.4.

#### 7.2.3.2 Benchmarks

The benchmarks used for the SWD circuit evaluation are selected from a set of relatively large combinational designs. All designs have been synthesized with MIG [35] for a straightforward mapping with the gate primitives shown in Fig. 7.8. The ten benchmarks selected are shown in Table 7.5. These benchmarks have varying input and output number of bits (I/O bits), which is critical in order to quantify the impact of the CMOS peripheral circuitry that enables digital I/O to the SWD circuits. The list includes three 64-bit adders (BKA264, HCA464, CSA464), three 32- and 64-bit multipliers (DTM32, WTM32, DTM64—Dadda tree and Wallace tree), a Galois-Field multiplier (GFMUL), a 32-bit MAC module (MAC32), a 32-bit divider (DIV32), and a cyclic redundancy check XOR tree (CRC32). All benchmarks (except DIV32 and CRC32) were generated using the *Arithmetic module generator* [62].

#### 7.2.3.3 Circuit Estimations

The specifications in Table 7.4 are used to calculate the results presented in Table 7.6. It's important to note that in these results energy and power metrics of SWD are calculated including the interconnection capacitances for each benchmark. This means that contrary to [48], here a more **realistic** sum of capacitances is accounted to calculate the minimum energy and power consumption of the SWD circuits. To quantify the benefits of SWD circuits, the same benchmarks were executed using a state-of-the-art CMOS technology of 10 nm feature size (hereafter named N10) [63]. All N10 reference results are provided post-synthesis by *Synopsys Design Compiler*. Table 7.6 includes the area metric for both technologies, the energy calculated to be consumed in the SWD circuits, the delay metric, and the power consumption metric.

**Table 7.5** Benchmark designs with I/O bits and MIG synthesis results ordered by size [48]

| Codename | Input bits | Output bits | MIG size | MIG depth |
|---|---|---|---|---|
| CRC32 | 64 | 32 | 786 | 12 |
| BKA264 | 128 | 65 | 1030 | 12 |
| GFMUL | 34 | 17 | 1269 | 17 |
| CSA464 | 256 | 66 | 2218 | 18 |
| HCA464 | 256 | 66 | 2342 | 19 |
| WTM32 | 64 | 64 | 7654 | 49 |
| MAC32 | 96 | 65 | 8524 | 58 |
| DTM32 | 64 | 64 | 9429 | 35 |
| DIV32 | 64 | 128 | 26,001 | 279 |
| DTM64 | 128 | 128 | 34,485 | 43 |

First, we observe that for all benchmarks the SWD circuits give smaller area (on average 3.5× smaller). This is based on two main factors: (1) the Majority synthesis in conjunction with the MAJ SWD gate yield great results, and (2) the output voltage assumed doesn't require bulky output SAs. Second, we note that for all benchmarks the SWD circuits are much slower than the reference circuits (on average 12× slower). This is due to the large ME cell switching delay (0.42 ns for INV/MAJ operation—Table 7.4) which is accumulated according to the longest path of the MIG netlists. However, due to the low energy consumption of both the SWD gates and the SA design, the power consumption metrics are in large favor of the SWD circuits for all the benchmarks (on average 51× lower).

Table 7.7 contains two important product metrics which help compare the two technologies, one is the product of area and energy (A·E—divided by 1000 for ease of presentation) and the other is the area, delay, and energy product (A·D·E—again divided by 1000). A·E serves as an indicator of the low-power application benefits of this technology, where the performance (delay) is the critical metric. The second product metric A·D·E combines all aspects of circuit evaluation. The energy consumption of the N10 reference benchmarks is not directly given by the synthesis tool, so it's calculated as the product of delay and power (from Table 7.6).

Figure 7.9 depicts the results of Table 7.7. On average in both product metrics, the SWD circuits outperform the N10 counterparts. Consider the A·E product, except one benchmark (BKA264), SWD technology produces smaller and less energy-consuming designs. However, when accounting for the long SWD delays with the A·D·E product, the benefits of the SWD technologies hold only for the two deepest benchmarks (CSA464 and DIV32). This means that SWD circuits outperform N10 ones only in the cases of large and complex benchmarks where CMOS circuit performance is not easily optimized (note the quite large delays of 1.78 ns and 14 ns for CSA464 and DIV32, respectively—Table 7.6).

These results compel us to characterize SWD (with CMOS overhead circuitry) as a technology extremely adept for ultralow-power applications, where latency is a secondary objective. SWD circuits perform in a way that CMOS circuits are not

**Table 7.6** Summary of benchmarking results

| Name | Area ($\mu m^2$) | | | | Energy (fJ) | | | Delay (ns) | | Power ($\mu W$) | |
|---|---|---|---|---|---|---|---|---|---|---|---|
| | SWD core | CMOS SA | SWD total | N10 | SWD core | CMOS SA | SWD total | SWD | N10 | SWD | N10 |
| CRC32 | 27.61 | 1.54 | 29.14 | 95.88 | 45.73 | 86.40 | 132.13 | 5.07 | 0.22 | 26.06 | 304.30 |
| BKA264 | 36.48 | 3.12 | 39.60 | 118.55 | 63.32 | 175.50 | 238.82 | 5.07 | 0.21 | 47.10 | 133.92 |
| GFMUL | 44.09 | 0.82 | 44.91 | 162.98 | 76.31 | 45.90 | 122.21 | 7.17 | 0.16 | 17.04 | 433.92 |
| CSA464 | 78.42 | 3.17 | 81.59 | 240.26 | 152.55 | 178.20 | 330.75 | 7.59 | 1.78 | 43.58 | 663.17 |
| HCA464 | 82.71 | 3.17 | 85.88 | 262.63 | 162.06 | 178.20 | 340.26 | 8.01 | 0.29 | 42.48 | 594.28 |
| WTM32 | 264.96 | 3.07 | 268.04 | 1163.37 | 635.03 | 172.80 | 807.83 | 20.61 | 0.58 | 39.20 | 3571.90 |
| MAC32 | 295.25 | 3.12 | 298.37 | 1372.83 | 727.32 | 175.50 | 902.82 | 24.39 | 0.66 | 37.02 | 3872.10 |
| DTM32 | 326.31 | 3.07 | 329.38 | 1183.64 | 822.32 | 172.80 | 995.12 | 14.73 | 0.52 | 67.56 | 3667.50 |
| DIV32 | 899.04 | 6.14 | 905.18 | 3347.73 | 3009.03 | 345.60 | 3354.63 | 117.21 | 14.00 | 28.62 | 5346.10 |
| DTM64 | 1192.69 | 6.14 | 1198.83 | 3459.32 | 4373.85 | 345.60 | 4719.45 | 18.09 | 0.63 | 260.89 | 12,793.10 |
| Averages | 324.76 | 3.34 | 328.09 | 1140.72 | 1006.75 | 187.65 | 1194.40 | 22.79 | 1.91 | 60.95 | 3138.03 |

**Table 7.7** Summary of benchmarking products

| | A·E (/1000) | | | A·D·E (/1000) | | |
|---|---|---|---|---|---|---|
| Name | SWD | N10 | Impr. (×) | SWD | N10 | Impr. (×) |
| CRC32 | 3.9 | 6.4 | 1.7 | 6.8 | 1.4 | 0.2 |
| BKA264 | 9.5 | 3.3 | 0.4 | 12.7 | 0.7 | 0.1 |
| GFMUL | 5.5 | 11.3 | 2.1 | 24.6 | 1.8 | 0.1 |
| CSA464 | 27.0 | 283.6 | 10.5 | 94.5 | 504.8 | 5.3 |
| HCA464 | 29.2 | 45.3 | 1.5 | 111.5 | 13.1 | 0.1 |
| WTM32 | 216.5 | 2410.2 | 11.1 | 3508.0 | 1397.9 | 0.4 |
| MAC32 | 269.4 | 3508.4 | 13.0 | 5292.9 | 2315.5 | 0.4 |
| DTM32 | 327.8 | 2257.3 | 6.9 | 3989.7 | 1173.8 | 0.3 |
| DIV32 | 3036.5 | 250,562.2 | 82.5 | 319,246.9 | 3,507,871.1 | 11.0 |
| DTM64 | 5657.8 | 27,880.9 | 4.9 | 94,854.9 | 17,565.0 | 0.2 |
| Averages | 958.3 | 28,696.9 | **13.5** | 42,714.3 | 353,084.5 | **1.8** |

able to even if their optimized only for power consumption. Just their innate leakage power would be enough in large designs to exceed the power consumption of their SWD equivalents.

### *7.2.4 Discussion*

In Sect. 7.2.2, we presented several experimental advancements toward the realization of the SWD concept, and in Sect. 7.2.3 we showcased the potential of the SWD with circuit evaluations. In order for these projections to become a reality, there should be several more steps implemented at the experimental level. The two main benefits the SWD concept has to offer are smaller area and lower energy than CMOS. In order for the first to be realized, more experimental work is needed for studying the behavior of exchange spin waves, which due to their short wavelength would have the ability to propagate in narrow and short waveguides (less than 100 nm wide). For SWD to deliver their low energy potential, the most crucial component to be experimentally verified is the ME cell spin wave generation and detection. Having experimental proof that the operational ME cell stack consisting of thin layers (not bulk piezoelectric as in [55]) can be integrated next to (or on top of) a ferromagnetic waveguide *and* that this cell would produce (and detect) well-controlled spin waves would be ideal. However, there are many challenges remaining for an ME cell realization. Such challenges include (but are not limited to) stacking a piezoelectric layer with other layers, maintaining its piezoelectric properties. Additionally, an ME cell should be optimized for spin wave detection, so that the read-out voltage is more than a few mV [9].

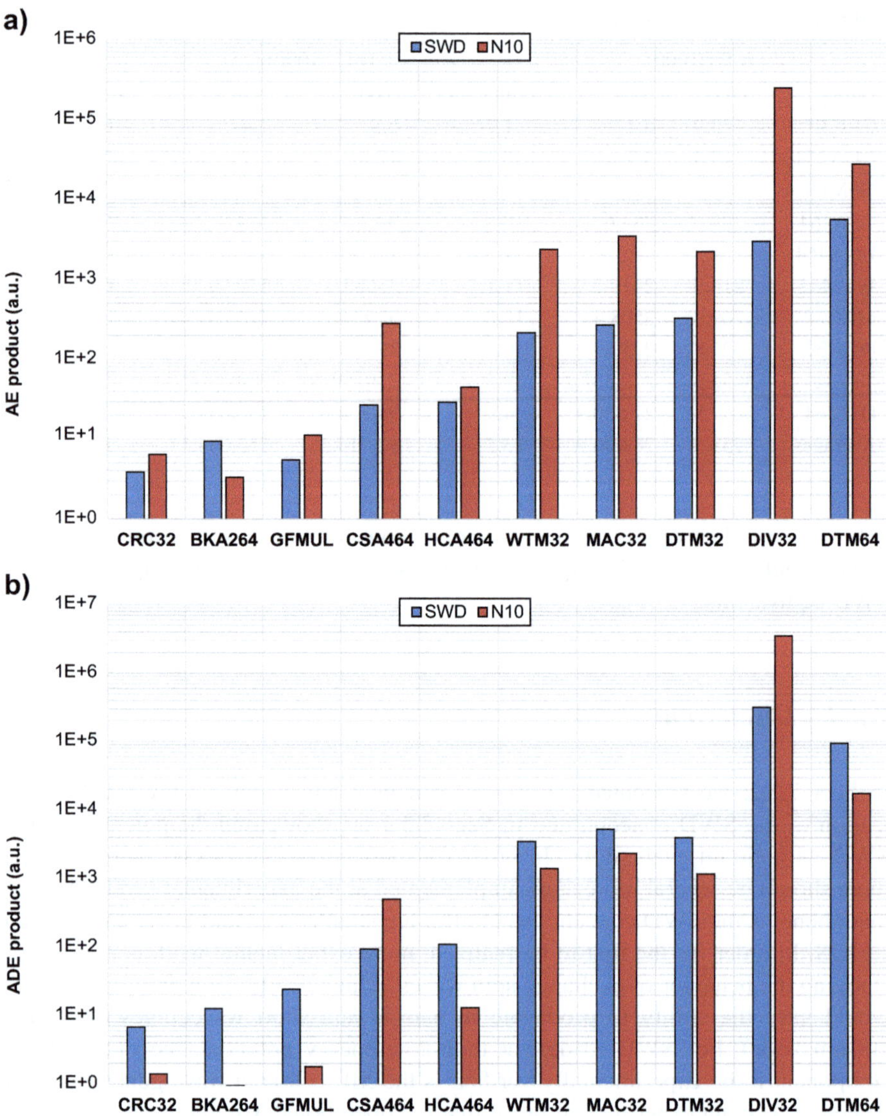


**Fig. 7.9** Product metrics for all benchmarks, ordered according to benchmark size. (**a**) Area-Energy product. (**b**) Area-Delay-Energy product

In conclusion, the SWD circuit can be promising and be very useful as CMOS technology is reaching its limits, especially for low-power applications. Paving the way for the realization of this concept has started but there is a lot for improvement and necessary advancements.

## 7.3 Spin Torque Majority Gate

The concept of Spin Torque Majority Gate (STMG) was introduced by Nikonov et al. in 2011 [6]. Before introducing the working principle of STMG, two key spintronics notions will be explained: the Spin Transfer Torque and the Tunnel Magnetoresistance, which are the write and read mechanisms of STMG.

### *7.3.1 Working Principle of STMG*

#### 7.3.1.1 Device Description

The STMG consists of a perpendicularly magnetized free layer shared by four Magnetic Tunnel Junctions. The logic state ('0' or '1') is represented by the orientation of the free layer magnetization ('UP' or 'DOWN'), as illustrated in Fig. 7.10. The input magnetic states are written by STT via the three input Magnetic Tunnel Junctions. The output magnetization state is detected by the fourth MTJ via Tunnel Magnetoresistance (TMR). The cross shape of the free layer has a main advantage: It should allow for easy cascading by utilizing the output arm of the cross as an input arm for the next gate. Several types of cross were simulated [64]. However, the "simple cross" remained the most reliable.

It is important to note that the current is perpendicular to the plane of the free layer. In practice, a voltage is applied between the top and the bottom electrodes.

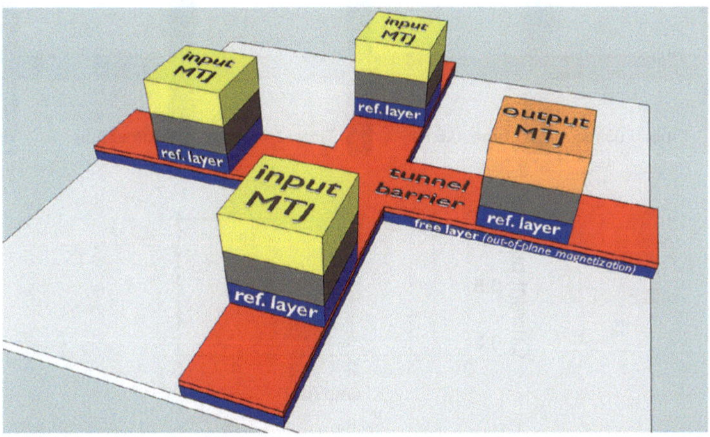


**Fig. 7.10** Schematic view of a Spin Torque Majority Gate. The red layer is the oxide tunnel barrier. The reference layer and the free layer (both in blue) have perpendicular magnetic anisotropy. The reference layer induces the spin polarization, and the free layer carries information. The input MTJs convert information from charge to spin while the output MTJ converts it from spin to charge

Depending on the voltage polarity, the current flows either downward or upward, creating a torque that pushes the magnetization either up or down, representing a '1' or a '0'. Contrary to other concepts of DW logic [65–67], here, no current is injected in-plane. The vertical current flows in the areas defined by the input MTJs. Thus, the spin torque is exerted only at the inputs, while the rest of the free layer is mainly driven by the exchange interaction. Since exchange is a short-range interaction, the MTJs have to be close enough to each other for correct STMG operation. How close? This question will be addressed in the following section.

#### 7.3.1.2 Micromagnetic Simulations and Analytical Model

Extensive micromagnetic simulations were performed to simulate the magnetization dynamics of the free layer [68]. The size and the main material parameters were varied for every possible input combination. The device is a functional majority gate if the majority of inputs is '1' lead to an output state UP (i.e., '1'), and if the majority is '0' lead to an output state DOWN (i.e., '0').

An example of simulation is shown in Fig. 7.11. The initial state is pointing UP (red). A negative voltage (i.e., '0') is applied to two input MTJs, pushing the magnetization down. A positive voltage (i.e., '1') is applied to the third MTJ, holding the magnetization up. These input signals, sent for 2 ns, are followed by relaxation time of 4 ns. At the end of the simulation, the magnetization under the output MTJ has switched to a down state (blue), as expected.

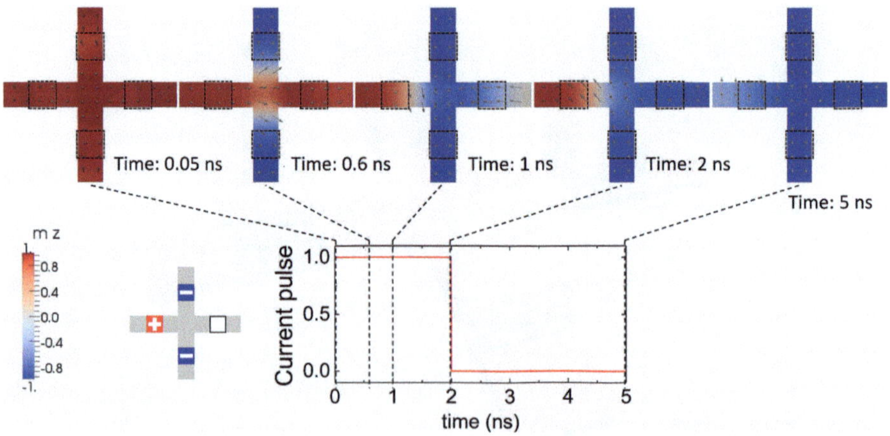


**Fig. 7.11** Micromagnetic simulation of a Spin Torque Majority Gate having a strip width of 10 nm and typical material parameters of CoFeB. At the top: simulation snapshots. The color represents the magnetization orientation; red: up; blue: down. Here, the combination of inputs induces a down state in the top and bottom arms of the cross and an up state in the left arm. At the end of the pulse, the majority state down has been transferred to the output arm. This output state remains stable after turning off the current

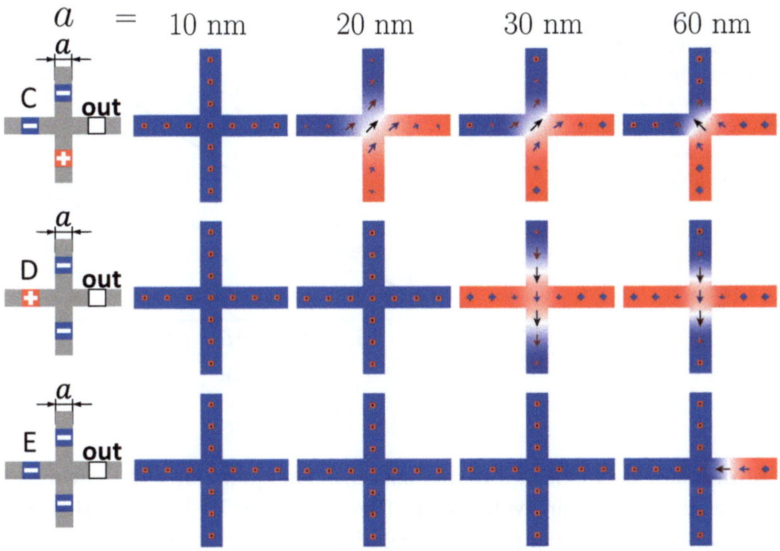


**Fig. 7.12** From [68]. Final magnetic states of the STMG free layer for four different sizes, for three combinations of inputs that are supposed to switch the output arm. For a strip width of 10 nm, no failure is observed

For simplicity, all the simulations were started from an initial UP state. Therefore, it is expected that, for a majority of '1', the output does not switch, and that it switches for majority '0'. The expected behavior has been confirmed by simulation for all the trivial combinations that do not induce any switching. However, several failures have been observed when output switching is expected. In some cases, the failure can be easily explained by a current density being too small or a pulse duration being too short. However, in other cases, failure is observed even at large pulse duration and amplitude. This is illustrated in Fig. 7.12 for the combinations "C", "D" and "E" that are supposed to switch the output. Interestingly, the failures always disappear below a critical size, confirming the essential role of the short-range exchange interaction. "E" (last line of Fig. 7.12) has the largest critical size, while "C" (first line of Fig. 7.12) is the most critical input combination. In the latter, a domain wall (shown in white) is pinned at the center of the cross, along the diagonal. In magnetism, "domain wall" refers to the transition region between two magnetic domains. Here the two magnetic domains are pointing up (in red) and down (in blue), while the domain wall is in-plane (in white).

The results of the micromagnetic simulations for the input combination "C" have been summarized in the phase diagram of Fig. 7.13. The failure region corresponds to a final state with a domain wall pinned at the center of the cross. The working region corresponds to a switched output. In the simulations, the width $a$ of the cross has been varied, as well as the exchange parameter $A_{ex}$ and the anisotropy constant $K_{eff}$. For a given size, it was found that $\sqrt{A_{ex}/K_{eff}}$ is a relevant parameter

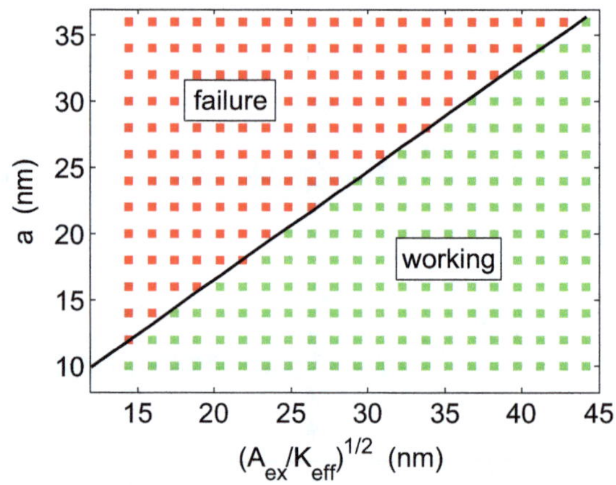


**Fig. 7.13** From [68]. Phase diagram of a STMG of aspect ratio $k = 5$ obtained from micromagnetic simulations. Switched (*working*) and non-switched (*failure*) output states are given as a function of the strip width $a$ and $\sqrt{A_{ex}/K_{eff}}$

that discriminates between failure and success. This parameter is known as being proportional to the domain wall width. Therefore, Fig. 7.13 reveals that majority operation is determined by a particular relation between the size of the device and the width of the domain wall. More specifically, for an aspect ratio $k = 7$, STMG is functional if $\sqrt{A_{ex}/K_{eff}} < 1.21a$.

Further investigation showed that STMG is very likely to fail when the domain wall is energetically stable at the center of the cross. In contrast, if the domain wall is unstable, the device exhibits majority operation, provided that the pulse of current is sufficiently large and long. Thus, STMG functionality is determined by the energy landscape.

Based on this conclusion, an analytical model was developed to derive the magnetic energy of the domain wall state [69] along the diagonal of the cross. Describing the domain wall as a function of two parameters, its position $x_0$ and its width $\Delta$, the total energy was obtained.

$$E = 2t\,\zeta\,\left(A_{ex} + K_{eff}\Delta^2\right) + cst, \tag{7.18}$$

where $t$ is the thickness of the free layer, $cst$ is a constant, and $\zeta$ is given by

$$\zeta = \frac{3d}{\Delta} + \ln\left(\frac{1 + e^{\frac{2x_0}{\Delta}}}{e^{\frac{d}{\Delta}} + e^{\frac{2x_0}{\Delta}}}\right) + \ln\left(\frac{1 + e^{-\frac{2x_0}{\Delta}}}{e^{\frac{d}{\Delta}} + e^{-\frac{2x_0}{\Delta}}}\right) - \frac{2d}{\Delta}\left(\frac{e^{\frac{2x_0}{\Delta}}}{e^{\frac{L}{\Delta}} + e^{\frac{2x_0}{\Delta}}} + \frac{e^{-\frac{2x_0}{\Delta}}}{e^{\frac{L}{\Delta}} + e^{-\frac{2x_0}{\Delta}}}\right) \tag{7.19}$$

where $d = \sqrt{2}a$ and $L = ka/\sqrt{2}$. The function $\zeta$ reveals the major differences with the common 1D model of domain wall. Here, the center of the cross acts like a pinning site. Moreover, the effect of the finite length $L$ is included in the last term.

The dependence of the energy with respect to the domain wall position $x_0$ is directly given by $\zeta$. Figure 7.14 shows $\zeta$ as a function of $x_0$ for several domain wall widths $\Delta$. For $\Delta = 10$ nm, the domain wall is clearly a minimum of the energy in

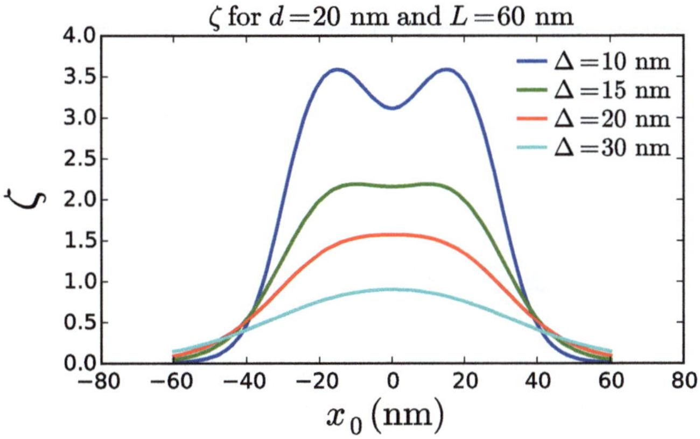


**Fig. 7.14** $\zeta$ as a function of the domain wall position for several domain wall widths. The lateral length $L$ and the distance $d$ correspond to a cross of aspect ratio $k = 6$ and arm width $a = 14$ nm

**Table 7.8** The operating condition expressed as a function of $a$ (arm width) and, equivalently, as a function of $ka$ (total length of the cross)

| | $k = 5$ | $k = 7$ | $k = 9$ |
|---|---|---|---|
| $\sqrt{\frac{A_{ex}}{K_{eff}}} >$ | $0.95\,a$ | $1.27\,a$ | $1.57\,a$ |
| $\sqrt{\frac{A_{ex}}{K_{eff}}} >$ | $0.190\,ka$ | $0.181\,ka$ | $0.174\,ka$ |

$x_0 = 0$. In other words, it is pinned at the center of the cross, along its diagonal, which leads to STMG failure. In contrast, for $\Delta = 20$ and 30 nm, the domain wall state is not in a minimum, which means that it cannot be pinned at the center. As mentioned previously, in that case, a pulse of current sufficiently large and long leads to the expected output. The case of $\Delta = 15$ nm is uncertain: The domain wall is in a shallow energy minimum in $x_0 = 0$, but it can be overcome when the STT is applied. For reliable STMG, this state should also be avoided.

The analytical model is valid for any aspect ratio $k$. The condition for the domain wall not being an energy minimum has been solved numerically at several values of $k$. The results are summarized in Table 7.8. These results are in very good agreement with the micromagnetic simulations at $k = 7$, confirming the validity of the analytical model. Interestingly, the ratio of the total length $ka$ and the domain width determines the operating condition. In summary, the domain wall width should be larger than about $0.2\,ka$ to be unstable at the center, leading to functional STMG.

### 7.3.2 *Circuit Outlook of STMGs*

As mentioned in Sect. 7.2.3, it is important to evaluate each emerging technology and identify potential advantages and drawbacks of their circuit implementation. The following section introduces the results of such benchmarking calculations

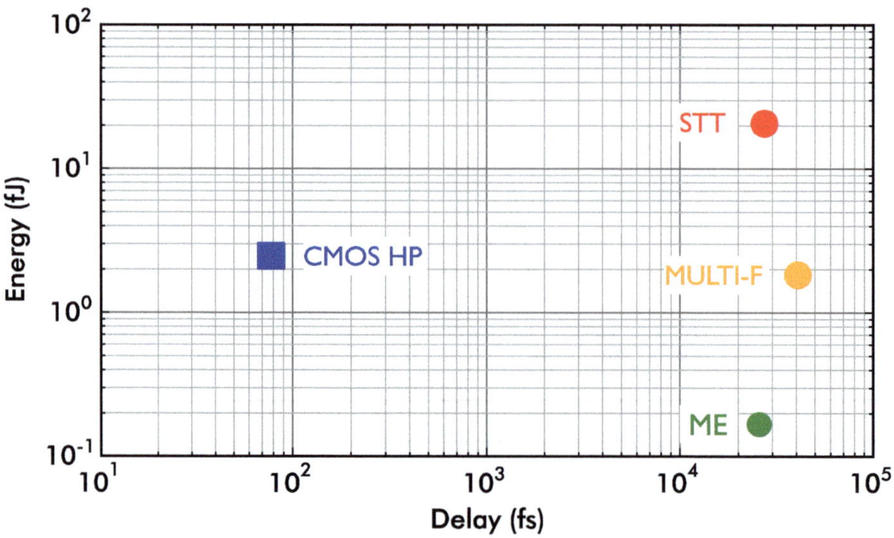


**Fig. 7.15** Energy over delay, for a 32-bit adder, all data from [4, 70]. CMOS HP is the CMOS High-Performance implementation, STT [4] is the original proposal of STMG implementation, MULTI-F [4] assumes use of multi-ferroic input and output elements, ME [70] assumes use of magnetoelectric input and output elements

along with the requirements STMG technology has to fill in order to fully exploit its potential.

#### 7.3.2.1 Benchmarking

STMGs have been benchmarked several times [4, 5, 70] versus CMOS and other beyond-CMOS technologies. A summary of the benchmarking results presented over the years is shown in Fig. 7.15. Energy and delay of a 32-bit full adder are used as metrics to compare CMOS High-Performance (CMOS HP) implementations to different flavors of STMG.

The first version of STMG shown in Fig. 7.15 (STT) is the one that uses MTJs and STT for generating the inputs. This version has been the original proposal [6] and the one studied in this chapter so far. We can clearly see that the circuit modeling of this version produces a result which is inferior to CMOS by one order of magnitude in energy and two orders of magnitude in delay. However, in [4], an alternative version of STMGs was modeled, which used voltage-controlled multi-ferroic elements for signal generation (Fig. 7.15 (MULTI-F)). These elements consume less energy and produce an 11× more energy-efficient result.

Lastly, Nikonov and Young presented in [70] a model of an STMG technology that utilizes Magnetoelectric cells (ME) as inputs and outputs. Taking this into account, the targeted 32-bit full adder can be implemented with an order of magnitude improvement in energy compared to CMOS HP. From the results in Fig. 7.15, two statements can be made: (a) the appropriate application of STMGs is on designs that target low-energy operation and not high-performance, and (b) the input/output elements of STMG circuit should be voltage-controlled and as energy-efficient as possible to maximize the benefits of the technology.

#### 7.3.2.2 Discussion

With the aforementioned results in mind, we can define a set of requirements for efficient implementation of STMGs from a circuit perspective. To be a serious contender to CMOS, STMG-based circuits should be reliable and consume less energy, but they should also meet the need of an application that exploits its intrinsic non-volatility. More specifically, the following points should be addressed.

1. **Energy-efficient generation and detection of domains**
   In the original concept, proposed in [6], MTJs are used to generate the input domains by STT and detect the output domain by TMR. These two mechanisms require current to flow through a tunnel barrier, which leads to a substantial energy consumption, especially at the inputs where the current density is larger. Instead, domains could be nucleated using Voltage-Controlled Magnetic Anisotropy, magnetoelectric effect or Spin-Orbit Torques, for instance. These effects have been actively studied in recent years as they are promising alternatives to STT.
2. **Energy-efficient domain propagation**
   The majority domain should propagate as fast as possible between the inputs and the output. This is critical for delay but also for energy, as the input signal must be activated until the end of the operation. In the present concept of STMG, the domains are switched via the exchange interaction that couples the STT-driven spins to their neighbors. The efficiency of this method is not very well known but it could certainly be increased by a more direct coupling between the input signal and the magnetization to switch. Improving the domain propagation would also enable easier cascading of the gates. All in all, the STMG could be operated with two independent mechanisms: One that would switch the inputs and another one that would assist the propagation of the majority domain. Thus, both could be optimized independently without trade-off.
3. **Wider operating range**
   The STMG can operate only when a domain wall is not stable inside the cross. This is a restrictive condition that implies small anisotropy and small size, hence small thermal stability. A device that would allow a domain wall could have a much wider operating range and would give much more flexibility for circuit design.
4. **Use of non-volatility**
   Having a magnetic domain as the information carrier lends itself to inherit non-volatility at each gate output. In order to maximize the benefits of STMG,

this non-volatility has to be exploited by the circuit design. A common way of doing this is to utilize non-volatility to reduce static/leakage energy consumption [71]. This aspect of STMG has not been addressed yet but should yield significant advantages compared to CMOS and other volatile emerging technologies.

## References

1. J. Hutchby, G. Bourianoff, V. Zhirnov, J. Brewer, IEEE Circuits Devices Mag. **18**, 28 (2002)
2. G. Moore, Electronics **38**, 114 (1965)
3. V. Zhirnov, R. Cavin, J. Hutchby, G. Bourianoff, Proc. IEEE **9**, 1934 (2003)
4. D.E. Nikonov, I.A. Young, IEEE Proc. **101**(12), 2498 (2013)
5. K. Bernstein, R.K. Cavin, W. Porod, A. Seabaugh, J. Welser, Proc. IEEE **98**, 2169 (2010)
6. D.E. Nikonov, G.I. Bourianoff, T. Ghani, IEEE Electron Device Lett. **32**(8), 1128 (2011)
7. M. Manfrini, J.V. Kim, S. Petit-Watelot, W. Van Roy, L. Lagae, C. Chappert, T. Devolder, Nat. Nanotechnol. **9**(20), 121 (2014)
8. S. Dutta, S.C. Chang, N. Kani, D.E. Nikonov, S. Manipatruni, I.A. Young, A. Naeemi, Sci. Rep. **5**, 9861 (2015)
9. A. Khitun, K.L. Wang, J. Appl. Phys. **110**(3), 034306 (2011)
10. D. Griffiths, *Introduction to Quantum Mechanics*. Pearson International Edition (Pearson Prentice Hall, Upper Saddle River, 2005)
11. P. Dirac, *The Principles of Quantum Mechanics*. International Series of Monographs on Physics (Clarendon, Oxford, 1981)
12. W. Pauli, Z. Phys. **31**(1), 765 (1925)
13. C. Kittel, *Introduction to Solid State Physics* (Wiley Eastern Pvt Limited, New York, 1966)
14. D. Griffiths, *Introduction to Electrodynamics* (Prentice Hall, Upper Saddle River, 1999)
15. B. Cullity, C. Graham, *Introduction to Magnetic Materials* (Wiley, New York, 2011)
16. S. Chikazumi, *Physics of Ferromagnetism*. International Series of Monographs on Physics (Oxford University Press, Oxford, 2009)
17. J. Slonczewski, J. Magn. Magn. Mater. **159**(1–2), L1 (1996)
18. M. Julliere, Phys. Lett. A **54**(3), 225 (1975)
19. W.H. Butler, X.G. Zhang, T.C. Schulthess, J.M. MacLaren, Phys. Rev. B **63**(5), 054416 (2001). http://link.aps.org/doi/10.1103/PhysRevB.63.054416
20. S. Yuasa, T. Nagahama, A. Fukushima, Y. Suzuki, K. Ando, Nat. Mater. **3**(12), 868 (2004). https://doi.org/10.1038/nmat1257. http://www.ncbi.nlm.nih.gov/pubmed/15516927
21. S. Yuasa, D.D. Djayaprawira, J. Phys. D Appl. Phys. **40**(21), R337 (2007). https://doi.org/10.1088/0022-3727/40/21/R01. http://stacks.iop.org/0022-3727/40/i=21/a=R01?key=crossref.5e4ad43797a155e306a576b1744a6d26
22. S. Datta, B. Das, Appl. Phys. Lett. **56**(7), 665 (1990)
23. Y.A. Bychkov, E.I. Rashba, J. Phys. C Solid State Phys. **17**(33), 6039 (1984)
24. G. Schmidt, D. Ferrand, L.W. Molenkamp, A.T. Filip, B.J. van Wees, Phys. Rev. B **62**, R4790 (2000)
25. S. Bandyopadhyay, M. Cahay, Appl. Phys. Lett. **85**(10), 1814 (2004)
26. P. Chuang, S.C. Ho, L.W. Smith, F. Sfigakis, M. Pepper, C.H. Chen, J.C. Fan, J.P. Griffiths, I. Farrer, H.E. Beere, G.A.C. Jones, D.A. Ritchie, T.M. Chen, Nat. Nano **10**(1), 35 (2015)
27. R.P. Cowburn, M.E. Welland, Science **287**(5457), 1466 (2000)
28. G. Csaba, A. Imre, G. Bernstein, W. Porod, V. Metlushko, IEEE Trans. Nanotechnol. **1**(4), 209 (2002)
29. S. Breitkreutz, J. Kiermaier, I. Eichwald, C. Hildbrand, G. Csaba, D. Schmitt-Landsiedel, M. Becherer, IEEE Trans. Magn. **49**(7), 4464 (2013)
30. B. Behin-Aein, D. Datta, S. Salahuddin, S. Datta, Nat. Nano **5**(4), 266 (2010)

31. S. Manipatruni, D.E. Nikonov, I.A. Young, Appl. Phys. Rev. **5**, 014002 (2016)
32. S. Natarajan, M. Agostinelli, S. Akbar et al., *2014 IEEE International Electron Devices Meeting* (2014), pp. 3.7.1–3.7.3
33. L. Amarú, P.E. Gaillardon, S. Mitra, G. De Micheli, Proc. IEEE **103**(11), 2168 (2015)
34. J. Von Neumann, Non-linear capacitance or inductance switching, amplifying, and memory organs. US Patent 2815488 (1957)
35. L. Amarú, P.E. Gaillardon, G. De Micheli, *Proceedings of Design Automation Conference (DAC)*, 2015
36. I.T.R. for Semiconductors. Executive summary (2013)
37. O. Zografos, L. Amarú, P.E. Gaillardon, P. Raghavan, G.D. Micheli, *2014 17th Euromicro Conference on Digital System Design (DSD)* (2014), pp. 691–694
38. K. Kong, Y. Shang, R. Lu, IEEE Trans. Nanotechnol. **9**(2), 170 (2010)
39. B. Hillebrands, K. Ounadjela, *Spin Dynamics in Confined Magnetic Structures I*. Topics in Applied Physics (Springer, Berlin, 2001)
40. Y. Xu, D. Awschalom, J. Nitta, *Handbook of Spintronics* (Springer, Dordrecht, 2015)
41. D. Stancil, A. Prabhakar, *Spin Waves: Theory and Applications* (Springer, Berlin, 2009)
42. A. Gurevich, G. Melkov, *Magnetization Oscillations and Waves* (Taylor & Francis, Boca Raton, 1996)
43. B.A. Kalinikos, A.N. Slavin, J. Phys. C: Solid State Phys. **19**, 7013 (1986)
44. A. Khitun, M. Bao, K.L. Wang, IEEE Trans. Magn. **44**(9), 2141 (2008)
45. T. Wu, A. Bur, K. Wong et al., J. Appl. Phys. **109**(7), 07D732 (2011)
46. O. Zografos, P. Raghavan, Y. Sherazi, A. Vaysset, F. Ciubatoru, B. Sorée, R. Lauwereins, I. Radu, A. Thean, in *2015 International Conference on IC Design Technology (ICICDT)* (2015), pp. 1–4
47. O. Zografos, P. Raghavan, L. Amarú, B. Sorée, R. Lauwereins, I. Radu, D. Verkest, A. Thean, *2014 IEEE/ACM International Symposium on Nanoscale Architectures (NANOARCH)* (2014), pp. 25–30
48. O. Zografos, B. Sorée, A. Vaysset, S. Cosemans, L. Amarú, P.E. Gaillardon, G.D. Micheli, R. Lauwereins, S. Sayan, P. Raghavan, I.P. Radu, A. Thean, *2015 IEEE 15th International Conference on Nanotechnology (IEEE-NANO)* (2015), pp. 686–689
49. S.O. Demokritov, B. Hillebrands, A.N. Slavin, Phys. Rep. **348**(6), 441 (2001)
50. S.O. Demokritov, A.A. Serga, A. André, V.E. Demidov, M.P. Kostylev, B. Hillebrands, A.N. Slavin, Phys. Rev. Lett. **93**(4), 047201 (2004)
51. F. Ciubotaru, T. Devolder, M. Manfrini, C. Adelmann, I.P. Radu, Appl. Phys. Lett. **109**(1) (2016)
52. F. Ciubotaru, O. Zografos, G. Talmelli, C. Adelmann, I. Radu, T. Fischer, et al., Spin waves for interconnect applications, in *Interconnect Technology Conference (IITC), 2017 IEEE International* (IEEE, 2017), pp. 1–4
53. V.E. Demidov, S. Urazhdin, R. Liu, B. Divinskiy, A. Telegin, S.O. Demokritov, Nat. Commun. **7**, 10446 (2016)
54. A.V. Chumak, P. Pirro, A.A. Serga, M.P. Kostylev, R.L. Stamps, H. Schultheiss, K. Vogt, S.J. Hermsdoerfer, B. Laegel, P.A. Beck, B. Hillebrands, Appl. Phys. Lett. **95**(26) (2009)
55. S. Cherepov, P. Khalili Amiri, J.G. Alzate et al., Appl. Phys. Lett. **104**(8), 082403 (2014)
56. T. Schneider, A.A. Serga, B. Leven, B. Hillebrands, R.L. Stamps, M.P. Kostylev, Appl. Phys. Lett. **92**(2), 022505 (2008)
57. S. Klingler, P. Pirro, T. Brächer, B. Leven, B. Hillebrands, A.V. Chumak, Appl. Phys. Lett. **105**(15), 152410 (2014)
58. S. Klingler, P. Pirro, T. Brächer, B. Leven, B. Hillebrands, A.V. Chumak, Appl. Phys. Lett. **106**(21), 212406 (2015)
59. G. Csaba, A. Papp, W. Porod, R. Yeniceri, *2015 45th European Solid State Device Research Conference (ESSDERC)* (2015), pp. 101–104

60. O. Zografos, M. Manfrini, A. Vaysset, B. Sorée, F. Ciubotaru, C. Adelmann, R. Lauwereins, P. Raghavan, I.P. Radu, Sci. Rep. **7**(1), 12154 (2017). https://doi.org/10.1038/s41598-017-12447-8
61. A. Khitun, D.E. Nikonov, K.L. Wang, J. Appl. Phys. **106**(12), 123909 (2009)
62. T.U. Aoki Laboratory. Arithmetic module generator. http://www.aoki.ecei.tohoku.ac.jp/arith/
63. J. Ryckaert, P. Raghavan, R. Baert et al., *2014 IEEE Proceedings of the Custom Integrated Circuits Conference (CICC)* (2014), pp. 1–8
64. D.E. Nikonov, S. Manipatruni, I.A. Young, J. Appl. Phys. **115**(17), 2014 (2014)
65. D.M. Bromberg, D.H. Morris, L. Pileggi, J.G. Zhu, IEEE Trans. Magn. **48**(11), 3215 (2012). https://doi.org/10.1109/TMAG.2012.2197186
66. J.A. Currivan, Y. Jang, M.D. Mascaro, M.A. Baldo, C.A. Ross, IEEE Magn. Lett. **3**, 3 (2012). https://doi.org/10.1109/LMAG.2012.2188621
67. J.A. Currivan-Incorvia, S. Siddiqui, S. Dutta, E.R. Evarts, J. Zhang, D. Bono, C.A. Ross, M.A. Baldo, Nat. Commun. **7**, 10275 (2016). http://dx.doi.org/10.1038/ncomms10275. http://www.nature.com/doifinder/10.1038/ncomms10275
68. A. Vaysset, M. Manfrini, D.E. Nikonov, S. Manipatruni, I.A. Young, G. Pourtois, et al., Toward error-free scaled spin torque majority gates. AIP Adv. **6**(6), 065304 (2016). https://doi.org/10.1063/1.4953672
69. A. Vaysset, M. Manfrini, D.E. Nikonov, S. Manipatruni, I.A. Young, I.P. Radu, et al. Operating conditions and stability of spin torque majority gates: analytical understanding and numerical evidence. J. Appl. Phys. **121**(4), 043902 (2017). https://doi.org/10.1063/1.4974472
70. D.E. Nikonov, I.A. Young, IEEE J. Explor. Solid-State Comput. Devices Circuits **1**, 3 (2015)
71. M. Natsui, D. Suzuki, N. Sakimura, R. Nebashi, Y. Tsuji, A. Morioka, T. Sugibayashi, S. Miura, H. Honjo, K. Kinoshita et al., IEEE J. Solid-State Circuits **50**(2), 476 (2015)

# Index





R. O. Topaloglu, H.-S. P. Wong (eds.), *Beyond-CMOS Technologies for Next Generation Computer Design*, https://doi.org/10.1007/978-3-319-90385-9

**T**

MIX
Papier aus verantwortungsvollen Quellen
Paper from responsible sources
FSC® C105338

If you have any concerns about our products,
you can contact us on
**ProductSafety@springernature.com**

In case Publisher is established outside the EU,
the EU authorized representative is:
**Springer Nature Customer Service Center GmbH**
**Europaplatz 3, 69115 Heidelberg, Germany**

Printed by Libri Plureos GmbH
in Hamburg, Germany